AN
INTRODUCTION
TO THE
THEORY OF

STATIONARY
RANDOM FUNCTIONS

AN
INTRODUCTION
TO THE
THEORY OF

STATIONARY
RANDOM FUNCTIONS

A. M. YAGLOM

*Institute of Atmospheric Physics*
*Academy of Sciences, U.S.S.R.*

*Revised English Edition*
*Translated and Edited by*

Richard A. Silverman

DOVER PUBLICATIONS, INC.
NEW YORK

Published in Canada by General Publishing Company, Ltd., 30 Lesmill Road, Don Mills, Toronto,
Ontario.
Published in the United Kingdom by Constable
and Company, Ltd., 10 Orange Street, London WC 2.


This Dover edition, first published in 1973, is an
unabridged and unaltered republication of the work
originally published by Prentice-Hall, Inc. in 1962.


*International Standard Book Number: 0-486-64688-2*
*Library of Congress Catalog Card Number:73-86439*

Manufactured in the United States of America
Dover Publications, Inc.
31 East 2nd Street
Mineola, N.Y. 11501

# AUTHOR'S PREFACE TO
# THE RUSSIAN EDITION

The basic aim of this monograph is to give a simple and at the same time mathematically rigorous treatment of the problem of extrapolating and filtering stationary random functions (both sequences and processes). The material stems from two sets of lectures given at Moscow State University, one for a group of graduate students in the Mechanics and Mathematics Department, and another for a seminar under the direction of E. B. Dynkin. In order to keep the presentation as elementary as possible, emphasis has been put on the simplest special case, where the spectral density is a rational function (in $e^{i\lambda}$ for sequences, and in $\lambda$ for processes). However, Chapter 8 contains a brief survey of the results obtained by A. N. Kolmogorov in his deep study of the general case.

Part 1 of the book is devoted to a rather complete presentation of the spectral theory of stationary random functions. This sophisticated theory, originating in A. Y. Khinchin's paper in the *Mathematische Annalen* (vol. 109, p. 604, 1934), is now the basis for almost all research on the subject. Part 1 should be of independent interest, aside from its connection with the theory of extrapolation and filtering, presented in Part 2.

The reader is assumed to have at his command little more than the rudiments of probability theory and complex variable theory. However, to understand Part 2, he will need to know the most elementary facts about the geometry of Hilbert space. The reader who is not familiar with this material may have to take certain statements on faith.

Frequent discussions with A. N. Kolmogorov have had a considerable influence on all my work in the field of random functions, and I have also profited greatly from conversations with A. M. Obukhov.  While writing this monograph, I have received substantial help from A. S. Monin, which has enabled me to finish the work much sooner than would otherwise have been possible.  A. N. Kolmogorov and A. M. Obukhov have both read the manuscript, making many valuable suggestions.  I am delighted to take this occasion to express my sincere gratitude to these three colleagues.

1952                                                            A. M. Y

# AUTHOR'S PREFACE TO THE REVISED ENGLISH EDITION

The original version of this monograph was published a decade ago as a long review article in the Russian journal *Uspekhi Matematicheskikh Nauk* (vol. 7, no. 5, 1952). At that time, no book specifically devoted to the mathematical theory of stationary random functions was available, either in Russian or in any other language, and it was my intention to remedy, at least partially, this gap in the literature.  I also wanted to popularize, for as large an audience as possible, the theory of linear extrapolation and filtering of stationary sequences and processes, due to A. N. Kolmogorov and N. Wiener, a theory which is both of intrinsic mathematical interest and of great practical importance.  This task seemed to me even more worthwhile for another reason: On the one hand, Kolmogorov's basic paper, containing complete proofs of his profound results in this field, had been published in a journal with a rather limited circulation (*Bulletin Moscow State University*, vol. 2, no. 6, 1941), while the methods used in the paper are complicated and accessible only to students with an extensive mathematical background.  On the other hand, Wiener's celebrated report, written during the war, had just appeared in book form at about this time, and it immediately acquired a reputation among engineers of being extraordinarily abstruse, whereas most mathematicians, unaccustomed to its heuristic level of rigor and engineering terminology, had great difficulty in understanding it.  However, I had found that it was quite feasible to give a simple and entirely rigorous treatment of the problem of extrapolating and filtering stationary random functions, for the case of rational spectral densities.  In fact, a recent article by S. Darlington (*Bell System Technical Journal*, vol. 37, p. 1221, 1958) advocates the same method as the most suitable approach for engineers encountering the subject for the first time.

During the last decade, a number of interesting books on the theory of random functions have been published, many of which are cited in the bibliographies at the end of this volume.  Nevertheless, it seems to me that my book, which, without sacrificing mathematical rigor, stresses the physical meaning of results rather than delving into logical subtleties, will be of interest to beginning mathematicians, as well as to physicists and engineers who actually deal with random functions in practice. However, despite the fact that the book treats only elementary aspects of the theory, the presentation of some topics in the Russian original was in danger of becoming out of date, because of subsequent advances in the field, on all levels.  Therefore, the translator, Dr. R. A. Silverman, decided to "rejuvenate" the book, by adding new material in keeping with the present development of the subject (especially in Chapters 3 and 8, which have a survey character), by enlarging and modernizing the Bibliography, by supplying numerous footnotes containing explanatory comments and references to recent work, etc.  I myself have also made numerous improvements and additions, working from a copy of the manuscript sent to me by the translator in ample time for revision. Moreover, I have read through all the galley proofs, making sure that I approved of the final version.  Thus, in my opinion, the English edition of the book reflects the contemporary state of research in the theory of stationary random functions.

Of course, in a book of this size, it has been impossible to discuss all aspects of a theory which has progressed so rapidly, and it is inevitable that some important results are mentioned only in passing, or are not touched upon at all. However, it seemed to me that failure to include a discussion of the recent theory of generalized random processes would have been regrettable, since this theory greatly simplifies the treatment of some important topics, e.g., random processes with stationary increments. Moreover, the concept of a generalized random process is very "physical," and hence quite accessible to those readers for whom the book is primarily intended. Thus, I thought it advisable to write an elementary exposition of the theory of generalized random processes, expressly for the English edition, and I accordingly sent the translator an outline which he expanded into the first appendix. In addition, Dr. D. B. Lowdenslager contributed a second appendix, containing a brief survey of some recent developments (mostly pertaining to extrapolation and filtering of multidimensional stationary random functions), with appropriate supplementary references. This appendix should also enhance the value of the book.

I am pleased that my book is appearing in a special series of Russian translations, since this will serve to bring it to the attention of many new readers. I would like to express my appreciation to the Prentice-Hall Publishing Company, to Dr. D. B. Lowdenslager, and especially to Dr. R. A. Silverman for his careful translation and painstaking effort to eliminate typographical errors in the Russian original, for helping me write Appendix I, and in general, for doing everything in his power to improve the English edition of the book.

1962                                                    A. M. Y.

# TRANSLATOR'S PREFACE

The present volume, the fourth in a new series of Russian translations under my editorship, has long been regarded as a classic presentation of a subject of great theoretical and practical interest. It has been a great pleasure to work with Professor Yaglom in preparing a revised English edition of his book. I would like to thank him for his kind words above, and for his indefatigable cooperation during a four-month period of heavy correspondence. Professor Yaglom has already described both his aims in writing the monograph, and the ways in which the English edition differs from the Russian original. Thus, at this point, I would only like to call attention to a few stylistic matters:

1. The system of references in "letter-number form" used in the main Bibliography is almost self-explanatory. For example, K9 refers to the ninth paper (or book) whose (first) author's surname begins with the letter K, where the entire Bibliography is arranged in alphabetical order, and in chronological order as well, whenever there are several papers by the same author.

2. The main Bibliography contains only items cited in the text, and the Supplementary Bibliography contains only items cited in the two appendices. It is not claimed that the bibliographies are exhaustive, especially as regards papers on applied topics.

3. Sections marked by asterisks, and also individual passages lying between asterisks, contain more advanced or detailed material, which can be omitted without loss of continuity. However, no attempt has been made to indicate such passages in Chapter 8, which is essentially a survey chapter.

1962                                                    R. A. S.

# CONTENTS

# INTRODUCTION

The mathematical theory of probability deals with phenomena *en masse*, i.e., observations (experiments, trials) which can be repeated many times under similar conditions. The principal concern of classical probability theory is with numerical characteristics of the phenomenon being studied, i.e., quantities which take various values depending on the result of the observation. Such quantities are called *random variables*. The following are all examples of random variables: (a) The number of points which appear when a die is tossed; (b) The number of calls which arrive at a telephone station during a given time interval; (c) The lifetime of an electric bulb, reckoned from the time it is produced to the time it finally burns out; (d) The error made in measuring a physical quantity, when a given device is used. In probability theory, a random variable $\xi$ is regarded as specified if one knows its *distribution function*

$$F(x) = \mathbf{P}\{\xi < x\}, \tag{0.1}$$

where the symbol $\mathbf{P}\{\cdot\}$ will always be used to denote the probability of occurrence of the relation which appears between the braces.

In the case where the result of the observation consists of $n$ different numbers $\xi_1, \xi_2, \ldots, \xi_n$, these numbers can be regarded as a single *multidimensional random variable*

$$\xi = (\xi_1, \xi_2, \ldots, \xi_n).$$

Probability theory is also habitually concerned with such multidimensional random variables. These random variables are specified by *multidimensional distribution functions*

$$F(x_1, x_2, \ldots, x_n) = \mathbf{P}\{\xi_1 < x_1, \xi_2 < x_2, \ldots, \xi_n < x_n\}, \tag{0.2}$$

which are, of course, functions of several variables.

However, in recent years, in many physical and engineering applications,

1

2 INTRODUCTION

it has become necessary to consider not only one-dimensional and multidimensional random variables, but also *random functions*, i.e., functions which are specified by the results of an observation, and which can take different values when the observation is repeated many times. One of the earliest and most typical examples of such a situation is to be found in the theory of Brownian motion (see C2, E1, L1, L4, L5, L6, V3, W2),[1] where each coordinate of the Brownian particle is a random function (depending on time). A phenomenon which is closely related to Brownian motion is that of fluctuations in electrical circuits (see B4, B13, D2, M2, R2, V1). Strictly speaking, the voltage across any conductor and the current in the conductor are always random functions of time, since the thermal motion of the electrons produces uncontrollable fluctuations in the voltage and current, known as "thermal noise." The role of noise in electrical circuits has become particularly important, now that circuits involving vacuum tubes have been extensively developed. A vacuum tube is always a source of considerable noise, both because of fluctuations in the number of electrons passing through the tube during identical time intervals (the "shot effect") and because of fluctuations in cathode emission intensity ("flicker noise"). In radio receivers, one always observes not only noise arising in the electrical circuits themselves, but also random changes in the level of the received signals (an effect known as "fading"), due to scattering of the radio waves caused by inhomogeneities in the refractive index of the atmosphere, and due to the influence of random electrical discharges (meteorological and industrial noise). Finally, radars and servomechanisms (which have undergone considerable development in recent years) are found to have their own special sources of noise (see e.g., J2, Chaps. 6 to 8, and L3). All this explains why the development of electrical engineering, especially as applied to problems of radio communication and radar, has served as a powerful stimulus to the proliferation in recent years of research on the theory of random functions.

Electrical noise is certainly the most important, but hardly the only example of a random function. As other examples of random functions, we can cite the pressure, the temperature and the three components of the velocity vector at a point in a turbulent flow (in particular, at a point in the atmosphere). These random functions depend on four variables, the three spatial coordinates and time. Another example of a random function is encountered in the manufacture of thread: Due to a whole series of uncontrollable factors, the thread never turns out to be completely homogeneous; instead, the diameter of a cross section of the thread is a random function of the position coordinate of the cross section. Similarly, random functions arise in a variety of other problems, pertaining to the most diverse branches of technology. Moreover, there is a whole category of random functions stemming

---

[1] The reference scheme is explained in the Translator's Preface.

from geophysical and related investigations.   For example, the set of monthly values of the Wolf number characterizing solar activity (the number of sunspots) can be regarded as a random function depending on a discrete argument, i.e., the number of the month.[2]

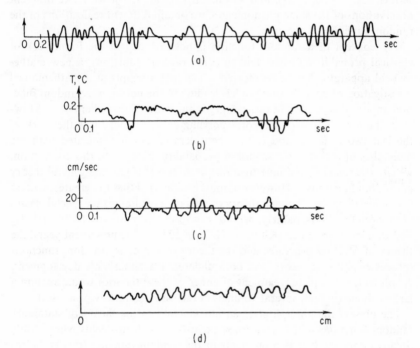

FIG. 1. Observed values of random functions. (a) Fading of the intensity of radio signals received in a radar (J2). (The units of intensity are not indicated); (b) Fluctuations of air temperature at a point in the atmosphere (K20); (c) Fluctuations of the difference in wind velocity at two points of the atmosphere separated by a distance of 8 cms (O2); (d) Variation in the diameter of a manufactured thread (F2). (The units are not indicated.)

To illustrate the appearance of the observed values of random functions, we show in Figure 1 the curves obtained in measuring the following quantities: (a) The intensity of radio signals received in a radar; (b) The air temperature at a given point of the atmosphere at different instants of time; (c) The difference in wind velocity at two nearby points of the atmosphere, and (d) The diameter of a manufactured thread at different points along the

<hr />

[2] We can apply all the results to be derived below to random functions of this type, if the mathematical expectation is replaced everywhere by the time average (see e.g., K4, W6).

thread.   We call attention to the qualitative similarity of these four curves, which pertain to completely different phenomena.   Such observed curves will be called *realizations* (or *sample functions*) of the corresponding random functions.   We shall denote realizations of a random function $\xi(t)$ by superscripts, e.g., $\xi^{(1)}(t)$, $\xi^{(2)}(t)$, and so on.   It is clear that if we make different observations of the same phenomenon, we obtain different realizations of the random function in question.

The concept of a random function is not included in the framework of classical probability theory, and to study random functions, a new mathematical apparatus has to be created.   The first attempts at a mathematical investigation of probabilistic models leading to the notion of a random function appeared very early in this century in the work of Bachelier (B1).   However, the formulation of the corresponding general theory was the work of the last two decades, and is to a considerable extent associated with the researches of the Moscow school of probability theory.   In this connection, we cite the fundamental investigations of Slutski (S10) on the general theory of random functions.   However, it must be admitted that the greatest success was achieved in developing a theory of the two most important special classes of random functions, namely, *Markov processes* [Kolmogorov (K8), in 1931] and *stationary processes* [Khinchin (K6), in 1934].   In subsequent years, the theory of Markov processes and the theory of stationary random functions (processes and sequences) have both undergone a remarkable development. At present, a complete treatment of either of these theories would require a large volume (in fact, several volumes in the case of Markov processes).

The present monograph gives an introduction to the theory of stationary random functions.   It is concerned exclusively with problems whose study involves only the first two moments of the random function (this is the so-called *correlation theory*).   Part 1 is devoted to a presentation of the basic facts of the general correlation theory of stationary random functions.   A central position is occupied by the problem of the spectral representation of such functions.   Without doubt, this topic represents the most important part of the general theory.   We shall pay considerable attention to the physical meaning of the mathematical concepts analyzed in Part 1, thereby clarifying a variety of ideas essential to a better understanding of subsequent material.

The subject matter of Part 2 is somewhat more specialized, being devoted to a particular aspect of the correlation theory which is of great importance in the applications.   We refer to the problem of *extrapolating* (*predicting*) and *filtering* stationary random sequences and processes.   The basic contribution to the solution of this problem is due to Kolmogorov (K12, K13), who was the first to formulate the problem clearly and give a general formula for the mean square error of the best linear extrapolation of an arbitrary stationary sequence.   Kolmogorov's investigations were extended by his pupil

Zazukhin (Z2), who studied extrapolation of multidimensional sequences, and by Krein (K24) and Karhunen (K3), who studied extrapolation of stationary processes.

In recent years, the practical significance of this whole class of ideas has become particularly apparent, due in large measure to Wiener's famous book (W4), which was written with a view to the interests of readers specializing in the field of radio engineering.[3, 4] Wiener considers problems similar to those treated by Kolmogorov (K12, K13), but from a theoretical point of view, his book dilutes the content of Kolmogorov's general investigations. In fact, instead of general stationary sequences and processes, Wiener is primarily concerned with a much narrower class of sequences and processes, which have spectral densities of a very special form. However, for this narrower class of sequences and processes, Wiener succeeds in finding not only the mean square error, but more generally an expression (actually, a simple one) for the operator which performs the extrapolation. This greatly simplifies the problem of making practical use of the results obtained. Moreover, besides the theory of linear extrapolation, Wiener also considers the theory of linear filtering of stationary random functions. The latter theory, although containing little of novelty from a theoretical standpoint, is very important in engineering applications.

In proving the basic results pertaining to the extrapolation of stationary random functions in the papers K3, K12, K13, K24, Z2, use is made of advanced techniques from functional analysis and from the theory of functions of a complex variable. Thus, a study of these papers requires that the reader have a very extensive mathematical background. In Wiener's book W4, the same kind of results are obtained by solving certain integral equations of a special type. This approach is rather complicated, and moreover does not give a rigorous proof of the formulas obtained, since both in formulating and solving the integral equations in question, one has to make extensive use of mathematically meaningless expressions (of the kind involving divergent integrals).

Nevertheless, for the class of random functions considered in Wiener's book, it is possible to give a rigorous treatment of the theory of extrapolation and filtering, which leans on the most rudimentary notions of the geometry of Hilbert space, and which uses only the simplest facts from the theory of functions of a complex variable (of the kind involving residue theory). Such

---

[3] Actually, Wiener did not succeed in writing a treatment of the subject which is accessible to readers who are primarily engineers. For an attempt to explain the content of Wiener's book in terms which are familiar to radio engineers, see B10.

[4] In this connection, reference should also be made to Solodovnikov's paper S11 and to his comprehensive book S12. We also call attention to the books by Davenport and Root (D2), Laning and Battin (L2), and Wainstein and Zubakov (W1).

a treatment is given in Part 2 of the present monograph.[5] In order to be as elementary as possible, at each step we begin by analyzing concrete examples, and only afterwards do we turn to a consideration of the general case. The reader who is particularly interested in the theory of extrapolation and filtering, and who is familiar with the basic facts of the correlation theory of stationary random functions, as can be found, for example, in the paper by Khinchin (K6) or in Gnedenko's book (G3, Secs. 57, 58) can proceed at once to read Part 2. He will occasionally find references in Part 2 to formulas in Part 1, which he can look up as the occasion warrants.

We conclude each part of the present monograph with a chapter devoted to further development of the questions touched upon earlier. In these chapters, we enumerate (citing the appropriate literature) a series of important facts from the theory of stationary random functions, which do not appear in the main part of the book. In particular, in Chap. 8, Sec. 36, we present the basic results of the general theory of extrapolation and filtering of stationary random functions, due in most part to Kolmogorov (K12, K13). This treatment contains some examples illustrating the theory, but no proofs are given. For a detailed presentation of Kolmogorov's theory, containing complete proofs, we refer the interested reader to Doob's treatise (D6, Chap. 12).

---

[5] The possibility of such a simplified treatment of the theory was pointed out by Doob (D4, D5) for the easiest special case, i.e., extrapolation of a stationary sequence one step ahead.

# Part 1

# THE GENERAL THEORY
# OF STATIONARY
# RANDOM FUNCTIONS

# 1

---

## BASIC PROPERTIES
## OF STATIONARY
## RANDOM FUNCTIONS

---

## I. Definition of a Random Function

Let $T$ be an arbitrary set of elements $t, s, \ldots$. By a *random function* of an argument $t$ in $T$ (synonymously, a random function *on* $T$), we mean a function $\xi(t)$ whose values are random variables. Thus, a random function on $T$ is a family of random variables $\xi(t), \xi(s), \ldots$ corresponding to all elements $t, s, \ldots$ in the set $T$. If the set $T$ contains only a finite number of elements, then the random function $\xi(t)$ is a finite family of random variables. In this case, we can regard $\xi(t)$ as a single multidimensional random variable. Such random variables are studied in classical probability theory, and they are specified by using multidimensional distribution functions.[1] If the set $T$ is infinite, then $\xi(t)$ is an infinite family of random variables. Such families of random variables are not studied in classical probability theory, and hence the specification of $\xi(t)$ requires a special definition.

Following Slutski, *we shall regard the random function $\xi(t)$ as being specified if for each element $t$ in the set $T$ we are given the distribution function*

$$F_t(x) = \mathbf{P}\{\xi(t) < x\} \tag{1.1}$$

*of the quantity $\xi(t)$, if for each pair of elements $t_1, t_2$ in the set $T$, we are given the distribution function*

$$F_{t_1, t_2}(x_1, x_2) = \mathbf{P}\{\xi(t_1) < x_1, \xi(t_2) < x_2\} \tag{1.2}$$

---

[1] See formula (0.2) of the Introduction.

*of the two-dimensional random variable* $\xi(t_1, t_2) = (\xi(t_1), \xi(t_2))$, *and so on, and in general if for any n elements* $t_1, t_2, \ldots, t_n$ *in the set T we are given the distribution function*

$$F_{t_1, t_2, \ldots, t_n}(x_1, x_2, \ldots, x_n) = \mathbf{P}\{\xi(t_1) < x_1, \xi(t_2) < x_2, \ldots, \xi(t_n) < x_n\} \quad (1.3)$$

*of the n-dimensional random variable*

$$\xi(t_1, t_2, \ldots, t_n) = (\xi(t_1), \xi(t_2), \ldots, \xi(t_n)).$$

The distribution functions (1.3) must obviously satisfy the following two conditions:

(a) The *symmetry condition*, according to which

$$F_{t_{j_1}, t_{j_2}, \ldots, t_{j_n}}(x_{j_1}, x_{j_2}, \ldots, x_{j_n}) = F_{t_1, t_2, \ldots, t_n}(x_1, x_2, \ldots, x_n), \quad (1.4)$$

where $j_1, j_2, \ldots, j_n$ is any permutation of the indices $1, 2, \ldots, n$;

(b) The *compatibility condition*, according to which

$$F_{t_1, t_2, \ldots, t_m, t_{m+1}, \ldots, t_n}(x_1, x_2, \ldots, x_m, \infty, \ldots, \infty)$$
$$= F_{t_1, t_2, \ldots, t_m}(x_1, x_2, \ldots, x_m) \quad (1.5)$$

for any $t_{m+1}, \ldots, t_n$ if $m < n$.

The converse is also true: Any system of distribution functions (1.3) satisfying the conditions (1.4) and (1.5) can be regarded as defining some random function on $T$.

There exist other methods of specifying a random function. Thus, it is often convenient to define a random function by an analytic formula, containing parameters which are random variables. For example, one can consider polynomials (ordinary or trigonometric) with random coefficients. We shall occasionally use this method of specifying a random function.

A few words are in order concerning still another method of specifying a random function, a method which is closely associated with the way a random function arises in an actual physical context. As already emphasized in the Introduction, in order to apply probabilistic methods in the first place, we must have an experiment which can be repeated many times under similar conditions and which can lead to different outcomes. The set $\Omega$ of all possible outcomes $\omega$ of such an experiment (the set of "elementary events") plays a basic role in the contemporary axiomatic formulation of probability theory (see K17). In fact, by a *random variable* $\xi$ is meant a quantity which takes different numerical values for different outcomes of an experiment. In other words, a random variable is a numerical function of the point $\omega$ of the set $\Omega$. Therefore, it would be more accurate to write $\xi(\omega)$ instead of $\xi$. (This was not done anywhere above, since in probability theory, the dependence on $\omega$ is traditionally suppressed.) Since we have defined a *random function* $\xi(t)$ *on* $T$ to be a family of random variables corresponding to all the elements $t$ of the set $T$, from our new point of view, a random function is a function $\xi(\omega, t)$ of the two variables $\omega$ and $t$.

If we fix the value of the argument $\omega$ in $\xi(\omega, t)$, we have a real function $\xi(\omega, t) = \xi^{(\omega)}(t)$ of the variable $t$, depending on the parameter $\omega$. Thus, to each outcome $\omega$ of our experiment, there corresponds a definite real function of the variable $t$. This function is called a *realization* (or a *sample function*) of the random function (cf. Introduction). Thus, a random function $\xi(\omega, t)$ can be regarded either as a family of random variables $\xi_t(\omega)$ depending on the parameter $t$, or as a family of realizations $\xi^{(\omega)}(t)$, depending on the parameter $\omega$. In the latter case, to specify the random function we have to give the probability (or probability density) of occurrence of the various realizations. This leads to the definition of a particular measure $P$ (the *probability measure*) on the function space of realizations, and by specifying this measure, we specify the random function.

This approach to the concept of a random function, originating in Kolmogorov's work (K17) on the foundations of probability theory, has been very fruitful in the further development of the theory. It can be shown that specifying a random function by defining a probability measure on the set of realizations of the function includes as special cases all the other ways of specifying a random function discussed above,[2] and moreover has many advantages over the other ways. However, a more detailed treatment of these matters would require the use of comparatively advanced techniques of function theory and set theory, and since the use of such techniques is not essential in any of our subsequent considerations, we confine ourselves to the remarks just made, referring the interested reader elsewhere (see e.g., D6).

## 2. Stationarity

From now on, the variable $t$ will only take real values, and will be interpreted as the *time*. As for the set $T$, it will often consist of the whole set of real numbers $-\infty < t < \infty$, in which case the random function $\xi(t)$ will be called a *(random) process*, while on other occasions, $T$ will consist only of the set of all integers

$$\ldots, -2, -1, 0, 1, 2, \ldots,$$

in which case $\xi(t)$ will be called a *random sequence*. The term *random function* will be reserved for situations in which we wish to consider random processes and random sequences simultaneously.

*The random function $\xi(t)$ will be called stationary if all the finite-dimensional distribution functions* (1.3) *defining $\xi(t)$ remain the same if the whole group of points $t_1, t_2, \ldots, t_n$ is shifted along the time axis, i.e., if*

$$F_{t_1+\tau,\, t_2+\tau,\, \ldots,\, t_n+\tau}(x_1, x_2, \ldots, x_n) = F_{t_1,\, t_2,\, \ldots,\, t_n}(x_1, x_2, \ldots, x_n) \qquad (1.6)$$

---

[2] In this regard, Kolmogorov's "fundamental theorem" is particularly relevant (K17, p. 29).

*for any n, $t_1, t_2, \ldots, t_n$ and $\tau$.* In particular, this implies that for a stationary random function, all the one-dimensional distribution functions (1.1) have to be identical [i.e., $F_t(x)$ cannot depend on $t$], all the two-dimensional distribution functions (1.2) can only depend on the difference $t_2 - t_1$, and so on.

The physical meaning of the concept of stationarity is clear. It means that $\xi(t)$ describes the time variation of a numerical characteristic $\xi$ of an event such that none of the observed macroscopic factors influencing the occurrence of the event changes in time. Thus, for example, the current and voltage fluctuations ("noise") in an electrical circuit represent stationary processes, provided that the circuit has a "stationary regime," i.e., that none of the circuit parameters (including the external e.m.f.) changes in time. In the same way, a component of the velocity of a turbulent flow will be a stationary random function if we are dealing with a steady flow, the diameter of a cross section of a thread will be a homogeneous random function [3] of the position coordinate of the cross section if the conditions under which the thread is manufactured do not change in time, and so on. In general, the random functions encountered in practical applications can very often be regarded as stationary, to a high degree of accuracy. This fact explains the great practical importance of the theory of stationary random functions.

Finally, we note that in order to regard a random function $\xi(t)$ as stationary, it is not really necessary that the external characteristics of the phenomenon giving rise to the random function be invariant in time. For practical purposes, it is sufficient that these external characteristics be constant during the entire period for which $\xi(t)$ is observed and also during certain adjacent time intervals, which are long enough to allow any transient processes to die out.

## 3. Moments. The Correlation Function

To give a complete description of a stationary random function, we have to know all its finite-dimensional distribution functions (1.3), which according to (1.6) depend only on the differences $t_j - t_1$ ($j = 2, \ldots, n$). However, in most actual physical and mechanical problems, it is excessively complicated and costly to measure these distribution functions experimentally, and they are too cumbersome to be used in practice. Therefore, in developing a theory of stationary random functions, it is natural to restrict oneself (at least, initially) to a study of properties determined by just the simplest numerical characteristics of the multidimensional distributions (1.3). The *moments* of

---

[3] Suppose that the real variable $t$ on which a random function depends is different from the time. Then, a random function satisfying the condition (1.6) is said to be *homogeneous* (rather than *stationary*). Moreover, one often refers to stationary random functions as random functions which are *homogeneous in time*.

a probability distribution serve as simple numerical characteristics of the distribution, and are widely used in probability theory. In the case of the distribution (1.3), the moments take the form

$$\mu_{m_1 m_2 \dots m_n}(t_1, t_2, \dots, t_n) = \mathbf{E}\xi^{m_1}(t_1)\xi^{m_2}(t_2)\dots\xi^{m_n}(t_n)$$
$$= \int_{-\infty}^{\infty}\int_{-\infty}^{\infty}\cdots\int_{-\infty}^{\infty} x_1^{m_1} x_2^{m_2} \dots x_n^{m_n}\, dF_{t_1, t_2, \dots, t_n}(x_1, x_2, \dots, x_n) \qquad (1.7)$$

where the symbol $\mathbf{E}$ denotes the *averaging operator*, i.e., $\mathbf{E}\eta$ denotes the *mathematical expectation* of the random variable $\eta$.

The integral in the right-hand side of (1.7) is an $n$-fold *Stieltjes integral* (W5, Sec. 2.5). We recall that the ordinary Stieltjes integral

$$\int_a^b f(x)\, dF(x)$$

is defined as the limit of the quantity

$$\sum_{k=1}^{n} f(x_k')[F(x_k) - F(x_{k-1})],$$

as $\max |x_k - x_{k-1}| \to 0$, where

$$a = x_0 < x_1 < \cdots < x_{n-1} < x_n = b, \qquad x_{k-1} \leqslant x_k' \leqslant x_k.$$

The improper integral

$$\int_{-\infty}^{\infty} f(x)\, dF(x)$$

is defined in the usual way as the limit

$$\lim_{a \to -\infty,\, b \to \infty} \int_a^b f(x)\, dF(x).$$

Multiple Stieltjes integrals are defined similarly. For example,

$$\int_a^b \int_c^d f(x, y)\, dF(x, y)$$

is the limit of the quantity

$$\sum_{k=1}^{n}\sum_{l=1}^{m} f(x_k', y_l')[F(x_k, y_l) - F(x_{k-1}, y_l) - F(x_k, y_{l-1}) + F(x_{k-1}, y_{l-1})]$$

as $\max |x_k - x_{k-1}| \to 0$, $\max |y_l - y_{l-1}| \to 0$, where

$$a = x_0 < x_1 < \cdots < x_{n-1} < x_n = b, \qquad x_{k-1} \leqslant x_k' \leqslant x_k,$$
$$c = y_0 < y_1 < \cdots < y_{m-1} < y_m = d, \qquad y_{l-1} \leqslant y_l' \leqslant y_l,$$

and the improper integral

$$\int_{-\infty}^{\infty} \int_{-\infty}^{\infty} f(x, y)\, dF(x, y)$$

is defined as the limit

$$\lim_{\substack{a \to -\infty, b \to \infty \\ c \to -\infty, d \to \infty}} \int_a^b \int_c^d f(x, y)\, dF(x, y).$$

In the case where the probability distribution (1.3) has a density

$$f_{t_1, t_2, \ldots, t_n}(x_1, x_2, \ldots, x_n) = \frac{\partial^n F_{t_1, t_2, \ldots, t_n}(x_1, x_2, \ldots, x_n)}{\partial x_1 \partial x_2 \ldots \partial x_n},$$

the integral (1.7) can be written in the form

$$\mu_{m_1 m_2 \ldots m_n} = \int_{-\infty}^{\infty} \int_{-\infty}^{\infty} \cdots \int_{-\infty}^{\infty} x_1^{m_1} x_2^{m_2} \ldots x_n^{m_n} f_{t_1, t_2, \ldots, t_n}(x_1, x_2, \ldots, x_n)\, dx_1\, dx_2 \ldots dx_n.$$

For our present purposes, it is quite sufficient to consider only this simplest case, without using Stieltjes integrals. However, later we shall encounter situations where the use of Stieltjes integrals is indispensable (see e.g., Sec. 9).

The simplest of the moments (1.7) is the first moment

$$\mu_1(t) = m(t) = \mathbf{E}\xi(t) = \int_{-\infty}^{\infty} x\, dF_t(x), \tag{1.8}$$

the *mean value* of the random function $\xi(t)$. If $\xi(t)$ is stationary, then the distribution function $F_t(x)$ does not depend on $t$, and then the mean value of $\xi(t)$ is obviously just a numerical constant:

$$\mathbf{E}\xi(t) = m. \tag{1.9}$$

The mean value is a very important characteristic of a random function $\xi(t)$, but it only describes the coarsest properties of $\xi(t)$. A much more precise description of $\xi(t)$ is given by its second moment

$$\mathbf{E}\xi(t)\xi(s) = \mu_{11}(t, s) = B(t, s). \tag{1.10}$$

The function $B(t, s)$ is called the *correlation function* of $\xi(t)$, and the theory which takes into account only those properties of random functions which are determined by their first and second moments is called the *correlation theory* (of random functions).[4] If $\xi(t)$ is stationary, then obviously the correlation function $B(t, s)$ depends only on the difference $t - s$, i.e.,

$$\mathbf{E}\xi(t)\xi(s) = B(t - s). \tag{1.11}$$

Thus, in the correlation theory, stationary random functions are characterized by one constant $m$ and one function $B(\tau)$ of a single variable $\tau$, which takes integral values for random sequences and arbitrary real values for random processes.

---

[4] Other terminology is often encountered in the literature. For example, the function $B(t, s)$, or else the *centered* function $B(t, s) - m(t)m(s)$, is often called the *covariance* (*function*), and the term correlation function, or *autocorrelation function* (cf. footnote 1, p. 78) is reserved for a *stationary* covariance. What we call correlation theory is often called *second-order theory*, or the theory of *second-order random functions*.

The fact that the multidimensional distribution functions (1.3) play no role in the correlation theory suggests that we modify the definition of stationarity used in this theory. In fact, in the correlation theory, it is natural to call a random function stationary provided that its mean value (1.8) is a constant and that its correlation function (1.10) depends only on $t - s$. Random functions satisfying these two conditions are said to be *stationary in the wide sense* (or *stationary in Khinchin's sense*).[5] In the rest of this book, the word *stationary* will always be meant in this sense. It is clear that in general, the condition (1.6) may not hold for functions which are stationary in the wide sense. (See e.g., the example in K6 of a stationary process with a given correlation function.) However, we note that in practice, one almost never encounters random functions which are stationary in the wide sense but which are not stationary in the sense of the definition given in Sec. 2.

It should be emphasized that the mean value and the correlation function obviously do not specify the random function $\xi(t)$ uniquely. Therefore, the correlation theory cannot replace a complete theory of random functions which uses the multidimensional distributions (1.3). However, at present, the correlation theory of stationary random functions is the only theory sufficiently developed to be broadly applicable in practice. Moreover, the practical value of the correlation theory is still growing because of the fact that the random functions encountered in practice very often turn out to be *normal* (or *Gaussian*), which means that all their finite-dimensional distribution functions (1.3) are multidimensional normal (or Gaussian). (For a discussion of normal distributions, see e.g., C5, G3.) For normal random functions $\xi(t)$, the mean value and the correlation function completely specify $\xi(t)$, i.e., they determine all the distribution functions (1.3). Therefore, in principle, the correlation theory can answer any question pertaining to a normal random function $\xi(t)$. However, as we shall see below, even in the case of random functions $\xi(t)$ which are not normal, the correlation theory can answer a whole series of questions which are important in the applications. It should also be noted that for normal random functions, the concepts of stationarity in the wide sense and of stationarity in the *strict sense* (i.e., in the sense of the definition given in Sec. 2) are exactly the same: A normal random function always has finite moments $m(t)$ and $B(t, s)$, and if $m(t)$ is independent of $t$ and $B(t, s)$ depends only on $t - s$, it can be shown that all the multidimensional distribution functions, which in this case are normal, satisfy the condition (1.6).

Finally, we make a further remark concerning normal random functions.

---

[5] We note that the statement of these two conditions presupposes that the random function $\xi(t)$ has a finite mean value and a finite correlation function, i.e., that for every value of $t$, $\xi(t)$ is a random variable with finite variance. In this sense, the new definition of stationarity is actually somewhat less general than the old definition, since with the old definition, the first and second moments may not exist. Henceforth, it will be assumed tacitly that the mean value and the correlation function of $\xi(t)$ are finite.

In a great many mechanical and physical problems, the probability distributions (1.3) can be regarded as normal, either because of the *central limit theorem* of probability theory (C5, G3, L8) [for example, this is the case in studying various phenomena associated with the "shot effect" (K7)] or because of the normality of Maxwell's distribution of molecular velocities [as an example of this case, we cite the theory of "thermal noise" (M2)]. For criteria which permit one to verify the normality of random functions, see the paper by Bunimovich and Leontovich (B14). We note in passing that sometimes the normality of the distribution (1.3) is taken for granted in physical problems without adequate justification, often even without any need to do so.[6]

## 4. The Ergodic Theorem

In the applications of probability theory, one ordinarily deals with phenomena which repeat themselves many times. Hence, as the mean value of a random variable $\xi$ characterizing an actual observed phenomenon, we can take the arithmetic mean of all the observed values $\xi^{(j)}$ of $\xi$. If we follow this procedure, then to determine the mean value $m(t)$ and the correlation function $B(t, s)$ of a random function $\xi(t)$, we must first take a large number $N$ of realizations of $\xi(t)$, written $\xi^{(1)}(t), \xi^{(2)}(t), \ldots, \xi^{(N)}(t)$, and then calculate the arithmetic mean (with respect to $j$) of $\xi^{(j)}(t)$ for every value of $t$, or the arithmetic mean of $\xi^{(j)}(t)\xi^{(j)}(s)$ for every pair of values of $t$ and $s$. However, in practice, observation of a random function and the subsequent processing of the data usually turn out to be quite complicated, and therefore it is very desirable to be able to get along with as small a number of realizations as possible. Indeed, the practical value of the correlation theory of *stationary* random functions is to a considerable extent due to the fact that if $\xi(t)$ is stationary, its mean value $m$ and its correlation function $B(\tau)$ can usually be easily calculated by using just one realization of $\xi(t)$, of the kind illustrated by Figure 1.

The possibility of calculating these characteristics of a stationary random function from a single realization is a consequence of the fact that the so-called *ergodic theorem*[7] (or *law of large numbers*) is applicable to stationary random functions (or at least to those usually encountered in practice).

---

[6] A random function may be accurately normal as far as its first-order distribution is concerned, but may not be normal as far as its second and higher-order distributions are concerned. An important example is furnished by the turbulent velocity field (see e.g., B3, Chap. 8). Another example is constructed in S5.

[7] The term *ergodic theorem* is borrowed from statistical mechanics. In probability theory, theorems of the same type come under the heading of the *law of large numbers*. In order to avoid confusion, we note that the term *ergodic theorem* often refers to a result considerably more general than the result given here (which is sufficient for our limited purposes). Cf. footnote 10, p. 21.

According to the ergodic theorem, the mathematical expectation of both the quantity $\xi(t)$ and the quantity $\xi(t + \tau)\xi(t)$, obtained by averaging the corresponding quantities over the whole space $\Omega$ of experimental outcomes, can be replaced by the *time averages* of the same quantities.   More precisely, *if $\xi(t)$ is a stationary random sequence (here t is an integer) satisfying certain very general conditions to be indicated below (which are almost always met in practice), then, with a suitable definition of the limit of a sequence of random variables, the following limiting relations are valid:*

$$m = \mathbf{E}\xi(t) = \lim_{N \to \infty} \frac{1}{N + 1} \sum_{t=0}^{N} \xi(t), \tag{1.12}$$

$$B(\tau) = \mathbf{E}\xi(t + \tau)\xi(t) = \lim_{N \to \infty} \frac{1}{N + 1} \sum_{t=0}^{N} \xi(t + \tau)\xi(t). \tag{1.13}$$

Similarly, *if $\xi(t)$ is a stationary random process (here t is a continuous variable), satisfying certain quite general conditions, then*

$$m = \mathbf{E}\xi(t) = \lim_{T \to \infty} \frac{1}{T} \int_0^T \xi(t) \, dt, \tag{1.14}$$

$$B(\tau) = \mathbf{E}\xi(t + \tau)\xi(t) = \lim_{T \to \infty} \frac{1}{T} \int_0^T \xi(t + \tau)\xi(t) \, dt, \tag{1.15}$$

*where the integrals in* (1.14) *and* (1.15) *are defined as the limits of the corresponding approximating sums.*   (Here, we again use the concept of the limit of a sequence of random variables.)

In probability theory, there are several different "reasonable" definitions of the limit of a sequence of random variables (see e.g., L8, pp. 151, 157). To make the statement of the ergodic theorem explicit, we have to choose one of these definitions.   In the correlation theory, the following definition is the most appropriate : *The random variable $\xi$ is the limit of the sequence of random variables $\xi_1, \xi_2, \ldots, \xi_n, \ldots$ if*

$$\lim_{n \to \infty} \mathbf{E}|\xi - \xi_n|^2 = 0, \tag{1.16}$$

*i.e., if for any $\varepsilon > 0$, there exists $N = N(\varepsilon)$ such that*

$$\mathbf{E}|\xi - \xi_n|^2 < \varepsilon \quad \text{if} \quad n > N. \tag{1.17}$$

A limit defined in this way is usually called a *mean square limit*, or a *limit in the mean (square)*, and $\xi_n$ is said to *converge to $\xi$ in the mean (square)*.   Henceforth, by the limit of a sequence of random variables, we shall always understand a limit of this kind.

Using *Chebyshev's inequality* [8]

$$\mathbf{P}\{|\xi_n - \xi| \geqslant \varepsilon\} \leqslant \frac{\mathbf{E}|\xi_n - \xi|^2}{\varepsilon^2},$$

we see that (1.16) implies that the relation

$$\lim_{n\to\infty} \mathbf{P}\{|\xi_n - \xi| \geqslant \varepsilon\} = 0 \qquad (1.18)$$

holds for any $\varepsilon > 0$. Equation (1.18) means that given any $\varepsilon > 0$ and $\eta > 0$, there exists $N_0 = N_0(\varepsilon, \eta)$ such that

$$\mathbf{P}\{|\xi_n - \xi| < \varepsilon\} > 1 - \eta \quad \text{for} \quad n > N_0. \qquad (1.19)$$

The relation (1.19) makes it particularly clear why it is natural to call the random variable $\xi$ satisfying (1.16) the *limit* of the sequence $\xi_1, \xi_2, \ldots,$ $\xi_n, \ldots$. A random variable $\xi$ satisfying (1.18) is usually called the *limit in probability* of the sequence of random variables $\xi_1, \xi_2, \ldots, \xi_n, \ldots$, and the result just proved can be stated as follows: *If $\xi$ is the limit in the mean of the sequence $\xi_1, \xi_2, \ldots, \xi_n, \ldots$, then $\xi$ is also the limit in probability of $\xi_1, \xi_2, \ldots,$ $\xi_n, \ldots$*

According to (1.19), the relation (1.12) means that for any $\varepsilon > 0$ and $\eta > 0$, and for all $N$ greater than some $N_0$, we have

$$P\left\{\left|\frac{1}{N+1}\sum_{t=0}^{N}\xi(t) - m\right| < \varepsilon\right\} > 1 - \eta. \qquad (1.20)$$

This is the usual formulation of the law of large numbers as applied to dependent random variables $\xi(t)$. Formula (1.20) gives us grounds for believing that for sufficiently large $N$, we have

$$m \approx \frac{1}{N+1}\sum_{t=0}^{n}\xi^{(1)}(t), \qquad (1.21)$$

where $\approx$ denotes approximate equality, and $\xi^{(1)}(t)$ is a realization of the stationary sequence $\xi(t)$, i.e., a set of observed values of the sequence. Similarly, the relation (1.13) allows us to make an approximate calculation of the correlation function $B(\tau)$ of a stationary sequence, by using the formula

$$B(\tau) \approx \frac{1}{N+1}\sum_{t=0}^{N}\xi^{(1)}(t+\tau)\xi^{(1)}(t). \qquad (1.22)$$

---

[8] Chebyshev's inequality follows at once from

$$\mathbf{E}|\xi_n - \xi|^2 = \int_{-\infty}^{\infty} x^2 \, dF(x) \geqslant \left(\int_{-\infty}^{-\varepsilon} + \int_{\varepsilon}^{\infty}\right) x^2 \, dF(x) \geqslant \varepsilon^2 \left(\int_{-\infty}^{-\varepsilon} + \int_{\varepsilon}^{\infty}\right) dF(x)$$
$$= \varepsilon^2 \mathbf{P}\{|\xi_n - \xi| \geqslant \varepsilon\},$$

where $F(x)$ is the distribution function of the random variable $\xi_n - \xi$.

In order to make use of the formulas (1.14) and (1.15), pertaining to a stationary process $\xi(t)$, we first replace the integrals in (1.14) and (1.15) by their approximating sums, noting that

$$\int_0^T \xi(t)\, dt = \lim_{N \to \infty} \frac{T}{N} \sum_{k=1}^{N} \xi\left(k\,\frac{T}{N}\right),$$

$$\int_0^T \xi(t + \tau)\xi(t)\, dt = \lim_{N \to \infty} \frac{T}{N} \sum_{k=1}^{N} \xi\left(k\,\frac{T}{N} + \tau\right)\xi\left(k\,\frac{T}{N}\right).$$

(1.23)

It then follows from (1.14), (1.15) and (1.23) that the mean value $m$ and the correlation function $B(\tau)$ of the stationary random process $\xi(t)$ can be calculated approximately by using the formulas

$$m \approx \frac{1}{N} \sum_{k=1}^{N} \xi^{(1)}(k\Delta), \qquad B(\tau) \approx \frac{1}{N} \sum_{k=1}^{N} \xi^{(1)}(k\Delta + \tau)\xi^{(1)}(k\Delta), \qquad (1.24)$$

where $\Delta$ is a small time interval, and $N$ is chosen so as to make $N\Delta = T$ sufficiently large. In practice, $\Delta$ should be chosen in such a way that the function $\xi^{(1)}(t)$ does not change appreciably during time intervals of length $\Delta$, while the numbers $N$ appearing in (1.21), (1.22) and (1.24) should be such that a further increase in the number of terms has only a slight influence on the value of the arithmetic mean being calculated.

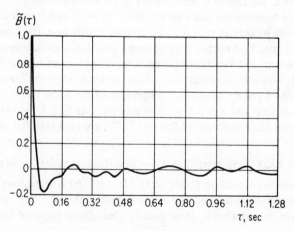

FIG. 2. The normalized correlation function for fading of radio signals (J2).

The tedious calculations of correlation functions based on formulas (1.22) and (1.24) can be greatly simplified by using computing machines specially designed to perform such calculations. In recent years, many such machines have been described in the literature (see e.g., B11, D1, S3, S6, S12). As an example of these calculations, in Figure 2 we show the *normalized* correlation function

$$\tilde{B}(\tau) = \frac{B(\tau)}{B(0)}$$

for fading of radio signals, calculated from the curve given in Figure 1(a), by using formula (1.24) with $\Delta = 0.016$ sec.

*We now discuss the conditions which have to be imposed on a stationary random function $\xi(t)$ in order to guarantee the validity of the ergodic theorem just stated. Consider the *centered* correlation function

$$R(\tau) = \mathbf{E}[\xi(t + \tau) - m][\xi(t) - m] = B(\tau) - m^2, \qquad (1.25)$$

which differs only by the factor $R(0)$ from the *correlation coefficient* (see e.g., G3, p. 200) of the random variables $\xi(t + \tau)$ and $\xi(t)$. As was first shown by Slutski (S9) [see also D6, O3], formulas (1.14) and (1.12) hold, with our previous definition of the limit of a sequence of random variables, if and only if

$$\lim_{T \to \infty} \frac{1}{T} \int_0^T R(\tau) \, d\tau = 0 \qquad (1.26)$$

and

$$\lim_{N \to \infty} \frac{1}{N + 1} \sum_{\tau=0}^{N} R(\tau) = 0, \qquad (1.27)$$

respectively.

In practice, the function $R(\tau)$ usually approaches zero as $\tau \to \infty$, since the dependence between the random variables $\xi(t + \tau)$ and $\xi(t)$ usually becomes progressively weaker as $\tau \to \infty$. Of course, in this case, the condition (1.26) or (1.27) is met.[9] Another case of some practical significance occurs when $R(\tau)$ is the sum of a function approaching zero at $\infty$ and of several periodic terms (e.g., sinusoidal oscillations), as for instance, when the random function $\xi(t)$ contains a purely harmonic component of the form $\xi_0 e^{i\lambda t}$, where $\xi_0$ is a random variable and $\lambda$ is a fixed real number (see Sec. 8 below). It is easy to verify that the conditions (1.26) and (1.27) are also satisfied in this case.

In order for the relation (1.15) or (1.13) to hold as well, the function

$$R_1(\tau) = \mathbf{E}[\xi(t + \tau_0 + \tau)\xi(t + \tau) - B(\tau_0)][\xi(t + \tau_0)\xi(t) - B(\tau_0)] \qquad (1.28)$$

must satisfy the condition (1.26) or (1.27). [If $\xi(t)$ is stationary in the wide sense, then the requirement that $R_1(\tau)$ depend only on $\tau$ has to be imposed as an extra assumption.] It is usually clear from physical considerations

---

[9] A proof of (1.14) under the assumption that

$$\lim_{\tau \to \infty} R(\tau) = 0 \qquad (1.26')$$

is essentially contained in the considerations given in Sec. 60 of Leontovich's book (L4), and a quite explicit proof is given in Sec. 8 of Bunimovich's book (B13). By making the last part of these arguments a bit more precise, we can replace the condition (1.26') by the condition (1.26), thereby proving Slutski's ergodic theorem.

that the dependence between the random variables $\xi(t+\tau_0+\tau)\xi(t+\tau)$ and $\xi(t+\tau_0)\xi(t)$ "wears off" as $\tau \to \infty$, so that

$$\lim_{\tau \to \infty} R_1(\tau) = 0.$$

In this case, the ergodic theorem for the second moments (1.15) and (1.13) is certainly valid.

It should be noted that in general, $R_1(\tau)$ involves fourth-order moments of the original function $\xi(t)$, and hence, in the general case, the conditions for the validity of the ergodic theorem for second moments cannot be expressed in terms of the correlation function. It is only in the special (but very important) case where the random function $\xi(t)$ is normal that its fourth moments can be expressed in terms of $B(\tau)$ and $m$. We can assume for simplicity that $m = 0$, since if $m \neq 0$ in practice, the formulas (1.15) and (1.13) are usually applied to the random function $\xi(t) - m$, which has mean value zero. Then $R(\tau)$ reduces to $B(\tau)$, and straightforward calculations show that if $\xi(t)$ is normal, the conditions (1.26) and (1.27) involving $R_1(\tau)$ are equivalent to the conditions

$$\lim_{T\to\infty} \frac{1}{T} \int_0^T |R(\tau)|^2 \, d\tau \equiv \lim_{T\to\infty} \frac{1}{T} \int_0^T |B(\tau)|^2 \, d\tau = 0 \qquad (1.29)$$

and

$$\lim_{N\to\infty} \frac{1}{N+1} \sum_{\tau=0}^{N} |R(\tau)|^2 \equiv \lim_{N\to\infty} \frac{1}{N+1} \sum_{\tau=0}^{N} |B(\tau)|^2 = 0, \qquad (1.30)$$

respectively. Thus, in the case where $\xi(t)$ is a normal stationary random function of time, with mean value zero, the conditions (1.29) and (1.30) are necessary and sufficient for the validity of (1.15) and (1.13). It can also be shown that if these conditions are met, then *all* higher moments of $\xi(t)$ equal the corresponding time averages (see G4, M1).[10]   It is clear that if

$$\lim_{\tau \to \infty} R(\tau) = 0 \qquad (1.26')$$

(cf. footnote 9) then the conditions (1.29) and (1.30) are met, but if $R(\tau)$ contains periodic components, then (1.29) and (1.30) are no longer valid.[11]

---

[10] In fact, in this case, if $\Phi(\xi_1, \xi_2, \ldots, \xi_n)$ is any measurable function such that

$$\mathbf{E}|\Phi(\xi(\tau_1), \xi(\tau_2), \ldots, \xi(\tau_n))| < \infty,$$

then the time average of $\Phi(\xi(t + \tau_1), \xi(t + \tau_2), \ldots, \xi(t + \tau_n))$ equals the corresponding mathematical expectation. Random functions which have this property are often said to be *ergodic*.

[11] By using the spectral representation of the correlation function $R(\tau)$, as in Secs. 10 and 12 below, we can formulate the conditions (1.26) and (1.27) as continuity conditions for the corresponding spectral distribution function $F(\lambda)$ at the point $\lambda = 0$, while (1.29) and (1.30) become continuity conditions for $F(\lambda)$ on the whole line $-\infty < \lambda < \infty$ and the whole interval $-\pi \leqslant \lambda \leqslant \pi$, respectively.

The existence of the limits appearing in (1.12) and (1.14) was proved by Khinchin (K6), in the early stages of the development of the theory of stationary random functions, and his proof imposes no restrictions on $\xi(t)$. A more precise result of this type is the famous ergodic theorem of Birkhoff and Khinchin (see G3, Sec. 59, and K9), which also deals with the existence of time averages, but starts from a stricter definition of convergence of a sequence of random variables (i.e., *almost sure convergence*). For further ergodic theorems using this kind of convergence, see B5, D3, D6, L8.*

## 5. Differentiation and Integration of a Random Process. Properties of the Correlation Function

Since the mean value $m$ of a stationary random function $\xi(t)$ can be easily measured, then, instead of $\xi(t)$ itself, it is legitimate to consider the new random function $\xi(t) - m$, sometimes called the "fluctuation" of $\xi(t)$, which has mean value zero. Henceforth we shall assume that all stationary random functions $\xi(t)$ under consideration have mean value zero. Thus, the only characteristic of $\xi(t)$ which we shall use will be the correlation function $B(\tau)$.

If $\xi(t)$ is a random sequence, the argument of the correlation function $B(\tau)$ is an integer, i.e., $B(\tau)$ is a sequence of numbers

$$\ldots, B(-2), B(-1), B(0), B(1), B(2), \ldots.$$

If $\xi(t)$ is a random process, the argument of $B(\tau)$ takes all real values. In the latter case, we shall consider only $\xi(t)$ such that $B(\tau)$ is a continuous function. [It should be noted that a sufficient condition for $B(\tau)$ to be continuous for *all* $\tau$ is that $B(\tau)$ be continuous at the point $\tau = 0$.[12]] It follows at once from the formula

$$\mathbf{E}|\xi(t + h) - \xi(t)|^2 = 2\,B(0) - 2\,B(h),$$

that if $B(\tau)$ is continuous, the random process $\xi(t)$ is *mean square continuous* (or *continuous in the mean*) in the sense that

$$\lim_{h \to 0} \mathbf{E}|\xi(t + h) - \xi(t)|^2 = 0$$

for all $t$, and hence,

$$\lim_{h \to 0} \mathbf{P}\{|\xi(t + h) - \xi(t)| \geqslant \varepsilon\} = 0$$

for all $t$ and any $\varepsilon > 0$.

---

[12] According to Schwarz's inequality (applied to the operator $\mathbf{E}$),

$$|\mathbf{E}\eta\zeta|^2 \leqslant \mathbf{E}|\eta|^2\,\mathbf{E}|\zeta|^2,$$

where $\eta$ and $\zeta$ are any two random variables. If we now let $\eta = \xi(\tau + h) - \xi(\tau), \zeta = \xi(0)$, we find that

$$|B(\tau + h) - B(\tau)|^2 \leqslant 2\,B(0)[B(0) - B(h)],$$

so that $B(\tau)$ is continuous everywhere if it is continuous at $\tau = 0$.

By using the correlation function, we can give simple conditions for the existence of the derivative and integral of a stationary random process $\xi(t)$. We shall say that the random process $\xi(t)$ is (*mean square*) *differentiable* at the point $t$, if given any sequence of numbers $h_1, h_2, \ldots, h_n, \ldots$ converging to zero, the sequence of random variables

$$\frac{\xi(t + h_j) - \xi(t)}{h_j}$$

converges (in the mean) to a unique random variable, which we call the *derivative* of $\xi(t)$ at the point $t$, and denote by $\xi'(t)$. It is not hard to show that a stationary random process $\xi(t)$ is differentiable for any $t$ provided that its correlation function $B(\tau)$ has a continuous second derivative with respect to $\tau$, in which case $\xi'(t)$ is also a continuous stationary process, with correlation function $B^{[1]}(\tau) = -B''(\tau)$. In fact, if $B(\tau)$ has a continuous second derivative, then it can easily be verified that

$$\lim_{h \to 0, \, h' \to 0} \mathbf{E} \left| \frac{\xi(t + h) - \xi(t)}{h} - \frac{\xi(t + h') - \xi(t)}{h'} \right|^2 = 0,$$

and moreover

$$\lim_{h \to 0} \mathbf{E} \left[ \frac{\xi(t + \tau + h) - \xi(t + \tau)}{h} \frac{\xi(t + h) - \xi(t)}{h} \right] = -B''(\tau),$$

which implies that $-B''(\tau)$ is the correlation function of the derivative $\xi'(t)$. It should also be noted that the correlation function $B(\tau)$ has a continuous second derivative for any $\tau$ if it has a continuous second derivative at $\tau = 0$.[13]

The integral

$$\int_a^b f(t)\xi(t) \, dt, \tag{1.31}$$

where $f(t)$ is a numerical function and $\xi(t)$ is a stationary process, is defined as the limit (in the mean) of the corresponding approximating sum.[14] The improper integral

$$\int_{-\infty}^{\infty} f(t)\xi(t) \, dt$$

is defined in the usual way as the limit of (1.31) as $a \to -\infty$ and $b \to \infty$. Then, it is not hard to show that a sufficient condition for the existence of the

---

[13] To see this, repeat the argument in footnote 12, with $\eta = \xi'(\tau + h) - \xi'(\tau)$ and $\zeta = \xi'(0)$.

[14] I.e., the sum

$$\sum_{j=1}^{n} f(t_k')\xi(t_k')(t_k - t_{k-1})$$

where $a = t_0 < t_1 < \cdots < t_{n-1} < t_n = b$, $t_{k-1} \leqslant t_k' \leqslant t_k$.

integral (1.31), where the limits of integration are permitted to be infinite, is that the double integral

$$\int_a^b \int_a^b B(t - s)f(t)f(s) \, dt \, ds \tag{1.32}$$

exists, where (1.32) is the mathematical expectation of the square of (1.31). The argument resembles that given in connection with the existence of the derivative of $\xi(t)$.

The following properties of the correlation function $B(\tau)$ are simple consequences of the definition (1.11):

(a)                                                        $B(0) > 0,$

(b)                                               $B(-\tau) = B(\tau),$ $\qquad$ (1.33)

(c)                                               $|B(\tau)| \leqslant B(0).$

[We assume that $\xi(t)$ does not vanish identically.] Properties (a) and (b) are obvious, and property (c) follows from the inequality

$$\mathbf{E}[\xi(t + \tau) \pm \xi(t)]^2 = \mathbf{E}[\xi^2(t + \tau) \pm 2\,\xi(t + \tau)\xi(t) + \xi^2(t)] \geqslant 0,$$

which implies that $B(0) \geqslant \pm B(\tau)$. Formula (1.11) also implies that the inequality

$$\sum_{j, k=1}^n B(\tau_j - \tau_k)a_j a_k \geqslant 0 \tag{1.34}$$

holds for any $n$ real numbers $a_1, a_2, \ldots, a_n$ and any $\tau_1, \tau_2, \ldots, \tau_n$ (integers or arbitrary real numbers, depending on whether we are dealing with a sequence or a process). This follows at once from the obvious relation

$$\sum_{j, k=1}^n B(\tau_j - \tau_k)a_j a_k = \sum_{j, k=1}^n \mathbf{E}\xi(\tau_j)\xi(\tau_k)a_j a_k$$

$$= \mathbf{E}\left[\sum_{j=1}^n \xi(\tau_j)a_j\right]^2 \geqslant 0.$$

A function $B(\tau)$ for which (1.34) holds for any $n, a_1, \ldots, a_n, \tau_1, \ldots, \tau_n$ is said to be *nonnegative definite* (or often simply *positive definite*). Thus, we have just shown that a correlation function must be nonnegative definite. The converse result is also true, i.e., every nonnegative definite function of a real (or integral) argument is the correlation function of a stationary random process (or sequence). We omit the proof due to Khinchin and Kolmogorov (K6),[15] which can be found, for example, in G3, Sec. 57.

---

[15] In K6, Khinchin considers only stationary random processes (i.e., it is assumed that $\tau$ is an arbitrary real number), but the proof given carries over without modification to the case of stationary sequences (see e.g., W7).

## 6. Complex Random Functions. Geometric Interpretation

Before considering examples of stationary random functions, we discuss two topics which are essential for our subsequent work:

1. So far, we have been dealing exclusively with real random variables and real random functions. However, in theoretical studies it is often more convenient to consider the more general case of *complex random variables* and *complex random functions*. A complex random variable $\xi$ can be defined as the sum

$$\xi = \eta + i\zeta,$$

where $\eta$ and $\zeta$ are real random variables. Similarly, a complex random function $\xi(t)$ can be defined as the sum

$$\xi(t) = \eta(t) + i\zeta(t),$$

where $\eta(t)$ and $\zeta(t)$ are real random functions.

A complex random function $\xi(t)$ is specified by the set of distribution functions of the $2n$ real random variables

$$\eta(t_1), \ldots, \eta(t_n), \zeta(t_1), \ldots, \zeta(t_n)$$

for all possible values of $n$, $t_1, t_2, \ldots, t_n$. The function $\xi(t)$ is said to be *stationary* if all these distribution functions are invariant under arbitrary shifts of the time origin, i.e., if they do not change when the same arbitrary number $\tau$ is added to all the arguments $t_1, \ldots, t_n$. It is clear that in this case the functions $\eta(t)$ and $\zeta(t)$, the real and imaginary parts of $\xi(t)$, are real-valued stationary random functions.

The mean value $m$ of a complex stationary random function $\xi(t)$ is obviously a (complex) constant, and without loss of generality, we can restrict ourselves to the case of functions $\xi(t)$ for which

$$E\xi(t) = m = 0.$$

As for the second moments of $\xi(t)$, the most important for our purposes is the following expression, which in the stationary case is a function of $\tau$, but not of $t$:

$$\begin{aligned} B(\tau) &= E\xi(t + \tau)\overline{\xi(t)} \qquad\qquad\qquad\qquad\qquad (1.35)\\ &= E[\eta(t + \tau)\eta(t) + \zeta(t + \tau)\zeta(t)] - iE[\eta(t + \tau)\zeta(t) - \zeta(t + \tau)\eta(t)]. \end{aligned}$$

(The overbar denotes the complex conjugate.) We call the function $B(\tau)$ the *correlation function* of the complex stationary random function $\xi(t)$. In particular, if $\xi(t)$ is real, this definition obviously reduces to the definition (1.11). $B(\tau)$ is a complex function satisfying the first and third of the relations (1.33). However, instead of the second relation (1.33) and the inequality (1.34), which hold in the real case, we now have

$$B(-\tau) = \overline{B(\tau)} \qquad\qquad\qquad\qquad\qquad (1.36)$$

and

$$\sum_{j,\,k=1}^{n} B(\tau_j - \tau_k) a_j \bar{a}_k \geqslant 0, \tag{1.37}$$

where $\tau_1, \tau_2, \ldots, \tau_n$ are arbitrary integers or real numbers [depending on whether $\xi(t)$ is a sequence or a process], and $a_1, a_2, \ldots, a_n$ are arbitrary *complex* numbers.  As before, a complex function $B(\tau)$ for which (1.37) holds is said to be *nonnegative definite* (or simply *positive definite*).

From now on, we shall assume that the stationary random functions under consideration are complex, with zero mean values.  However, as a rule we shall choose examples involving only real random functions, since they are the kind of random functions encountered in most applications.

2. In studying random variables and random functions, it is frequently very convenient to use geometric notions and geometric language.  First, we consider random variables associated with an experiment which only has a finite number $n$ of different outcomes (e.g., the toss of a die).  Let $p_1, p_2, \ldots, p_n$ denote the probabilities of these $n$ outcomes.  Then, a random variable $\xi$ is a quantity which takes certain numerical values $\xi^{(1)}, \xi^{(2)}, \ldots, \xi^{(n)}$ depending on the outcome of the experiment.  Quantities of this kind can be regarded as $n$-dimensional vectors with components $\xi^{(1)}, \xi^{(2)}, \ldots, \xi^{(n)}$.  The set of all random variables associated with our experiment is then an $n$-dimensional vector space $H_n$, which is real or complex, depending on the nature of the random variables under consideration.

To make this geometric model more meaningful, we still have to introduce a metric in the space $H_n$, i.e., we have to define a "length" or *norm* $\|\xi\|$ of the vector $\xi$.  The most appropriate definition of the norm $\|\xi\|$ is

$$\|\xi\|^2 = \sum_{j=1}^{n} p_j [\xi^{(j)}]^2, \tag{1.38}$$

or in the complex case

$$\|\xi\|^2 = \sum_{j=1}^{n} p_j |\xi^{(j)}|^2.$$

In other words, we let the probabilities $p_j$ give the units along the $n$ coordinate axes of our $n$-dimensional space.  Then, the *scalar product* $(\xi, \eta)$ of two vectors

$$\xi = (\xi^{(1)}, \xi^{(2)}, \ldots, \xi^{(n)}) \quad \text{and} \quad \eta = (\eta^{(1)}, \eta^{(2)}, \ldots, \eta^{(n)})$$

should be defined by

$$(\xi, \eta) = \sum_{j=1}^{n} p_j \xi^{(j)} \eta^{(j)}, \tag{1.39}$$

or in the complex case, by

$$(\xi, \eta) = \sum_{j=1}^{n} p_j \xi^{(j)} \overline{\eta^{(j)}}.$$

In probability theory language, (1.38) and (1.39) mean that the square of the norm of a vector and the scalar product of two vectors are given by the following mathematical expectations:

$$\|\xi\|^2 = \mathbf{E}\xi^2, \qquad (\xi, \eta) = \mathbf{E}\xi\eta. \tag{1.40}$$

Thus, the metric just introduced has a simple probabilistic meaning.

This geometric interpretation of random variables can be extended in a natural way to the case where the set of all outcomes of the experiment is infinite. Here, the random variables $\xi$ can again be thought of as vectors in a "space of random variables" $H$, which however now has an infinite number of dimensions. The vectors in $H$ can be added and multiplied by numbers (scalars), since random variables can be added and multiplied by numbers. The square of the length of a vector, and the scalar product of two vectors in $H$ are defined by the formulas

$$\|\xi\|^2 = \mathbf{E}|\xi|^2, \qquad (\xi, \eta) = \mathbf{E}\xi\bar{\eta}, \tag{1.41}$$

by analogy with (1.40). (However, we now consider the complex case from the outset.) It is easy to see that these definitions of the square of the length of a vector and the scalar product of two vectors have the usual properties of the same quantities in a finite-dimensional vector space. Moreover, the fact that the space $H$ is in general infinite-dimensional (i.e., is a *Hilbert space*) only slightly restricts the use of ordinary geometric notions as applied to $H$ (see e.g., A2). In particular, we note that in Hilbert space, just as in a finite-dimensional space, we can drop a unique perpendicular from any point $P$ to a given linear subspace, and the length of this perpendicular is the shortest distance between the point $P$ and points of the subspace (A2, pp. 8–10). This fact is very important for us, since, as we shall see later, the problem of extrapolating and filtering stationary random functions reduces to the problem of dropping a perpendicular from a point of $H$ to a linear subspace of $H$.

From our standpoint, a random sequence is a sequence of vectors in a Hilbert space $H$, while a random process is a one-parameter family of vectors in $H$, in other words, a *curve* in $H$.[16] According to (1.10), the correlation function

---

[16] Strictly speaking, in order to be able to talk about a *curve* in the ordinary sense of the word, we must also require that the following continuity condition hold: The distance between the points $\xi(t)$ and $\xi(t')$ of the curve must approach zero as $t' \to t$. It is easy to see that this condition is equivalent to the condition that the process $\xi(t)$ be continuous in the mean. In general, the statement that a sequence of random variables $\xi_1, \xi_2, \ldots, \xi_n, \ldots$ converges in the mean to a random variable $\xi$ has a very simple geometric interpretation, i.e., the distance between the point $\xi$ of the space $H$ and the variable point $\xi_n$ goes to zero as $n \to \infty$.

$B(t, s)$ is just the scalar product of the vectors $\xi(t)$ and $\xi(s)$, i.e., $B(t, s)$ is a simple geometric characteristic of the corresponding sequence or curve. Thus, in a certain sense, the correlation theory of random processes is equivalent to the theory of curves in a Hilbert space $H$. In particular, stationary random processes correspond to curves in $H$ for which the scalar product $(\xi(t), \xi(s))$ depends only on $t - s$. It is not hard to see that geometrically this means that there exists a one-parameter group of rotations in $H$ which carry the curves into themselves. (In three dimensions, only circles have this property.)

# 2

---

# EXAMPLES OF STATIONARY
# RANDOM FUNCTIONS.
# SPECTRAL REPRESENTATIONS

---

## 7. Examples of Stationary Random Sequences

*Example 1.* The simplest example of a stationary random sequence is a sequence $\eta(t)$, $t = \ldots, -2, -1, 0, 1, 2, \ldots$ of independent random variables with identical probability distributions. Let the mean value $E\eta(t)$ of the random variables $\eta(t)$ equal 0, and let their variance

$$E\eta^2(t) - [E\eta(t)]^2 = E\eta^2(t)$$

equal 1. Then, the correlation function of the sequence $\eta(t)$ equals

$$B(\tau) = E\eta(t + \tau)\overline{\eta(t)} = \begin{cases} 1 & \text{for } \tau = 0, \\ 0 & \text{for } \tau \neq 0. \end{cases} \tag{2.1}$$

In the general case, the elements $\eta(t)$ of a stationary sequence with mean value zero and correlation function (2.1) may not be independent. Formula (2.1) only implies that the random variables $\eta(t)$ are *uncorrelated*. For this reason, we shall refer to a sequence of the type considered here as a *sequence of uncorrelated random variables*.[1] From the standpoint of the Hilbert space $H$ of Sec. 6, a sequence of uncorrelated random variables is an orthonormal

---

[1] We note that in the special case of *normal* stationary sequences (Sec. 3), lack of correlation is equivalent to independence.

sequence of vectors, i.e., a sequence of mutually orthogonal vectors of length 1.

**Example 2.** Let

$$\xi(t) = \frac{\eta(t) + \eta(t - 1) + \cdots + \eta(t - m + 1)}{\sqrt{m}} \tag{2.2}$$

$$(t = -2, -1, 0, 1, 2, \ldots),$$

where $\eta(t)$ is a stationary sequence of uncorrelated random variables. It is clear that $\xi(t)$ is also a stationary random sequence, with mean value zero. Moreover, as is easily verified, in this case the correlation function is

$$B(\tau) = \mathbf{E}\xi(t + \tau)\overline{\xi(t)} = \begin{cases} \dfrac{m - |\tau|}{m} & \text{for } |\tau| \leqslant m, \\ 0 & \text{for } |\tau| > m, \end{cases} \tag{2.3}$$

i.e., as $\tau$ increases, $B(\tau)$ falls off linearly from 1 to 0.

**Example 3.** A generalization of the stationary sequence of Example 2 is given by the sequence

$$\xi(t) = \sum_{k=0}^{n} a_k \eta(t - k) \qquad (t = \ldots, -2, -1, 0, 1, 2, \ldots), \tag{2.4}$$

where the $a_k$ are given complex numbers, and $\eta(t)$ is the sequence of Example 1. The correlation function of this sequence has the form

$$B(\tau) = \mathbf{E}\xi(t + \tau)\overline{\xi(t)} = \sum_{\substack{0 \leqslant k \leqslant n \\ 0 \leqslant k - \tau \leqslant n}} a_k \bar{a}_{k-\tau}. \tag{2.5}$$

The number $n$ in (2.4) may also be infinite. Then, for the infinite sum

$$\xi(t) = \sum_{k=0}^{\infty} a_k \eta(t - k) \tag{2.6}$$

to be convergent, it is only necessary that

$$\sum_{k=0}^{\infty} |a_k|^2 < \infty, \tag{2.7}$$

and then (2.7) equals $\mathbf{E}|\xi(t)|^2$. Moreover, the index $k$ in the expressions (2.4) and (2.6) can also take negative values. We obtain the most general stationary sequence of the type considered here by setting

$$\xi(t) = \sum_{k=-\infty}^{\infty} a_k \eta(t - k), \quad \text{where} \quad \sum_{k=-\infty}^{\infty} |a_k|^2 < \infty. \tag{2.8}$$

In this case, we obviously have

$$B(\tau) = \mathbf{E}\xi(t + \tau)\overline{\xi(t)} = \sum_{k=-\infty}^{\infty} a_k \bar{a}_{k-\tau}. \tag{2.9}$$

Stationary sequences of the form (2.4), (2.6) or (2.8), obtained from a sequence of uncorrelated random variables in the way indicated, will be called *moving averages*. It is clear that in this way we can generate a very large class of stationary random sequences.[2]

## 8. Examples of Stationary Random Processes

Unfortunately, in the case of processes, we do not have a simple analog for a sequence of uncorrelated random variables $\eta(t)$, such as was used in Sec. 7 to give a simple construction of a variety of stationary sequences. (See, however, Sec. 14 below.)   Therefore, to construct examples of stationary processes, we must use other methods.

*Example 1.* Some interesting examples of stationary random processes can be obtained by considering analytic expressions containing random parameters.   The simplest of all such expressions is given by the formula

$$\xi(t) = \xi f(t), \tag{2.10}$$

where $\xi$ is a random variable, and $f(t)$ is a numerical function of $t$ ($-\infty < t < \infty$). Since we have agreed to consider only stationary processes for which $\mathbf{E}\xi(t) = 0$, and since $\mathbf{E}\xi f(t) = f(t)\mathbf{E}\xi$, we must require that

$$\mathbf{E}\xi = 0. \tag{2.11}$$

Moreover, for $\xi(t)$ to be stationary, it is necessary that the function

$$\mathbf{E}\xi(t + \tau)\overline{\xi(t)} = f(t + \tau)\overline{f(t)}\mathbf{E}|\xi|^2 \tag{2.12}$$

be independent of $t$, i.e., that $f(t + \tau)\overline{f(t)}$ be independent of $t$.   Thus, setting $\tau = 0$, we find that

$$|f(t)|^2 = r^2 = \text{const}, \qquad f(t) = re^{i\varphi(t)}, \tag{2.13}$$

where $r$ is a real number and $\varphi(t)$ is a real function of $t$.   Substituting (2.13) into the product $f(t + \tau)\overline{f(t)}$, we find that the difference $\varphi(t + \tau) - \varphi(t)$ must not depend on $t$.   If we assume that the function $\varphi(t)$ is differentiable, this implies that

$$\frac{d}{dt}[\varphi(t + \tau) - \varphi(t)] = 0, \qquad \varphi'(t + \tau) = \varphi'(t). \tag{2.14}$$

---

[2] The problem of how to characterize the class of sequences which can be represented in the form (2.6) or in the form (2.8), where $\eta(t)$ is a sequence of uncorrelated random variables, is solved in K12.   The solution uses the concept of the spectral distribution function of the stationary sequence (see Sec. 12 below), and is basic for the whole theory of extrapolation of stationary sequences, as developed by Kolmogorov.

Since $\tau$ must be arbitrary, this means that

$$\varphi'(t) = \lambda = \text{const}, \qquad \varphi(t) = \lambda t + \theta, \qquad (2.15)$$

so that we must have

$$f(t) = re^{i(\lambda t + \theta)}.$$

The numerical factor $re^{i\theta}$ can now be included in the random variable $\xi$, i.e., the product $\xi re^{i\theta}$ will be denoted simply by the single symbol $\xi$. Thus, finally, the random process (2.10) is stationary if and only if it has the form

$$\xi(t) = \xi e^{i\lambda t}, \qquad (2.16)$$

where $\xi$ is a complex random variable with mean value zero, and $\lambda$ is a real constant.

The random process (2.16) obviously describes a periodic oscillation of angular frequency $\lambda$, with random amplitude and random phase. The real part of each realization of (2.16) is a sinusoidal oscillation of the form

$$\eta^{(1)}(\tau) = a \sin (\lambda t + \psi),$$

where $a$ and $\psi$ vary from one realization to another. The correlation function of (2.16) has the form

$$B(\tau) = \mathbf{E}\xi(t + \tau)\overline{\xi(t)} = \mathbf{E}|\xi|^2 e^{i\lambda\tau} = be^{i\lambda\tau}, \qquad (2.17)$$

where $b = \mathbf{E}|\xi|^2$ is the mathematical expectation of the square of the amplitude, and is proportional to the average energy of the oscillations per unit time, i.e., to their average power.[3] We note that the correlation function $B(\tau)$ does not depend at all on the statistical characteristics of the phase of the oscillation.

If we regard the random process $\xi(t)$ as a function $\xi(\omega, t)$ of two variables, i.e., the element $\omega$ of the "space of elementary events" and the real argument $t$ (see Sec. 1), then the method used here to construct an example of a stationary random process can be regarded as completely analogous to the method of separation of variables, so widely used in mathematical physics for constructing particular solutions of partial differential equations. In fact, if we explicitly indicate the dependence on $\omega$, we can write (2.10) as

$$\xi(\omega, t) = \xi(\omega)f(t), \qquad (2.18)$$

where $\xi(\omega)$ is a (time-independent) random variable, and $f(t)$ is a numerical function of $t$. In mathematical physics, one often succeeds in representing a general solution as a superposition of particular solutions, obtained by the

---

[3] If $\xi(t)$ is a velocity undergoing random oscillations, then to within a constant factor, $b$ is the average kinetic energy, if $\xi(t)$ is a fluctuating current or voltage, then $b$ is proportional to the average electrical power, etc. We shall usually regard $\xi(t)$ as a current or voltage and $b$ as a power.

method of separation of variables. In the theory of stationary random processes, the principle of superposition can also be used to represent stationary processes of a general type (see Examples 2, 3 and Sec. 9 below).

*Example 2.* We can obtain further examples of stationary random processes by forming linear combinations of processes of the type (2.16). For example, consider the process

$$\xi(t) = \xi_1 e^{i\lambda_1 t} + \xi_2 e^{i\lambda_2 t}, \quad \lambda_1 \neq \lambda_2, \tag{2.19}$$

where $\xi_1$ and $\xi_2$ are random variables with mean zero. Clearly, we have $E\xi(t) = 0$, and moreover

$$\begin{aligned} E\xi(t + \tau)\overline{\xi(t)} &= E[\xi_1 e^{i\lambda_1(t+\tau)} + \xi_2 e^{i\lambda_2(t+\tau)}][\bar{\xi}_1 e^{-i\lambda_1 t} + \bar{\xi}_2 e^{-i\lambda_2 t}] \\ &= E|\xi_1|^2 e^{i\lambda_1 \tau} + E\xi_1 \bar{\xi}_2 e^{i(\lambda_1-\lambda_2)t+i\lambda_1\tau} \\ &\quad + E\bar{\xi}_1\xi_2 e^{-i(\lambda_1-\lambda_2)t+i\lambda_2\tau} + E|\xi_2|^2 e^{i\lambda_2\tau}. \end{aligned}$$

If the process (2.19) is to be stationary, the last expression must be independent of $t$. Because of the linear independence of the functions $e^{i(\lambda_1-\lambda_2)t}$, $e^{-i(\lambda_1-\lambda_2)t}$ and 1, this expression can be independent of $t$ only if

$$E\xi_1\bar{\xi}_2 = E\bar{\xi}_1\xi_2 = 0. \tag{2.20}$$

Thus, the random process (2.19) is stationary if and only if $\xi_1$ and $\xi_2$ are uncorrelated random variables with mean value zero. In this case, $\xi(t)$ will be a superposition of two uncorrelated (possibly independent) oscillations of different frequency, with random amplitudes and phases. The correlation function of (2.19) is

$$B(\tau) = E|\xi_1|^2 e^{i\lambda_1 \tau} + E|\xi_2|^2 e^{i\lambda_2 \tau} = b_1 e^{i\lambda_1 \tau} + b_2 e^{i\lambda_2 \tau} \quad (b_1 > 0, b_2 > 0), \tag{2.21}$$

where $b_1$ and $b_2$ are the average powers of the individual oscillations. Note that $B(\tau)$ is completely independent of the phases of these oscillations.

In this example, the process $\xi(t)$ can be real, whereas in Example 1, the process is obviously always complex. For $\xi(t)$ to be real, it is only necessary that $\lambda_2 = -\lambda_1 = -\lambda$, and $\xi_2 = \bar{\xi}_1$, e.g.,

$$\xi_1 = \tfrac{1}{2}(\eta - i\zeta), \quad \xi_2 = \tfrac{1}{2}(\eta + i\zeta).$$

In this case, the process (2.19) can be written in the form

$$\xi(t) = \eta \cos \lambda t + \zeta \sin \lambda t, \tag{2.22}$$

where $\xi$ and $\eta$ are real random variables. Then, (2.20) is equivalent to the two formulas

$$E\xi^2 = E\eta^2 = b, \quad E\xi\eta = 0, \tag{2.23}$$

and (2.21) becomes

$$B(\tau) = b \cos \lambda\tau. \tag{2.24}$$

*Example 3.* In the same way, we can construct processes of the form

$$\xi(t) = \sum_{k=1}^{n} \xi_k e^{i\lambda_k t}, \tag{2.25}$$

which are superpositions of $n$ periodic oscillations of different frequencies, where

$$\mathbf{E}\xi_1 = \mathbf{E}\xi_2 = \cdots \mathbf{E}\xi_n = 0.$$

Just as in Example 2, it can be shown that the process (2.25) is stationary if and only if

$$\mathbf{E}\xi_k \bar{\xi}_l = 0 \quad \text{for} \quad k \neq l. \tag{2.26}$$

In this case, the correlation function has the form

$$B(\tau) = \sum_{k=1}^{n} \mathbf{E}|\xi_k|^2 e^{i\lambda_k \tau} = \sum_{k=1}^{n} b_k e^{i\lambda_k \tau} \qquad (b_k > 0). \tag{2.27}$$

The coefficients $b_k$ specify the average powers of the separate harmonic oscillations appearing in (2.25). Setting $\tau = 0$ in (2.27), we obtain the formula

$$B(0) = \mathbf{E}|\xi(t)|^2 = \sum_{k=1}^{n} b_k, \tag{2.28}$$

which shows that in a superposition of uncorrelated periodic oscillations, the average power of the composite oscillation equals the sum of the average powers of the separate periodic components.

For the process (2.25) to be real, the number $n$ must be even, equal to $2m$, say, and the terms in the sum (2.25) must separate into $m$ pairs of complex conjugate terms. In this case, the process $\xi(t)$ can be rewritten in the form

$$\xi(t) = \sum_{j=1}^{m} (\eta_j \cos \lambda_j t + \zeta_j \sin \lambda_j t), \tag{2.29}$$

where, according to (2.26),

$$\mathbf{E}\eta_j \zeta_k = 0 \quad \text{for all} \quad j, k,$$
$$\mathbf{E}\eta_j \eta_k = \mathbf{E}\zeta_j \zeta_k = 0 \quad \text{for} \quad j \neq k,$$
$$\mathbf{E}\eta_j^2 = \mathbf{E}\zeta_j^2 = b_j$$

[cf. (2.23)]. The correlation function (2.27) then becomes

$$B(\tau) = \sum_{j=1}^{m} b_j \cos \lambda_j \tau. . \tag{2.30}$$

In formula (2.25), we can also set $n = \infty$. Then, if we are to be able to talk about a correlation function for the process $\xi(t)$, the series (2.28) must converge, i.e., we must have

$$\sum_{k=1}^{\infty} \mathbf{E}|\xi_k|^2 = \sum_{k=1}^{\infty} b_k < \infty. \tag{2.31}$$

If this condition holds, the series

$$\xi(t) = \sum_{k=1}^{\infty} \xi_k e^{i\lambda_k t}, \tag{2.32}$$

where $E\xi_k = 0$, $E\xi_k \bar{\xi}_l = 0$ for $k \neq l$ will converge (with the usual mean square interpretation for convergence of sequences of random variables), so that formula (2.32) is meaningful.   The correlation function $B(\tau)$ of the process (2.32) is [4]

$$B(\tau) = \sum_{k=1}^{\infty} b_k e^{i\lambda_k \tau}. \tag{2.33}$$

In the real case, we can write formulas (2.32) and (2.33) in a form similar to (2.29) and (2.30).

A stationary process of the form (2.25) or (2.32) is called a *process with a discrete spectrum*, and the set of numbers $\lambda_1, \lambda_2, \ldots, \lambda_n$ is called the *spectrum* of the process.   It is clear from formula (2.33) that in this case, the spectrum can be determined from the correlation function of the process.   In fact, the spectrum consists of the numbers $\lambda$ for which

$$\lim_{T \to \infty} \frac{1}{2T} \int_{-T}^{T} B(\tau) e^{-i\lambda \tau} \, d\tau \neq 0. \tag{2.34}$$

Moreover, the correlation function gives the mean values $b_k$ of the squares of the amplitudes $|\xi_k|$ of the harmonic components $\xi_k e^{i\lambda_k t}$ of the process $\xi(t)$. In the special case where it is known that a process $\xi(t)$ with a discrete spectrum is normal, it can be shown that each of the amplitudes $|\xi_k|$ of the separate harmonic components of $\xi(t)$ must have a so-called *Rayleigh distribution*, with a probability density of the form

$$p(x) = \frac{2x}{b_k} e^{-x^2/b_k} \qquad (x \geqslant 0),$$

while all the phases $\varphi_k = \arg \xi_k$ must be uniformly distributed from 0 to $2\pi$. Then, it is clear that the correlation function (2.33) will uniquely determine $\xi(t)$.

Formula (2.33) for the correlation function of a stationary process with a discrete spectrum was considerably generalized in Khinchin's important paper K6.   In fact, Khinchin proved that the correlation function of *any* stationary random process can be represented in the form of an integral

$$B(\tau) = \int_{-\infty}^{\infty} e^{i\lambda \tau} \, dF(\lambda), \tag{2.35}$$

---

[4] It was shown by Slutski (S9) that the converse is true, i.e., every stationary random process with a correlation function of the form (2.33) can be represented in the form (2.32), with $E\xi_k \bar{\xi}_l = 0$ for $k \neq l$.  This result of Slutski is a special case of the general spectral representation theorem which we shall discuss later.

where $F(\lambda)$ is a real nondecreasing bounded function, and conversely, every function of the form (2.35), where $F(\lambda)$ has the properties just specified, is the correlation function of a stationary random process.   [It is clear that (2.35) is a generalization of formula (2.33) derived above.]   This theorem of Khinchin plays a very important role in the general theory of stationary random functions, and is vital in most applications of the theory.   We shall discuss the theorem in detail in Sec. 10 below.   However, to get a clearer idea of the meaning of (2.35), it is best to consider first the spectral representation of stationary random processes themselves, although historically this representation was proved only after the appearance of Khinchin's result, and in fact its proof was based on this result.

## 9. Spectral Representation of Stationary Processes

Stationary processes with a discrete spectrum are far from being the only kind of stationary processes.   However, it turns out that any stationary random process can be obtained as the limit of a sequence of processes with discrete spectra, where these latter processes can even be chosen to have the special form (2.25).   In fact, it can be shown that given any stationary process $\xi(t)$, any $\varepsilon > 0$ (however small) and any $T > 0$ (however large), there exist random variables $\xi_1, \xi_2, \ldots, \xi_n$ which are uncorrelated in pairs (the number $n$ depends on $\varepsilon$ and $T$, of course) and real numbers $\lambda_1, \lambda_2, \ldots, \lambda_n$ such that

$$\mathbf{E}\left|\xi(t) - \sum_{k=1}^{n} \xi_k e^{i\lambda_k t}\right|^2 < \varepsilon \tag{2.36}$$

for any $t$ in the interval $-T \leqslant t \leqslant T$.   In particular, this implies that given any stationary process $\xi(t)$ and any $\varepsilon > 0, \eta > 0, T > 0$, there exist random variables $\xi_1, \xi_2, \ldots, \xi_n$ and real numbers $\lambda_1, \lambda_2, \ldots, \lambda_n$ such that $\mathbf{E}\xi_k\bar{\xi}_l = 0$ for $k \neq l$ and

$$\mathbf{P}\left\{\left|\xi(t) - \sum_{k=1}^{n} \xi_k e^{i\lambda_k t}\right| < \varepsilon\right\} > 1 - \eta \tag{2.36'}$$

for any $t$ for which $|t| \leqslant T$.   [Cf. the transition from (1.17) to (1.19) in Sec. 4.]

Formulas (2.36) and (2.36') show that every stationary random process $\xi(t)$ can be approximated arbitrarily closely by a linear combination of harmonic oscillations of the form (2.16).   Moreover, to improve the accuracy and reliability of this approximation [i.e., to decrease the numbers $\varepsilon$ and $\eta$ in formula (2.36')] and to increase the time during which it can be applied [i.e., to increase $T$], we must in general increase $n$, the number of harmonic components in the linear combination, and decrease all the differences $\lambda_{k+1} - \lambda_k$ between neighboring frequencies appearing in its spectrum.   If we make $\varepsilon$

arbitrarily small in formula (2.36), or if we make $\varepsilon$ and $\eta$ arbitrarily small in formula (2.36'), and if at the same time we make $T$ arbitrarily large, then in the general case the number of frequencies $\lambda_k$ lying in any interval $\Delta\lambda$ of values of $\lambda$ will become arbitrarily large. It turns out that when this limiting process is performed, the sum

$$\sum_{\Delta\lambda} \xi_k$$

of the uncorrelated random variables $\xi_k$ corresponding to frequencies $\lambda_k$ belonging to some interval $\Delta\lambda$ converges to a definite random variable (depending on $\Delta\lambda$, of course), which we denote by $Z(\Delta\lambda)$. This fact is the basic content of the spectral representation theorem for stationary processes, and allows us to represent the stationary process $\xi(t)$ in the form of the following *Fourier-Stieltjes integral*:

$$\xi(t) = \int_{-\infty}^{\infty} e^{i\lambda t} \, Z(d\lambda). \tag{2.37}$$

The integral (2.37) means the limit

$$\lim_{\Lambda \to \infty} \int_{-\Lambda}^{\Lambda} e^{i\lambda t} Z(d\lambda), \tag{2.38}$$

where

$$\int_{-\Lambda}^{\Lambda} e^{i\lambda t} Z(d\lambda) = \lim_{\max|\lambda_k - \lambda_{k-1}| \to 0} \sum_{k=1}^{n} e^{i\lambda_k' t} Z(\Delta_k \lambda). \tag{2.39}$$

In the right-hand side of (2.39), the summation is over all subintervals $\Delta_k\lambda = [\lambda_{k-1}, \lambda_k]$ appearing in the partition

$$-\Lambda = \lambda_0 < \lambda_1 < \cdots < \lambda_{n-1} < \lambda_n = \Lambda$$

of the interval $[-\Lambda, \Lambda]$, and $\lambda_k'$ is an arbitrary point in $\Delta_k\lambda$.

The function $Z(\Delta\lambda)$ associates a random variable with each interval $\Delta\lambda$, i.e., $Z(\Delta\lambda)$ is a *random interval function*. From the very definition of $Z(\Delta\lambda)$, it is clear that $Z(\Delta\lambda)$ has the following properties:

(a) $$EZ(\Delta\lambda) = 0$$

for all $\Delta\lambda$;

(b) $$Z(\Delta_1\lambda + \Delta_2\lambda) = Z(\Delta_1\lambda) + Z(\Delta_2\lambda)$$

if $\Delta_1\lambda$ and $\Delta_2\lambda$ are disjoint intervals (*additivity*);

(c) $$EZ(\Delta_1\lambda)\overline{Z(\Delta_2\lambda)} = 0$$

if $\Delta_1\lambda$ and $\Delta_2\lambda$ are disjoint intervals.

Instead of the random interval function $Z(\Delta\lambda)$, we can also consider the *random point function* [5]

$$Z(\lambda) = Z([-\infty, \lambda]).$$

Then, the integral representation (2.37) can be written in the more usual form

$$\xi(t) = \int_{-\infty}^{\infty} e^{i\lambda t}\, dZ(\lambda), \tag{2.40}$$

standing for the same limit as (2.38) and (2.39), which can now be written as

$$\int_{-\infty}^{\infty} e^{i\lambda t}\, dZ(\lambda) = \lim_{\Lambda \to \infty} \left\{ \lim_{\max|\lambda_k - \lambda_{k-1}| \to 0} \sum_{k=1}^{n} e^{i\lambda_k' t}[Z(\lambda_k) - Z(\lambda_{k-1})] \right\}. \tag{2.41}$$

(Cf. the definition of the ordinary Stieltjes integral given in Sec. 3.)   In the new notation, conditions (a) and (c) become

(a')                               $$EZ(\lambda) = 0$$

for all $\lambda$;

(c')      $$E[Z(\lambda_1 + \Delta\lambda_1) - Z(\lambda_1)]\overline{[Z(\lambda_2 + \Delta\lambda_2) - Z(\lambda_2)]} = 0$$

if the intervals $(\lambda_1, \lambda_1 + \Delta\lambda_1)$ and $(\lambda_2, \lambda_2 + \Delta\lambda_2)$ are disjoint.   Random functions satisfying the condition (c') are called *random functions with uncorrelated increments*.

In the case where the process $\xi(t)$ is real, the random interval function $Z(\Delta\lambda)$ is obviously such that $Z(\Delta^*\lambda) = \overline{Z(\Delta\lambda)}$, where $\Delta^*\lambda$ denotes the interval symmetric to $\Delta\lambda$ with respect to the point $\lambda = 0$.   Then, formula (2.40) can be rewritten in the form

$$\xi(t) = \int_0^{\infty} \cos \lambda t\, dZ_1(\lambda) + \int_0^{\infty} \sin \lambda t\, dZ_2(\lambda), \tag{2.42}$$

analogous to (2.29) where $Z_1(\lambda)$ and $Z_2(\lambda)$ are real random functions of the variable $\lambda$ (which takes only positive values) such that

$$E[Z_j(\lambda_1 + \Delta\lambda_1) - Z_j(\lambda_1)][Z_k(\lambda_2 + \Delta\lambda_2) - Z_k(\lambda_2)] = 0 \tag{2.43}$$

if $j \neq k$, or if $j = k$ and the intervals $(\lambda_1, \lambda_1 + \Delta\lambda_1)$, $(\lambda_2, \lambda_2 + \Delta\lambda_2)$ are disjoint, and

$$E[Z_1(\lambda + \Delta\lambda) - Z_1(\lambda)]^2 = E[Z_2(\lambda + \Delta\lambda) - Z_2(\lambda)]^2. \tag{2.44}$$

The representation of a stationary random process $\xi(t)$ in the form of the integral (2.40) [or (2.37), (2.42)], where $Z(\lambda)$ has the properties (a') and (c'), is called the *spectral representation* of $\xi(t)$.   The possibility of such a representation for an arbitrary stationary process was first shown by Kolmogorov

---

[5] By $[-\infty, \lambda]$ is meant the infinite half-line consisting of the points $\lambda'$ for which $-\infty < \lambda' \leqslant \lambda$.

(K10, K12), where the results were formulated in terms of the geometry of the Hilbert space $H$ (see Sec. 6), and were derived by using certain results from the spectral theory of operators. In recent years, an extensive literature has appeared which interprets and justifies this representation in a probability theory context (see e.g., B2, B8, C4, D6, G5, K2, L8, M1, R5).

*We now show the method which leads most directly to a proof the of spectral representation of $\xi(t)$. By analogy with the inversion formula for Fourier-Stieltjes integrals of numerical functions (see e.g., the inversion formula for characteristic functions in G3, Sec. 36), we can suppose that if the expansion (2.40) holds, then the function $Z(\lambda)$ is given in terms of $\xi(t)$ by the formula

$$Z(\lambda) = \lim_{T \to \infty} \frac{1}{2\pi} \int_{-T}^{T} \frac{e^{-i\lambda t} - 1}{-it} \xi(t)\, dt. \tag{2.45}$$

Now let $\xi(t)$ be a stationary random process, and define the random function $Z(\lambda)$ by formula (2.45). In order to make this formula meaningful, we still have to show that the integral from $-T$ to $T$ appearing in (2.45) exists (this is easily shown by using the results of Sec. 5) and that as $T \to \infty$, the random function of $T$ defined by this integral converges in the mean to a random variable $Z(\lambda)$. It is not hard to prove the last assertion by using Khinchin's formula (2.35), about which more will be said in Sec. 10.[6] In fact, it follows at once from (2.35) that for $T' > T$, we have

$$\mathbf{E} \left| \int_{-T'}^{T'} \frac{e^{-i\lambda t} - 1}{-2\pi i t} \xi(t)\, dt - \int_{-T}^{T} \frac{e^{-i\lambda t} - 1}{-2\pi i t} \xi(t)\, dt \right|^2$$

$$= \mathbf{E} \left| \int_{T < |t| < T'} \frac{e^{-i\lambda t} - 1}{-2\pi i t} \xi(t)\, dt \right|^2$$

$$= \int_{T < |t| < T'} \int_{T < |s| < T'} \frac{e^{-i\lambda t} - 1}{-2\pi i t} \frac{e^{i\lambda s} - 1}{2\pi i s} \mathbf{E}\xi(t)\overline{\xi(s)}\, dt\, ds \tag{2.46}$$

$$= \int_{T < |t| < T'} \int_{T < |s| < T'} \int_{-\infty}^{\infty} \frac{e^{-i\lambda t} - 1}{-2\pi i t} \frac{e^{i\lambda s} - 1}{2\pi i s} e^{i\lambda'(t-s)}\, dt\, ds\, dF(\lambda')$$

$$= \int_{-\infty}^{\infty} \left| \int_{T < |t| < T'} \frac{e^{-i\lambda t} - 1}{-2\pi i t} e^{i\lambda' t}\, dt \right|^2 dF(\lambda').$$

For simplicity, we shall assume temporarily that the function $F(\lambda)$ is continuous. Then, according to (2.46), the proof of the existence of the limit (2.45) is a consequence of the existence of the limit

$$\Psi(\lambda', \lambda) = \lim_{T \to \infty} \frac{1}{2\pi} \int_{-T}^{T} \frac{e^{i\lambda' t} - e^{i(\lambda' - \lambda)t}}{it}\, dt$$

$$= \frac{1}{\pi} \int_{0}^{\infty} \frac{\sin \lambda' t}{t}\, dt - \frac{1}{\pi} \int_{0}^{\infty} \frac{\sin (\lambda' - \lambda)t}{t}\, dt \tag{2.47}$$

---

[6] As already remarked, formula (2.40) was derived after Khinchin's formula (2.35) was proved (cf. end of Sec. 8). Moreover, most proofs of (2.40) rely on the possibility of representing $B(\tau)$ in the form (2.35).

involving only numerical functions, or, more precisely, a consequence of the fact that $\Psi(\lambda', \lambda)$ exists and is such that

$$\lim_{T \to \infty} \int_{-\infty}^{\infty} |\Psi(\lambda', \lambda) - \Psi_T(\lambda', \lambda)|^2 \, dF(\lambda) = 0, \qquad (2.48)$$

where

$$\Psi_T(\lambda', \lambda) = \frac{1}{2\pi} \int_{-T}^{T} \frac{e^{i\lambda' t} - e^{i(\lambda' - \lambda)t}}{it} \, dt.$$

However, it is well known that

$$\frac{1}{\pi} \int_0^{\infty} \frac{\sin \lambda t}{t} \, dt = \begin{cases} \frac{1}{2} & \text{for } \lambda > 0, \\ 0 & \text{for } \lambda = 0, \\ -\frac{1}{2} & \text{for } \lambda < 0, \end{cases} \qquad (2.49)$$

from which it follows that the limit (2.47) exists and equals

$$\Psi(\lambda', \lambda) = \begin{cases} 1 & \text{for } 0 < \lambda' < \lambda, \\ \frac{1}{2} & \text{for } \lambda' = \lambda > 0 \quad \text{or} \quad \lambda' = 0, \lambda > 0, \\ 0 & \text{for } \lambda' > 0, \lambda' > \lambda \quad \text{or} \quad \lambda' = \lambda = 0 \\ & \qquad \text{or} \quad \lambda' < 0, \lambda' < \lambda, \\ -\frac{1}{2} & \text{for } \lambda' = \lambda < 0 \quad \text{or} \quad \lambda' = 0, \lambda < 0, \\ -1 & \text{for } \lambda < \lambda' < 0, \end{cases} \qquad (2.50)$$

so that (2.48) holds for this $\Psi(\lambda', \lambda)$, as is easily verified. This means that (2.45) actually defines a random function $Z(\lambda)$.

The fact that $Z(\lambda)$ has the property (a') is obvious. Next, we verify that $Z(\lambda)$ has the property (c') and that it satisfies the relation (2.40). Using (2.45), (2.35) and (2.47), and arguing as we did in deriving (2.46), we can easily show that

$$\mathbf{E}[Z(\lambda_2) - Z(\lambda_1)]\overline{[Z(\lambda_4) - Z(\lambda_3)]}$$
$$= \int_{-\infty}^{\infty} \Psi(\lambda - \lambda_1, \lambda_2 - \lambda_1)\overline{\Psi(\lambda - \lambda_3, \lambda_3 - \lambda_4)} \, dF(\lambda). \qquad (2.51)$$

It follows from (2.50) that for $\lambda_2 > \lambda_1$

$$\mathbf{E}|Z(\lambda_2) - Z(\lambda_1)|^2 = F(\lambda_2) - F(\lambda_1) \qquad (2.52)$$

[cf. (2.62) below], and for $\lambda_1 < \lambda_2 \leqslant \lambda_3 < \lambda_4$

$$\mathbf{E}[Z(\lambda_2) - Z(\lambda_1)]\overline{[Z(\lambda_4) - Z(\lambda_3)]} = 0, \qquad (2.53)$$

i.e., $Z(\lambda)$ has the property (c'). Moreover, again using (2.45), (2.35) and (2.47), we find that

$$\mathbf{E}\{\xi(t)\overline{[Z(\lambda_2) - Z(\lambda_1)]}\} = \int_{-\infty}^{\infty} \Psi(\lambda_2 - \lambda, \lambda_2 - \lambda_1)e^{i\lambda t} \, dF(\lambda), \qquad (2.54)$$

so that for $\lambda_2 > \lambda_1$

$$\mathbf{E}\{\xi(t)\overline{[Z(\lambda_2) - Z(\lambda_1)]}\} = \int_{\lambda_1}^{\lambda_2} e^{i\lambda t} \, dF(\lambda). \qquad (2.55)$$

Then, from (2.55), (2.35) and (2.41) [the existence of the limit in (2.41) is easily proved by using (2.52) and (2.53)], we obtain

$$\mathbf{E}\left\{\xi(t)\overline{\int_{-\infty}^{\infty}e^{i\lambda s}\,dZ(\lambda)}\right\} = \lim_{\Lambda\to\infty}\int_{-\Lambda}^{\Lambda}e^{i\lambda(t-s)}\,dF(\lambda)$$

$$= \int_{-\infty}^{\infty}e^{i\lambda(t-s)}\,dF(\lambda) = B(t-s), \tag{2.56}$$

and it follows from (2.52), (2.53), (2.41) and (2.35) that

$$\mathbf{E}\left\{\int_{-\infty}^{\infty}e^{i\lambda t}\,dZ(\lambda)\overline{\int_{-\infty}^{\infty}e^{i\lambda s}\,dZ(\lambda)}\right\} = \int_{-\infty}^{\infty}e^{i\lambda(t-s)}\,dF(\lambda) = B(t-s). \tag{2.57}$$

It is now an easy matter to prove the relation

$$\mathbf{E}\left|\xi(t) - \int_{-\infty}^{\infty}e^{i\lambda t}\,dZ(\lambda)\right|^2 = 0, \tag{2.58}$$

equivalent to (2.40). This completes the proof of the existence of the spectral representation (2.40) in the case where the function $F(\lambda)$ is continuous.

If the function $F(\lambda)$ is not continuous, then, as before, we define the random function $Z(\lambda)$ by using (2.45) at the points of continuity of $F(\lambda)$, and we set

$$Z(\lambda) = \tfrac{1}{2}[Z(\lambda - 0) + Z(\lambda + 0)]$$

at the points of discontinuity of $F(\lambda)$, where $Z(\lambda - 0)$ and $Z(\lambda + 0)$ are defined in the usual way. It is not hard to verify that all the formulas (2.51) to (2.58) remain in force, so that our proof of the existence of the representation (2.40) goes through in this case too.

It should be noted that the function $Z(\lambda)$ is determined by (2.40) and by the properties (a') and (c') only to within an additional term which is a random variable with zero mathematical expectation, and hence we can add any such random variable to the right-hand side of (2.45).*

In the special case where the random interval function $Z(\Delta\lambda)$ is concentrated entirely on a discrete set of points $\lambda_1, \lambda_2, \lambda_3, \ldots$ [i.e., where $Z(\Delta\lambda) = 0$ for all intervals $\Delta\lambda$ that do not contain any of these points], the function $Z(\lambda)$ is a random "jump function," and the spectral representation (2.40) reduces to the representation (2.32) of the process $\xi(t)$ as a superposition of separate uncorrelated harmonic oscillations with random amplitudes and phases. In the general case, the points of discontinuity $\lambda_k$ of the function $Z(\lambda)$ [the "jumps" of $Z(\lambda)$] also correspond to purely periodic components of the process $\xi(t)$, of the form $\xi_k e^{i\lambda_k t}$, but then the process has a "continuous spectrum" as well, which corresponds to the part of the integral (2.40) that does not reduce to a sum of the form (2.32). However, even in the general case, the obvious interpretation of the spectral representation as a "decomposition" of the process into separate pairwise uncorrelated, periodic oscillations helps to clarify the meaning of the integral (2.40).

The real physical meaning of the spectral representation (2.40) is explained by the possibility of separating spectral components corresponding to different parts of the spectrum by using suitably chosen *filters*.   In engineering, a *filter* is a device which passes harmonic oscillations in a certain frequency range (the *pass band*), while suppressing oscillations with different frequencies.   The use of electrical filters is particularly widespread, and they play an important role in almost all radio engineering equipment.   Such a filter is an electrical circuit with four free terminals (a "four-pole"), of which two are called the *input* of the filter, and the other two the *output*.   If a voltage which is a harmonic oscillation of angular frequency $\lambda$ is applied at the filter input, then, depending on the frequency $\lambda$, the output voltage will either be almost exactly the same as the input voltage or else will be practically zero.   The basic types of filters used in practice are *low-pass* filters, passing all oscillations with frequencies less than a certain critical frequency $\lambda_0$ and suppressing all oscillations with frequencies greater than $\lambda_0$, *high-pass* filters, passing all oscillations with frequencies greater than a certain critical frequency and suppressing all oscillations with frequencies less than $\lambda_0$, and *band-pass* filters, passing only oscillations with frequencies $\lambda$ lying in a given interval (pass band) $\lambda_1 \leqslant \lambda \leqslant \lambda_2$.

The reader can find detailed information about these various kinds of filters, including circuit diagrams, in books on circuit theory (see e.g., K5, Chap. 11, and J1 for the elementary theory, and C1 for the advanced theory). In addition to electrical filters, extensive use of mechanical and acoustical filters (see e.g., O5) has become common in recent years.   Analytically, a filter can be characterized by its "weighting function" $s(\tau)$ appearing in the formula

$$\xi_2(t) = \int_0^\infty \xi_1(t - \tau)s(\tau)\, d\tau, \tag{2.59}$$

which gives the function $\xi_2(t)$ at the filter output in terms of the known function $\xi_1(t)$ at the filter input.   The form of the function $s(\tau)$ for a filter which can be regarded as a band-pass filter[7] can be found, for example, in Kolmogorov's article (K16).   Sometimes, the function $s(\tau)$ is called the "impulse response" of the filter.   Another common method of describing filters is to use the "transfer function" $S(\lambda)$ instead of $s(\tau)$, where $S(\lambda)e^{i\lambda t}$ is the filter output when the filter input is just $e^{i\lambda t}$.   It can be shown that $S(\lambda)$ is the Fourier transform of the function $s(\tau)$ [see e.g., B13, D2, J2].

Now let $\xi(t)$ be a real stationary random process with the spectral representation (2.40) or (2.42).   To be explicit, we assume that $\xi(t)$ corresponds to a fluctuating electrical voltage.   If we apply this voltage to the input of an

---

[7] It can be shown that an ideal band-pass filter is theoretically unrealizable.   However, filters can be constructed which approximate a band-pass filter to any preassigned degree of accuracy.

electrical filter with pass band $\lambda_1 \leqslant \lambda \leqslant \lambda_2$, then the output voltage will be a random process $\xi(\lambda_1, \lambda_2; t)$ with spectral representation

$$
\begin{aligned}
\xi(\lambda_1, \lambda_2; t) &= \int_{\lambda_1}^{\lambda_2} \cos \lambda t \, dZ_1(\lambda) + \int_{\lambda_1}^{\lambda_2} \sin \lambda t \, dZ_2(\lambda) \\
&= \int_{\lambda_1}^{\lambda_2} e^{i\lambda t} \, dZ(\lambda) + \int_{-\lambda_2}^{-\lambda_1} e^{i\lambda t} \, dZ(\lambda) = 2 \operatorname{Re} \int_{\lambda_1}^{\lambda_2} e^{i\lambda t} \, dZ(\lambda),
\end{aligned}
\tag{2.60}
$$

where $\operatorname{Re} X$ denotes the real part of $X$. The process (2.60) is the component of $\xi(t)$ corresponding to the spectral interval $[\lambda_1, \lambda_2]$. If the pass band $\Delta \lambda = [\lambda_1, \lambda_2]$ of the filter is taken to be sufficiently narrow, then we can approximate the corresponding spectral component

$$
\xi(\Delta \lambda; t) = 2 \operatorname{Re} \int_{\Delta \lambda} e^{i\lambda t} \, dZ(\lambda)
\tag{2.61}
$$

in any preassigned finite time interval $-T \leqslant t \leqslant T$, with any preassigned degree of accuracy, by a random harmonic oscillation

$$
2 \operatorname{Re} Z(\Delta \lambda) e^{i\lambda t},
$$

where $\lambda$ is an arbitrary point of $\Delta \lambda$. However, it is clear that in the case where $\xi(t)$ has a continuous spectrum, the component $\xi(\Delta \lambda; t)$ will not actually be strictly periodic for any $\Delta \lambda$. In fact, if we single out a particular realization of the process $\xi(\Delta \lambda; t)$, where $\Delta \lambda$ is small, and examine it during two time intervals which are sufficiently far apart, we can verify that although the realization is approximated in both intervals to very great accuracy by a sinusoidal oscillation of frequency $\lambda$, the amplitude and phase of these two "portions" of the oscillation will in general be different.

## 10. Spectral Representation of the Correlation Function

We now return to Khinchin's formula (2.35) for the correlation function $B(\tau)$. It is clear that this formula is an immediate consequence of the spectral representation (2.40) of the process $\xi(t)$ itself. In fact, replacing $\xi(t + \tau)$ and $\xi(t)$ in the formula

$$
B(\tau) = \mathbf{E}\xi(t + \tau)\overline{\xi(t)}
$$

by their expressions as limits of sums of the form (2.41), and then using property (c′) of the random function $Z(\lambda)$, we obtain

$$
B(\tau) = \int_{-\infty}^{\infty} e^{i\lambda t} \, dF(\lambda),
\tag{2.35}
$$

where

$$
F(\lambda + \Delta \lambda) - F(\lambda) = \mathbf{E}|Z(\lambda + \Delta \lambda) - Z(\lambda)|^2,
\tag{2.62}
$$

so that $F(\lambda)$ is a nondecreasing function. Thus, we have not only arrived at the formula (2.35), but we have also related the function $F(\lambda)$ appearing in (2.35) to the random function $Z(\lambda)$, a fact which will be essential in clarifying the meaning of (2.35) [see below].

In deriving (2.35), we made use of the definition of the Stieltjes integral given in Sec. 2. It is clear that in a similar fashion, we can derive the more general formula

$$\mathbf{E}\left\{\int_{-\infty}^{\infty} f(\lambda)\, dZ(\lambda) \overline{\int_{-\infty}^{\infty} g(\lambda)\, dZ(\lambda)}\right\} = \int_{-\infty}^{\infty} f(\lambda)\, \overline{g(\lambda)}\, dF(\lambda), \qquad (2.63)$$

where $F(\lambda)$ is defined by (2.62). (This formula will be needed later.) We also note that (2.62) determines $F(\lambda)$ only to within an arbitrary additive constant, which we choose so that $F(-\infty) = 0$ [see formula (2.67) below]. Of course, the derivation of formula (2.35) given here is not a proof, since we started from formula (2.40), whose proof uses (2.35)! We shall discuss the proof of (2.35) later in this section.

The function $F(\lambda)$ is called the *spectral distribution function* of the stationary random process $\xi(t)$. Because of (2.35), we have

$$B(0) = \mathbf{E}|\xi(t)|^2 = \int_{-\infty}^{\infty} dF(\lambda),$$

which means that

$$\int_{-\infty}^{\infty} dF(\lambda) = F(+\infty) - F(-\infty) < \infty. \qquad (2.64)$$

Thus, the spectral distribution function is a bounded function.

According to formula (2.35), the spectral distribution function $F(\lambda)$ of the process $\xi(t)$ can be easily determined from its correlation function. Conversely, if we know the correlation function $B(\tau)$ of the process $\xi(t)$, we can calculate its spectral distribution function. To do so, we have only to use the formula for inversion of Fourier-Stieltjes integrals [see e.g., G3, Sec. 36 and formula (2.45) of the preceding section]. For our purposes, we only need a special case of this inversion formula, pertaining to the most important situation encountered in practice, i.e., where the absolute value of $B(\tau)$ falls off so rapidly as $|\tau| \to \infty$ that

$$\int_{-\infty}^{\infty} |B(\tau)|\, d\tau < \infty. \qquad (2.65)$$

In this case, $B(\tau)$ can be represented as the Fourier integral

$$B(\tau) = \int_{-\infty}^{\infty} e^{i\lambda\tau} f(\lambda)\, d\lambda. \qquad (2.66)$$

Formula (2.66) is a special case of (2.35) and shows that if (2.65) holds, then the spectral distribution function $F(\lambda)$ can be written in the form

$$F(\lambda) = \int_{-\infty}^{\lambda} f(\lambda)\, d\lambda, \qquad (2.67)$$

where obviously

$$f(\lambda) = F'(\lambda), \tag{2.68}$$

so that the function $F(\lambda)$ is everywhere differentiable. The function $f(\lambda)$ is called the *spectral density* (*function*) of the process $\xi(t)$. Since $F(\lambda)$ is a nondecreasing function, the spectral density is nonnegative everywhere, i.e.,

$$f(\lambda) \geqslant 0.$$

In most applications of the theory, the use of the spectral density allows us to get along without Stieltjes integrals when considering correlation functions. However, in cases where the spectral representation (2.40) of the random process itself is also under consideration, the use of the Stieltjes integral is usually quite unavoidable. In all cases of interest in practice, the random function $Z(\lambda)$ turns out to be nondifferentiable, and then $Z(\lambda)$ certainly cannot be represented in a form similar to (2.67). This is understandable, since a comparison of (2.62) and (2.67) shows that in cases where a continuous nonvanishing spectral density $f(\lambda)$ exists, the mean square increment of the function $Z(\lambda)$ is of order $\Delta\lambda$ when $\Delta\lambda$ is small, i.e., as a rule, the increment of $Z(\lambda)$ itself is of order $\sqrt{\Delta\lambda}$, which is incompatible with the assumption that $Z(\lambda)$ is differentiable.

According to (2.66), the spectral density is the Fourier transform of $B(\tau)$. Hence, if the correlation function $B(\tau)$ is known, the spectral density $f(\lambda)$ can be obtained by using the usual formula for the inversion of a Fourier transformation :

$$f(\lambda) = \frac{1}{2\pi} \int_{-\infty}^{\infty} e^{-i\lambda\tau} B(\tau)\, d\tau. \tag{2.69}$$

Then, we can use (2.67) to find the spectral distribution function $F(\lambda)$ as well. Thus, formulas (2.69) and (2.67), taken together, are precisely the special case of the inversion formula for Fourier-Stieltjes integrals which is applicable when the condition (2.65) holds.

In the case of a real process $\xi(t)$, the spectral distribution function $F(\lambda)$ is an even function of $\lambda$, to within an additive constant, and then (2.35) can be written in the form

$$B(\tau) = \int_0^{\infty} \cos \lambda\tau\, dF_1(\lambda), \tag{2.70}$$

where

$$F_1(\lambda) = 2F(\lambda) + \text{const}, \tag{2.71}$$

or alternatively,

$$\begin{aligned}
F_1(\lambda + \Delta\lambda) - F_1(\lambda) &= 2\mathbf{E}|Z(\lambda + \Delta\lambda) - Z(\lambda)|^2 \\
&= \mathbf{E}[Z_1(\lambda + \Delta\lambda) - Z_1(\lambda)]^2 \\
&= \mathbf{E}[Z_2(\lambda + \Delta\lambda) - Z_2(\lambda)]^2
\end{aligned} \tag{2.72}$$

[cf. (2.42)]. Moreover, if the condition (2.65) holds, the function $B(\tau)$ can be represented in the form (2.66), where in the case of real $\xi(t)$, the function $f(\lambda)$ is even, i.e., $f(-\lambda) = f(\lambda)$. Since $B(\tau)$ is then also an even (real) function, formulas (2.35) and (2.69) can be rewritten in the form

$$B(\tau) = \int_0^\infty \cos \lambda\tau\, f_1(\lambda)\, d\lambda, \tag{2.73}$$

$$f_1(\lambda) = \frac{2}{\pi} \int_0^\infty \cos \lambda\tau\, B(\tau)\, d\tau \tag{2.74}$$

where

$$f_1(\lambda) = F_1'(\lambda) = 2f(\lambda). \tag{2.75}$$

We shall also call the functions $F_1(\lambda)$ and $f_1(\lambda)$ the *spectral distribution function* and the *spectral density (function)*, respectively, of the process $\xi(t)$; they differ

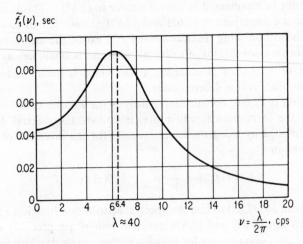

FIG. 3. The normalized spectral density for fading of radio signals (J2). (The ordinary frequency $\nu = \lambda/2\pi$ is plotted as abscissa, instead of the angular frequency $\lambda$.)

from the functions $F(\lambda)$ and $f(\lambda)$ introduced earlier only by the factor 2 and by the fact that they are defined only for $\lambda \geqslant 0$, instead of for $-\infty < \lambda < +\infty$. However, it should be noted that in this book, we shall as a rule use the previous formulas (2.35), (2.66) and (2.69), even when the processes are real.

As an example of the application of formula (2.74), we show in Figure 3 the normalized spectral density

$$\tilde{f}_1(\lambda) = \frac{f_1(\lambda)}{B(0)}$$

for fading of radio signals received in a radar. The function $\tilde{f}_1(\lambda)$ was found by applying formula (2.74) to the correlation function shown in Figure 2.

So far, we have not yet explained how one proves the formula (2.35), to which we have devoted so much attention.   The proof of (2.35) and also of the fact that every $B(\tau)$ which can be represented in the form (2.35), where $F(\lambda)$ is nondecreasing and bounded, is the correlation function of some stationary random process, was derived by Khinchin (K6) as a consequence of the following two facts, taken together:

(a) The class of functions which are correlation functions of stationary random processes coincides with the class of nonnegative definite functions of the variable $\tau$ (cf. the end of Sec. 5);

(b) A function $B(\tau)$ is nonnegative definite if and only if it can be represented in the form (2.35), where $F(\lambda)$ is nondecreasing and bounded.

Shortly before the appearance of Khinchin's paper (K6), the proposition (b) had been published by Bochner (B9), but it was also discovered independently by Khinchin, at about the same time.   Hence, this result is known as the *Bochner theorem*, or the *Bochner-Khinchin theorem*.   (A proof can be found, for example, in G3, Sec. 39 or L8, p. 207.)   It is clear that (a) and (b) immediately imply the two assertions made above concerning correlation functions.[8]

In our treatment, we preferred to point out first the possibility of the spectral representation (2.40) or (2.42) of the random process $\xi(t)$ itself, since this helps to explain the physical meaning of the spectral distribution function. Thus, if $\xi(t)$ is a real stationary random process, it follows from formulas given above that

$$F_1(\lambda + \Delta\lambda) - F_1(\lambda) = \mathbf{E}|\xi(\Delta\lambda; t)|^2, \qquad (2.76)$$

where $\xi(\Delta\lambda; t)$ is given by (2.61).   Therefore, the difference $F_1(\lambda + \Delta\lambda) - F_1(\lambda)$ equals the average power of the spectral components of the process corresponding to the frequency interval $[\lambda, \lambda + \Delta\lambda]$, i.e., to the average energy dissipated per unit time by the aggregate of all oscillations making up $\xi(t)$ which have frequencies between $\lambda$ and $\lambda + \Delta\lambda$.   This makes it clear that the spectral distribution function determines the *frequency distribution of the (average) power of the process*, and hence its derivative $f_1(\lambda)$ can also be called the *power spectral density*.   The functions $F(\lambda)$ and $f(\lambda)$, which in the real case differ from $F_1(\lambda)$ and $f_1(\lambda)$ only by an unimportant constant factor, have a similar meaning.   Formula (2.35) [or (2.37)] is completely analogous to formula (2.27) [or (2.30)] for the correlation function of the sum of a finite number of uncorrelated harmonic oscillations with random amplitudes and phases.   If we set $\tau = 0$ in (2.73), we find that

$$B(0) = \mathbf{E}|\xi(t)|^2 = \int_0^\infty dF_1(\lambda), \qquad (2.77)$$

---

[8] In K6, only the real case is considered, and hence formula (2.73) is derived instead of (2.35).   However, the argument given there can be extended in an obvious way to the complex case.

or if the condition (2.65) holds,

$$B(0) = \mathbf{E}|\xi(t)|^2 = \int_0^\infty f_1(\lambda) \, d\lambda, \tag{2.78}$$

which shows that the total power of the process $\xi(t)$ is obtained by adding up the powers of the separate harmonic components of $\xi(t)$.

The relation (2.76) allows us to make experimental measurements of the spectral distribution function $F_1(\lambda)$ by using a system of filters. In fact, if

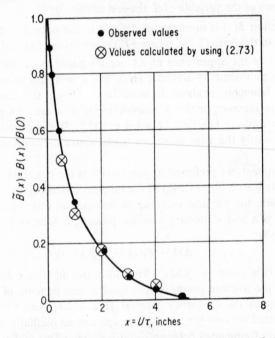

FIG. 4. The normalized correlation function for fluctuations of the longitudinal velocity component of a turbulent flow in a wind tunnel (T2). The abscissa is the time multiplied by the mean velocity $U$, and experimental results obtained for various values of $U$ are plotted as a single curve.[9]

we apply the random process $\xi(t)$, regarded as being a fluctuating electrical voltage or current, to the input of a low-pass filter, which only passes oscillations with frequencies less than $\lambda$, and if to the filter output we attach a wattmeter, which measures the average dissipated power, then the needle of the meter registers just the value $F_1(\lambda)$ [to within a constant of proportion-

---

[9] It should be noted that in practice, one usually calculates $B(x)$, defined as the mean value of the product of the *simultaneous* values of the velocity fluctuations at *two points* of the flow a distance $x$ apart along the axis of the wind tunnel. This replacement of $U\tau$ by $x$ is legitimate if the mean velocity $U$ is much larger than the fluctuational velocity $\xi(t)$.

ality].   Here, it is assumed that $F_1(0) = 0$, and we note that as a rule the time averaging of the instantaneous power will be carried out by the measuring apparatus itself, because of its inertia.   By varying $\lambda$, we obtain $F_1(\lambda)$ as a function of $\lambda$, and then by differentiating $F_1(\lambda)$, we find $f_1(\lambda)$.

Formulas (2.73) and (2.74) can be verified by direct experimental measurement of $F_1(\lambda)$ and $f_1(\lambda)$.   As an example, in Figures 4 and 5, borrowed from Taylor's paper (T2), we show the results of such calculations for the case of

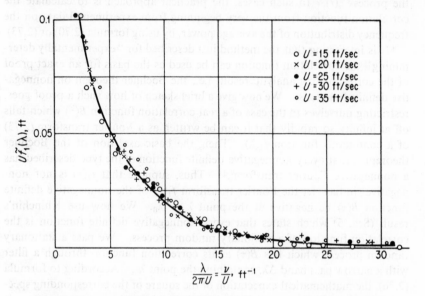

FIG. 5. The normalized spectral density of fluctuations of the longitudinal component of the turbulent velocity field in a wind tunnel (T2).   $U$ is the mean velocity of the flow.   The abscissa is the frequency divided by $U$, while the ordinate is the spectral density multiplied by $U$.   This allows the results of measurements made at different values of $U$ to be plotted as a single curve.

wind tunnel turbulence, where $\xi(t)$ represents the fluctuation of the longitudinal velocity (i.e., the velocity component along the axis of the wind tunnel) at a point in the turbulent flow.   Figure 4 shows the correlation function of the process $\xi(t)$, and Figure 5 shows the spectral density $f_1(\lambda)$ as measured directly by using a system of filters.[10]   In addition, on the curve in Figure 4, we indicate

---

[10] In order to be able to apply electrical filters, the velocity fluctuations $\xi(t)$ must first be converted into fluctuations of electrical current.   This conversion is carried out by using a *hot-wire anemometer*, a device whose basic element is a thin conducting wire located in the turbulent flow.   The change in velocity of the flow past the wire produces changes in the heat exchange of the wire, thereby changing its temperature and resistance, and finally the current in the wire.

a set of points representing the values of the correlation function calculated by using formula (2.73) and the data given in Figure 5. As we see, the agreement between observed and calculated values is completely satisfactory.

It should be noted that in practice, measurement of the spectral distribution function (or spectral density) by using a system of filters often turns out to be much simpler than determination of the correlation function by using formula (1.24), which requires that we record at least one realization of the process $\xi(t)$. In such cases, the practical approach is to calculate the correlation function from the very beginning from experimental data on the frequency distribution of the average power, by using formula (2.70) or (2.73).

*It is interesting that the method just described for "experimentally determining" the correlation function can be used as the basis for an exact proof of the corresponding analytic result, i.e., the Bochner theorem on nonnegative definite functions. We now give a brief sketch of how such a proof goes, restricting ourselves to the case of a real correlation function $B(\tau)$ which falls off at infinity so rapidly that it can be written as a Fourier transform (2.73) of a continuous function $f_1(\lambda)$. Then, the basic assertion of the Bochner theorem is that every nonnegative definite function of the type described has a nonnegative Fourier transform.[11] Thus, suppose that $f_1(\lambda)$ is not nonnegative; in fact, let the Fourier transform $f_1(\lambda)$ of the nonnegative definite function $B(\tau)$ be negative at the point $\lambda = \lambda_0$. We now use Khinchin's result (Sec. 5) which states that every nonnegative definite function is the correlation function of a stationary random process. We pass a stationary random process which has $B(\tau)$ as its correlation function through a filter with a narrow pass band $\Delta\lambda$, containing the point $\lambda_0$. According to formula (2.76), the mathematical expectation of the square of the corresponding spectral component $\xi(\Delta\lambda; t)$, a quantity which is inherently nonnegative, equals the integral over $\Delta\lambda$ of the Fourier transform $f_1(\lambda)$ of $B(\tau)$. Moreover, as $\Delta\lambda \to 0$, this integral equals $f_1(\lambda_0)\Delta\lambda$, to within terms of higher order in $\Delta\lambda$, since $f_1(\lambda)$ is continuous. This contradicts the assumption that $f_1(\lambda_0) < 0$. For a more detailed presentation of this proof, which in particular does not require that $B(\tau)$ fall off rapidly at infinity, we refer the reader to the original paper (B6).

In conclusion, we note that a filter can be defined purely mathematically as an operator transforming every stationary random process into another stationary random process. A band-pass filter with a given pass band can be defined by an analytic formula of the type (2.59). [See e.g., formulas (20)

---

[11] The converse assertion is proved quite simply, since if (2.66) holds and if $f(\lambda) \geqslant 0$, then obviously [cf. (1.37)] we have

$$\sum_{j,k=1}^{n} B(\tau_j - \tau_k)a_j\bar{a}_k = \sum_{j,k=1}^{n}\left[\int_{-\infty}^{\infty}e^{i\lambda(\tau_j-\tau_k)}f(\lambda)d\lambda\right]a_j\bar{a}_k = \int_{-\infty}^{\infty}\left|\sum_{j=1}^{n}e^{i\lambda\tau_j}a_j\right|^2 f(\lambda)d\lambda \geqslant 0.$$

[If $B(\tau)$ is real, then $f_1(\lambda) = 2f(\lambda)$ and $f(-\lambda) = f(\lambda)$.]

and (21) of K16.] Moreover, we note that for our purposes we can use an even simpler formula. In fact, the result of applying a filter to a process $\xi_1(t)$ can be written as

$$\xi_2(t) = \int_{-\infty}^{\infty} \xi_1(t - \tau)s(\tau)\, d\tau.$$

The supplementary condition

$$s(\tau) = 0 \quad \text{for} \quad \tau < 0,$$

imposed on $s(\tau)$ in K16, is the condition that the filter be physically realizable, and is not essential in the proof just outlined of Bochner's theorem. If

$$s(\tau) = \frac{1}{2\pi} \frac{e^{i\lambda_2\tau} - e^{i\lambda_1\tau}}{i\tau},$$

we obtain an ideal band-pass filter with pass band $[\lambda_1, \lambda_2]$. [Cf. formula (2.45).]*

## 11. Spectral Representation of the Derivative and Integral of a Stationary Process

In the theory of extrapolation and filtering of stationary random processes (see Part 2), we shall need formulas giving the spectral representations of the derivative of a stationary random process $\xi(t)$, and of integrals of the form

$$\int_0^{\infty} f(\tau)\xi(t - \tau)\, d\tau.$$

According to Sec. 5, the derivative $\xi'(t)$ is defined as the limit in the mean as $h \to 0$ of the random variable

$$\frac{\xi(t + h) - \xi(t)}{h}.$$

If the process $\xi(t)$ is specified by its spectral representation (2.40), then

$$\frac{\xi(t + h) - \xi(t)}{h} = \int_{-\infty}^{\infty} \frac{e^{i\lambda(t+h)} - e^{i\lambda t}}{h}\, dZ(\lambda). \qquad (2.79)$$

It is a simple consequence of the definition of the integral (2.40) as the limit of the sum (2.41) that the limit as $h \to 0$ in (2.79) can be taken under the integral sign, provided that the limit $\xi'(t)$ exists and that $\mathbf{E}|\xi'(t)|^2$ is finite. In this case, we have

$$\xi'(t) = \int_{-\infty}^{\infty} \lim_{h \to 0} \left[\frac{e^{i\lambda(t+h)} - e^{i\lambda t}}{h}\right] dZ(\lambda) = \int_{-\infty}^{\infty} e^{i\lambda t}\, i\lambda\, dZ(\lambda), \quad (2.80)$$

which is just the spectral representation of the derivative $\xi'(t)$ of the stationary random process $\xi(t)$.

It follows at once from (2.80) that the correlation function of the process $\xi'(t)$ can be represented in the form

$$B^{[1]}(\tau) = \mathbf{E}\xi'(t + \tau)\overline{\xi'(t)} = \int_{-\infty}^{\infty} e^{i\lambda\tau} \lambda^2 \, dF(\lambda), \qquad (2.81)$$

where $F(\lambda)$ is the spectral distribution function of $\xi(t)$.   Comparing (2.81) and (2.35), we find that

$$B^{[1]}(\tau) = -B''(\tau), \qquad (2.82)$$

a result which was derived earlier in Sec. 5.   According to (2.81), the spectral distribution function of the derivative $\xi'(t)$ equals

$$F^{[1]}(\lambda) = \int_{-\infty}^{\lambda} \lambda^2 \, dF(\lambda). \qquad (2.83)$$

If $\xi(t)$ has a spectral density, i.e., if $F(\lambda)$ can be represented in the form (2.67), then $\xi'(t)$ also has a spectral density, equal to

$$f^{[1]}(\lambda) = \lambda^2 f(\lambda). \qquad (2.84)$$

Since the spectral distribution function must always be bounded [see (2.64)], the formulas (2.83) and (2.84) make sense, provided that

$$\int_{-\infty}^{\infty} \lambda^2 \, dF(\lambda) < \infty. \qquad (2.85)$$

This condition guarantees that taking the limit as $h \to 0$ in (2.79) is legitimate.   In other words, (2.85) is the condition for the process $\xi(t)$ to be differentiable, which, according to (2.35), is equivalent to requiring the existence of $B''(0)$, the second derivative of the correlation function $B(\tau)$ at the point $\tau = 0$.

By differentiating the stationary process $\xi'(t)$, we obtain the second derivative $\xi''(t)$ of the process $\xi(t)$, and similarly, we can also form higher derivatives of $\xi(t)$.   It is easy to see that the $n$th derivative $\xi^{(n)}(t)$ exists if and only if

$$\int_{-\infty}^{\infty} \lambda^{2n} \, dF(\lambda) < \infty, \qquad (2.86)$$

a condition which is equivalent to requiring the existence of the derivative of order $2n$ of the function $B(\tau)$ at the point $\tau = 0$.   If the condition (2.86) is met, then the $n$th derivative $\xi^{(n)}(t)$ can be represented in the form

$$\xi^{(n)}(t) = \int_{-\infty}^{\infty} e^{i\lambda t} (i\lambda)^n \, dZ(\lambda), \qquad (2.87)$$

and the corresponding correlation function $B^{[n]}(\tau)$ equals

$$B^{[n]}(\tau) = \mathbf{E}\xi^{(n)}(t + \tau)\overline{\xi^{(n)}(t)} = \int_{-\infty}^{\infty} e^{i\lambda\tau} \lambda^{2n} \, dF(\lambda). \qquad (2.88)$$

Next, we consider the integral

$$\int_0^T f(\tau)\xi(t-\tau)\,d\tau, \tag{2.89}$$

where $f(\tau)$ is a numerical function and $\xi(t)$ is a stationary random process. For the time being, we assume that the upper limit of integration $T$ is finite. An integral of this kind is defined as the limit as $\max|\tau_k - \tau_{k-1}| \to 0$ of

$$\sum_{k=1}^n f(\tau_k')\xi(t-\tau_k')(\tau_k - \tau_{k-1}),$$

where

$$0 = \tau_0 < \tau_1 < \cdots < \tau_{n-1} < \tau_n = T \quad \text{and} \quad \tau_{k-1} \leqslant \tau_k' \leqslant \tau_k.$$

Using the spectral representation (2.40), we find that

$$\int_{-\infty}^\infty \left\{ \sum_{k=1}^n f(\tau_k')e^{i\lambda(t-\tau_k')}(\tau_k - \tau_{k-1}) \right\} dZ(\lambda)$$

$$= \int_{-\infty}^\infty e^{i\lambda\tau} \left\{ \sum_{k=1}^n f(\tau_k')e^{-i\lambda\tau_k'}(\tau_k - \tau_{k-1}) \right\} dZ(\lambda). \tag{2.90}$$

It is not hard to see that we can take the limit as $\max|\tau_k - \tau_{k-1}| \to 0$ under the integral sign in (2.90), provided only that the resulting integral represents a random variable $\eta$, say, for which $\mathbf{E}|\eta|^2 < \infty$. It follows that the integral (2.89) exists and is given by the formula

$$\int_0^T f(\tau)\xi(t-\tau)\,d\tau = \int_{-\infty}^\infty e^{i\lambda\tau}g(\lambda)\,dZ(\lambda), \tag{2.91}$$

where

$$g(\lambda) = \lim_{\max|\tau_k - \tau_{k-1}|\to 0} \sum_{k=1}^n f(\tau_k')e^{-i\lambda\tau_k'}(\tau_k - \tau_{k-1}) = \int_0^T f(\tau)e^{-i\lambda\tau}\,d\tau, \tag{2.92}$$

provided that

$$\int_{-\infty}^\infty |g(\lambda)|^2\,dF(\lambda) < \infty. \tag{2.93}$$

The condition (2.93) is obviously equivalent to the requirement that the integral

$$\int_0^T \int_0^T B(t-s)f(t)\overline{f(s)}\,dt\,ds$$

exist (see Sec. 5).

If we now let $T \to \infty$, it is easily seen that the above assertion remains valid for the resulting improper integral, so that we can choose $\infty$ as the upper limit of integration in (2.89) and (2.92). In particular, the fact that

$$\int_0^\infty e^{-(\alpha+i\lambda)\tau}\,d\tau = \frac{1}{\alpha + i\lambda} \qquad (\text{Re } \alpha > 0) \tag{2.94}$$

implies that

$$\int_0^\infty e^{-\alpha\tau}\,\xi(t-\tau)\,d\tau = \int_{-\infty}^\infty \frac{e^{i\lambda\tau}}{\alpha+i\lambda}\,dZ(\lambda), \tag{2.95}$$

a result which we shall find useful later.    [It is clear that the condition (2.93) always holds for $g(\lambda) = 1/(\alpha + i\lambda)$.]    Similarly, the more general formula

$$\frac{1}{(k-1)!}\int_0^\infty \tau^{k-1}\,e^{-(\alpha+i\lambda)\tau}\,d\tau = \frac{1}{(\alpha+i\lambda)^k} \qquad (\text{Re }\alpha > 0) \tag{2.94'}$$

implies that

$$\frac{1}{(k-1)!}\int_0^\infty \tau^{k-1}\,e^{-\alpha\tau}\,\xi(t-\tau)\,d\tau = \int_{-\infty}^\infty \frac{e^{i\tau\lambda}}{(\alpha+i\lambda)^k}\,dZ(\lambda). \tag{2.95'}$$

## 12. Spectral Representation of Stationary Sequences

So far, we have only discussed the spectral representation of stationary random processes.    However, it is clear that almost all the results of Secs. 8 to 11 can be carried over to the case of stationary random sequences as well, by making some changes which are hardly of a fundamental nature.    Thus, if as in Example 1 of Sec. 8 we construct a stationary sequence $\xi(t)$ of the form

$$\xi(t) = \xi f(t),$$

where $f(t)$ is a numerical function and $\xi$ is a random variable with $E\xi = 0$, we again arrive at a harmonic oscillation

$$\xi(t) = \xi e^{i\lambda t} \tag{2.96}$$

with random amplitude and phase.    The only difference is that since now $t$ takes only integral values, we have

$$e^{i(\lambda+2k\pi)t} = e^{i\lambda t}$$

for all $t$ if $k$ is an integer, and hence $\lambda$ is defined only to within an additive constant which is a multiple of $2\pi$.    This fact allows us to assume that in the case of stationary random sequences the angular frequency $\lambda$ always lies between $-\pi$ and $+\pi$.    Therefore, if we represent an arbitrary sequence $\xi(t)$ as a superposition of oscillations of the form (2.90), we have to take into account only oscillations with frequencies in the interval $[-\pi, \pi]$,    Thus, (2.40) is replaced by the formula

$$\xi(t) = \int_{-\pi}^\pi e^{i\lambda t}dZ(\lambda), \tag{2.97}$$

where the properties of $Z(\lambda)$ and the meaning of the integral are the same as in the case of stationary random processes.

The representation of the stationary random sequence $\xi(t)$ in the form (2.97) is called the *spectral representation* of $\xi(t)$. The proof of this representation can be found, for example, in Karhunen's paper (K2), which we have already cited in connection with the spectral representation of stationary processes (see Sec. 9). In fact, a more general result is proved in K2, which implies both (2.40) and (2.97) as special cases. Another possible way of proving (2.97) is to define $Z(\lambda)$ by the formula

$$Z(\lambda) = \frac{1}{2\pi} \left\{ \lambda \xi(0) - \sum_{t \neq 0} \frac{e^{-i\lambda t}}{it} \xi(t) \right\},$$

and then follow the argument given in Sec. 9 [formula (2.45) *et seq.*].

The correlation function $B(\tau)$ of a stationary sequence can always be represented in the form

$$B(\tau) = \int_{-\pi}^{\pi} e^{i\lambda \tau} \, dF(\lambda), \tag{2.98}$$

where $F(\lambda)$ is a nondecreasing function. If the absolute value of $B(\tau)$ falls off so rapidly as $|\tau| \to \infty$ that

$$\sum_{\tau = -\infty}^{\infty} |B(\tau)| < \infty, \tag{2.99}$$

then the function $F(\lambda)$ is differentiable, and can be represented in the form

$$F(\lambda) = \int_{-\pi}^{\lambda} f(\lambda) \, d\lambda,$$

where

$$f(\lambda) = F'(\lambda) \geqslant 0. \tag{2.100}$$

In this case, (2.98) can be replaced by the formula

$$B(\tau) = \int_{-\pi}^{\pi} e^{i\lambda \tau} f(\lambda) \, d\lambda. \tag{2.101}$$

It is clear from (2.101) that the quantities

$$\frac{1}{2\pi} B(-\tau) \qquad (\tau = 0, \pm 1, \pm 2, \ldots)$$

are the Fourier coefficients of the function $f(\lambda)$, so that the expansion of $B(\tau)$ in Fourier series takes the form

$$f(\lambda) = \frac{1}{2\pi} \sum_{\tau = -\infty}^{\infty} e^{-i\lambda \tau} B(\tau). \tag{2.102}$$

Obviously, (2.102) is the analog of formula (2.69) in the theory of stationary random processes.

The function $F(\lambda)$ is called the *spectral distribution function* of the stationary sequence $\xi(t)$, and the function $f(\lambda)$, if it exists, is called the *spectral density* (*function*) of $\xi(t)$.   The possibility of representing $B(\tau)$ as an integral of the form (2.98) can be proved in a way completely analogous to Khinchin's proof of formula (2.40) [see discussion in Sec. 10].   The only difference is that instead of the Bochner theorem, we must now use a result known as the *Herglotz lemma*, which states that a numerical sequence $B(\tau)$, $\tau = 0, \pm 1,$ $\pm 2, \ldots,$ is nonnegative definite if and only if it can be represented in the form (2.98), where $F(\lambda)$ is nondecreasing.   For a proof of this result, see e.g., L8, p. 207.

It should also be noted that conversely, any sequence $B(\tau)$, $\tau = 0, \pm 1,$ $\pm 2, \ldots,$ which can be represented as an integral of the form (2.98) where $F(\lambda)$ is nondecreasing, is the correlation function of a stationary random sequence.   (For a detailed proof, see W7.)   In particular, if the sequence $B(\tau)$ satisfies the condition (2.99), then to verify whether or not $B(\tau)$ is the correlation function of a stationary random sequence, it is only necessary to use (2.102) to calculate the function $f(\lambda)$, and then examine whether or not $f(\lambda)$ is nonnegative.   This fact makes it quite easy to construct examples of correlation functions of stationary sequences.

## 13. Examples of Correlation Functions of Stationary Sequences

In this and the next section, we shall give some examples of correlation functions of stationary random sequences and processes.   Subsequently, we shall make repeated use of these examples.

***Example 1.*** Let

$$B(0) = 1, \qquad B(\tau) = 0 \quad \text{for} \quad \tau \neq 0; \qquad (2.103)$$

then, according to formula (2.102), the corresponding spectral density is

$$f(\lambda) = \frac{1}{2\pi}. \qquad (2.104)$$

Since $f(\lambda)$ is nonnegative, (2.103) actually defines the correlation function of a stationary random sequence $\eta(t)$.   The sequence $\eta(t)$ is a stationary sequence of uncorrelated random variables, and has already been considered in Example 1 of Sec. 7.

We note that if we choose the spectral density $f(\lambda)$ to be a constant other than $1/2\pi$, we still have $B(\tau) = 0$ for $\tau \neq 0$, and hence, the corresponding stationary sequence still consists of uncorrelated random variables, the only difference being that the mathematical expectation of the square of the absolute values of these random variables is no longer 1.   Thus, stationary sequences of uncorrelated random variables are characterized by the fact that their spectral densities are constant over the whole interval $[-\pi, \pi]$.

***Example 2.*** Suppose we have a sequence of real numbers which fall off like the geometric progression

$$B(k) = Ca^k, \quad C > 0 \quad (k = 0, 1, 2, \ldots), \tag{2.105}$$

where $a$ is real and $|a| < 1$. If we assume that these numbers are the values of the correlation function of a stationary sequence, then by (1.33) we must have

$$B(-k) = B(k) = Ca^k,$$

so that

$$B(\tau) = Ca^{|\tau|}, \quad C > 0, \quad |a| < 1 \quad (\tau = 0, \pm 1, \pm 2, \ldots). \tag{2.106}$$

We now verify that (2.106) is actually a correlation function. Clearly, the condition (2.99) is satisfied in this case, and hence, according to (2.102),

$$f(\lambda) = \frac{1}{2\pi} \sum_{\tau = -\infty}^{\infty} B(\tau) e^{-i\tau\lambda} = \frac{C}{2\pi} \left\{ \sum_{k=-\infty}^{-1} a^{-k} e^{-ik\lambda} + \sum_{k=0}^{\infty} a^k e^{-ik\lambda} \right\}$$

$$= \frac{C}{2\pi} \left\{ \sum_{k=1}^{\infty} a^k e^{ik\lambda} + \sum_{k=0}^{\infty} a^k e^{-ik\lambda} \right\} = \frac{C}{2\pi} \left\{ \frac{ae^{i\lambda}}{1 - ae^{i\lambda}} + \frac{1}{1 - ae^{-i\lambda}} \right\}$$

$$= \frac{C}{2\pi} \left\{ \frac{a}{e^{-i\lambda} - a} + \frac{e^{i\lambda}}{e^{i\lambda} - a} \right\} = \frac{C}{2\pi} \frac{1 - a^2}{|e^{i\lambda} - a|^2}.$$

Thus, we finally have

$$f(\lambda) = \frac{C}{2\pi} \frac{1 - a^2}{|e^{i\lambda} - a|^2}, \tag{2.107}$$

so that $f(\lambda) \geqslant 0$. Therefore, formula (2.106) actually gives the correlation function of a stationary random sequence. It should be noted that this formula often closely fits experimental data on correlation coefficients of time series encountered in practice.

Next, we show that a sequence $\xi(t)$ with the correlation function (2.106) has a simple representation as a *moving average* of a sequence of uncorrelated random variables (cf. Example 3 of Sec. 7). In fact, let

$$\xi(t) = \sum_{k=0}^{\infty} a^k \eta(t - k), \tag{2.108}$$

where $a$ is real, $|a| < 1$, and $\eta(t)$ is a sequence of uncorrelated random variables. In the present case, the condition (2.7) is obviously satisfied, so that formula (2.108) defines a stationary random sequence. According to (2.9) and (2.108), if $\tau \geqslant 0$ the correlation function of this sequence equals

$$B(\tau) = \sum_{k=+\tau}^{\infty} a^k a^{k-\tau} = \sum_{l=0}^{\infty} a^l a^{l+\tau}$$

$$= \sum_{l=0}^{\infty} a^{2l+\tau} = \frac{a^\tau}{1 - a^2}.$$

For values of $\tau < 0$, we define $B(\tau)$ by the condition $B(-\tau) = B(\tau)$, so that

$$B(\tau) = \frac{1}{1 - a^2}\, a^{|\tau|}. \tag{2.109}$$

Thus, the sum (2.108) represents a stationary random sequence whose correlation function is of the form (2.106), where

$$C = \frac{1}{1 - a^2}.$$

It is clear that we obtain the general correlation function (2.106) by considering the sequence

$$\xi(t) = A \sum_{k=0}^{\infty} a^k \eta(t - k), \tag{2.110}$$

where

$$A = \sqrt{C(1 - a^2)}.$$

**Example 3.** Further examples of correlation functions of stationary random sequences are most easily obtained by calculating $B(\tau)$ from formula (2.101), after choosing $f(\lambda)$ to be some sufficiently simple nonnegative function. In particular, if we choose $f(\lambda)$ to be a nonnegative function which is rational in $e^{i\lambda}$, then the calculations based on (2.101) can be greatly simplified by using the residue theorem.

For example, let $a$ and $b$ be real, and consider the function

$$f(\lambda) = \frac{C}{2\pi} \frac{|e^{i\lambda} - b|^2}{|e^{i\lambda} - a|^2} = \frac{C}{2\pi} \frac{(e^{i\lambda} - b)(e^{-i\lambda} - b)}{(e^{i\lambda} - a)(e^{-i\lambda} - a)}, \tag{2.111}$$

where $|a| < 1$, $|b| < 1$.    Then

$$\begin{aligned}
B(k) &= \int_{-\pi}^{\pi} e^{ik\lambda} f(\lambda)\, d\lambda = \frac{C}{2\pi} \int_{-\pi}^{\pi} e^{ik\lambda} \frac{(e^{i\lambda} - b)(e^{-i\lambda} - b)}{(e^{i\lambda} - a)(e^{-i\lambda} - a)}\, d\lambda \\[2mm]
&= \frac{C}{2\pi i} \oint_{|z|=1} z^{k-1} \frac{(z - b)\left(\dfrac{1}{z} - b\right)}{(z - a)\left(\dfrac{1}{z} - a\right)}\, dz,
\end{aligned} \tag{2.112}$$

where $z = e^{i\lambda}$, and it follows from the residue theorem that $B(k)$ is the coefficient of $z^{-k}$ in the expansion of the rational function

$$f^*(z) = \frac{C(z - b)\left(\dfrac{1}{z} - b\right)}{(z - a)\left(\dfrac{1}{z} - a\right)} = \frac{C(z - b)(1 - bz)}{(z - a)(1 - az)} \tag{2.113}$$

in a Laurent series.   Expanding $f^*(z)$ in partial fractions, we obtain

$$f^*(z) = \frac{C(a - b)(1 - ab)}{a(1 - a^2)} \left( \frac{a}{z - a} + \frac{1}{1 - az} \right) + \frac{Cb}{a}. \qquad (2.114)$$

The function $1/(1 - az)$ is regular in the unit circle, and hence its series expansion contains only nonnegative powers of $z$.  On the other hand, the function $a/(z - a)$ is regular outside the unit circle, and has the series expansion

$$\frac{a}{z - a} = \frac{a}{z} \frac{1}{1 - \frac{a}{z}} = \sum_{n=1}^{\infty} \frac{a^n}{z^n}$$

for $|z| \geqslant 1$.   This makes it clear that for $k > 0$, the required coefficient of $z^{-k}$ is given by the formula

$$B(k) = \frac{C(a - b)(1 - ab)}{1 - a^2} a^{k-1}. \qquad (2.115)$$

Moreover, since the coefficient of $z^0$ (the constant term) in the expansion of $1/(1 - az)$ equals 1, we find that

$$B(0) = \frac{C(a - b)(1 - ab)}{a(1 - a^2)} + \frac{Cb}{a} = \frac{C(1 - 2ab + b^2)}{1 - a^2}. \qquad (2.116)$$

Thus, finally, in the case of the spectral density (2.111), the correlation function equals

$$B(\tau) = \begin{cases} \dfrac{C(1 - 2ab + b^2)}{1 - a^2} & \text{for} \quad \tau = 0, \\[3mm] \dfrac{C(a - b)(1 - ab)}{1 - a^2} a^{|\tau|-1} & \text{for} \quad \tau \neq 0, \end{cases} \qquad (2.117)$$

where we have used the fact that $B(-k) = B(k)$.

In particular, this result shows that a sequence with the spectral density (2.111) can be represented as the sum of two mutually uncorrelated sequences, one of which is a sequence of uncorrelated random variables, while the other is a sequence of the form (2.110).   In fact, suppose we write (2.117) in the form

$$B(\tau) = \begin{cases} \dfrac{C(a - b)(1 - ab)}{a(1 - a^2)} + \dfrac{Cb}{a} & \text{for} \quad \tau = 0, \\[3mm] \dfrac{C(a - b)(1 - ab)}{a(1 - a^2)} a^{|\tau|} & \text{for} \quad \tau \neq 0, \end{cases} \qquad (2.117')$$

and then let

$$\xi(t) = \xi_1(t) + \eta_1(t), \qquad (2.118)$$

where $\eta_1(t)$ is a sequence of uncorrelated random variables such that

$$\mathbf{E}|\eta_1(t)|^2 = \frac{Cb}{a},$$

whereas $\xi_1(t)$ is a stationary sequence which is not correlated with $\eta_1(t)$ and has a correlation function of the form (2.106), with $C$ replaced by

$$\frac{C(a - b)(1 - ab)}{a(1 - a^2)}.$$

Then, the correlation function of the sequence $\xi(t)$ equals

$$B(\tau) = \mathbf{E}\xi(t + \tau)\overline{\xi(t)} = \mathbf{E}[\xi_1(t + \tau) + \eta_1(t + \tau)][\overline{\xi_1(t) + \eta_1(t)}]$$
$$= \mathbf{E}\xi_1(t + \tau)\overline{\xi_1(t)} + \mathbf{E}\eta_1(t + \tau)\overline{\eta_1(t)},$$

and hence coincides with (2.117).   Of course, this result can be obtained directly from the spectral density without recourse to correlation functions, by writing the function (2.111) as

$$f(\lambda) = \frac{C}{2\pi} \frac{(e^{i\lambda} - b)(e^{-i\lambda} - b)}{(e^{i\lambda} - a)(e^{-i\lambda} - a)}$$
$$= \frac{Cb}{2\pi a} + \frac{C(a - b)(1 - ab)}{2\pi a} \frac{1}{(e^{i\lambda} - a)(e^{-i\lambda} - a)}.$$

Conversely, suppose that we have a sequence $\xi(t)$ which can be represented in the form (2.118), where $\xi_1(t)$ and $\eta_1(t)$ are mutually uncorrelated stationary sequences, such that $\eta_1(t)$ consists of uncorrelated random variables with

$$\mathbf{E}|\eta_1(t)|^2 = C_2,$$

while $\xi_1(t)$ has a correlation function of the form (2.106):

$$B_{\xi_1\xi_1}(\tau) = \mathbf{E}\xi_1(t + \tau)\overline{\xi_1(t)} = C_1 a^{|\tau|}, \quad |a| < 1. \qquad (2.119)$$

Then, it is a simple consequence of the formulas derived above that the spectral density $f(\lambda)$ of the sequence $\xi(t)$ can be written in the form (2.111), where $C$ and $b$ are determined by the system

$$\frac{Cb}{a} = C_2, \quad \frac{C(a - b)(1 - ab)}{a(1 - a^2)} = C_1. \qquad (2.120)$$

[We leave it to the reader to obtain the same result from formulas (2.111), (2.104) and (2.107) for the spectral densities.]   If we eliminate $C$, formula (2.120) leads to the following quadratic equation for the quantity $b$:

$$ab^2 - b[(1 + a^2) + \frac{C_1}{C_2}(1 - a^2)] + a = 0. \qquad (2.121)$$

It is not hard to verify that the solutions of (2.121) are real and unequal, and that one of them has absolute value less than 1.   [This follows immediately from the fact that the product of the roots of (2.121) equals 1.]  Determining $b$ from (2.121), we then immediately find $C$ from the first of the equations (2.120).

**Example 4.** Now let

$$f(\lambda) = \frac{C}{2\pi} \frac{1}{|e^{i\lambda} - a_1|^2 |e^{i\lambda} - a_2|^2}, \tag{2.122}$$

where $a_1$ and $a_2$ are real numbers such that $|a_1| < 1$, $|a_2| < 1$.  It is easy to see that this time $B(k)$ is the coefficient of $z^{-k}$ in the Laurent expansion of the function

$$
\begin{aligned}
f^*(z) &= \frac{Cz^2}{(z - a_1)(1 - a_1 z)(z - a_2)(1 - a_2 z)} \\
&= \frac{C}{(a_1 - a_2)(1 - a_1 a_2)} \\
&\times \left\{ \frac{a_1}{1 - a_1^2} \left( \frac{a_1}{z - a_1} + \frac{1}{1 - a_1 z} \right) - \frac{a_2}{1 - a_2^2} \left( \frac{a_2}{z - a_2} + \frac{1}{1 - a_2 z} \right) \right\}.
\end{aligned}
\tag{2.123}
$$

Expanding the functions $1/(1 - a_1 z)$ and $1/(1 - a_2 z)$ in series of powers of $z$, and the functions $1/(z - a_1)$ and $1/(z - a_2)$ in series of powers of $1/z$, we find that in this case

$$B(\tau) = \frac{C}{(a_1 - a_2)(1 - a_1 a_2)} \left\{ \frac{a_1}{1 - a_1^2} a_1^{|\tau|} - \frac{a_2}{1 - a_2^2} a_2^{|\tau|} \right\}. \tag{2.124}$$

## 14. Examples of Correlation Functions of Stationary Processes

**Example 1.**  In the case of functions of a continuous variable $\tau$, $0 \leqslant \tau < \infty$, the analog of the geometric progression (2.105) is the function

$$B(\tau) = Ca^\tau = Ce^{-\alpha\tau},$$

where $C > 0$, $a < 1$, so that

$$\alpha = \log \frac{1}{a} > 0.$$

Using the condition $B(-\tau) = B(\tau)$ to complete the definition of the correlation function for negative values of $\tau$, we have

$$B(\tau) = Ce^{-\alpha|\tau|}, \quad C > 0, \quad \alpha > 0. \tag{2.125}$$

We now verify that (2.125) is actually the correlation function of a stationary process.  According to formula (2.69), in the present case

$$
\begin{aligned}
f(\lambda) &= \frac{1}{2\pi} \int_{-\infty}^{\infty} Ce^{-\alpha|\tau| - i\lambda\tau}\, d\tau = \frac{C}{2\pi} \left\{ \int_{-\infty}^{0} e^{(\alpha - i\lambda)\tau}\, d\tau + \int_{0}^{\infty} e^{-(\alpha + i\lambda)\tau}\, d\tau \right\} \\
&= \frac{C}{2\pi} \left\{ \frac{1}{\alpha - i\lambda} + \frac{1}{\alpha + i\lambda} \right\} = \frac{C}{\pi} \frac{\alpha}{\alpha^2 + \lambda^2}.
\end{aligned}
$$

Thus we have

$$f(\lambda) = \frac{A}{\alpha^2 + \lambda^2}, \qquad A = \frac{C\alpha}{\pi}, \qquad (2.126)$$

so that $f(\lambda) > 0$, i.e., (2.125) is a correlation function for any $C > 0$ and $\alpha > 0$.

In practice, formulas (2.125) and (2.126) are very often used with success to describe the correlation functions and spectral densities of actual stationary processes, and hence in Figures 6 and 7 we illustrate the graphs of these functions. We note that the curve (2.125) falls off rapidly (in fact, exponentially) with $|\tau|$, and can already be considered virtually zero when the distance from the origin is only a few multiples of $\alpha^{-1}$. Thus, the quantity $\alpha^{-1}$ (with the dimension of time) characterizes the time needed for any correlation between the quantities $\xi(t)$ and $\xi(t + \tau)$ to die out.[12] Therefore, in order to

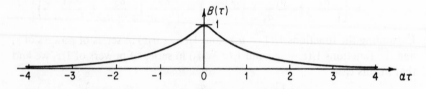

FIG. 6. Graph of the function $B(\tau) = e^{-\alpha|\tau|}$.

estimate the parameter $\alpha$, we need only know for how long appreciable correlation is preserved between $\xi(t)$ and the initial value $\xi(0)$. For this reason, the very simple formula (2.125) is often used in cases where the experimental data does not allow one to obtain more precise information about the character of the correlation function. Moreover, in order to simplify a problem, it is sometimes useful to replace the true correlation function by a function of the form (2.125), even when the form of the curve $B(\tau)$ is known rather accurately and does not actually have the form of the curve shown in Figure 6. For example, as a first approximation in solving certain problems in the theory of servomechanisms, we can replace the correlation function shown in Figure 2 on p. 19, which practically vanishes for $\tau > 0.1$ sec, by the function (2.125) with $\alpha$ approximately equal to 20–30 sec$^{-1}$. However, it should be noted that in many cases the fact that the correlation function has the form (2.125) can be justified on theoretical grounds, as will be shown on pp. 69–71.

We see from Figure 7, that the spectral density $f(\lambda)$ corresponding to the correlation function (2.125) has a maximum at zero, remains almost constant when $\lambda$ is small compared to $\alpha$, and then falls off rather slowly (like $\lambda^{-2}$). The value of $f(\lambda)$ is twice as small as its value at $\lambda = 0$ when $\lambda = \alpha$, five times

---

[12] The quantity $\alpha^{-1}$ is often called the "correlation time" of the process $\xi(t)$.

smaller when $\lambda = 2\alpha$, ten times smaller when $\lambda = 3\alpha$, and so on.  Thus, by increasing $\alpha$, we lengthen the interval in which $f(\lambda)$ is "approximately constant" and slow down the rate at which $f(\lambda)$ falls off with $\lambda$.  As we have seen, this is the opposite of how $B(\tau)$ depends on $\alpha$.

Of very great practical importance is the case where the correlation between the quantities $\xi(t)$ and $\xi(t + \tau)$ falls off extremely rapidly with $\tau$, so that $\alpha$ is very large.  For example, consider the Brownian motion of a small particle immersed in a fluid, and let $\xi(t)$ be the fluctuation in the number of collisions of molecules of the fluid with the particle.  Since under normal conditions, the particle in the fluid experiences about $10^{21}$ collisions per second, the values of $\xi(t)$ and $\xi(t + \tau)$ are practically independent for times $\tau$ of order $10^{-18}$ sec, say, and hence $\alpha$ is considerably larger than $10^{18}$ sec$^{-1}$.  The same

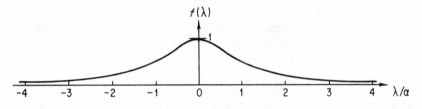

Fig. 7. Graph of the function $f(\lambda) = \dfrac{\alpha^2}{\alpha^2 + \lambda^2}$.

sort of behavior is manifested by the fluctuating e.m.f. (electromotive force) in a conductor, due to thermal motion of the electrons, or by the fluctuations of current flow in a vacuum tube, due to fluctuations in the number of electrons going from the cathode to the anode.  In all these cases, if we use a correlation function of the form (2.125) to describe the phenomenon, we must choose $\alpha$ to be extremely large.

As already remarked, if $\lambda$ is small compared to $\alpha$, the spectral density $f(\lambda)$ is practically constant and equals $f(0) = f_0$.  Consequently, in the case where $\alpha$ is very large, $f(\lambda)$ is constant over a very wide frequency range.  However, in practice, we are almost never interested in all possible values of $\lambda$.  In fact, in every particular case, only a certain restricted frequency range is of any importance.  Thus, if the spectral density is virtually constant over this frequency range, then in solving the given problem we can in general neglect the variation of $f(\lambda)$ and assume that

$$f(\lambda) = f_0 = \text{const.} \qquad (2.127)$$

It is clear that, strictly speaking, the spectral density (2.127) can never exist, for otherwise the quantity

$$\mathbf{E}|\xi(t)|^2 = \int_{-\infty}^{\infty} f(\lambda)\, d\lambda,$$

the total (average) power of the process, would be infinite [recall the condition (2.64) on p. 44]. Nevertheless, the idea of a random process with a constant spectral density turns out to be a very useful mathematical idealization[13] (similar to the idea of a point mass in mechanics, which contradicts the physically obvious fact that the density of any real object must be finite). The value of this idea is due to the fact that in the great majority of cases, if we carry out a calculation using the spectral density (2.126) and then in the final result let $A$ and $\alpha$ go to infinity in such a way that the ratio

$$f_0 = \frac{A}{\alpha^2} = \frac{C}{\pi\alpha}$$

remains constant, we ultimately obtain a meaningful result, which, of course, depends only on $f_0$, and not on $A$ and $\alpha$ separately. In all such cases, this limiting result can be used with considerable accuracy for all sufficiently large $\alpha$, and it is much simpler to obtain the final result by assuming from the outset that the random process under consideration has a spectral density of the form (2.127). In a certain sense, estimating the values of $\alpha$ for which we can begin to use the limiting result obtained with the assumption (2.127) is equivalent to estimating the "spectral bandwidth" which is important in the given problem. Here, we only indicate that the values of $\alpha$ corresponding to the three random processes just enumerated (i.e., fluctuations of the force acting on a Brownian particle, voltage fluctuations due to thermal noise in a conductor, and current fluctuations due to the shot effect) are actually large enough to justify the approximation (2.127) in all problems of practical interest.

In this connection, the following two remarks should be made:

(a) If $\alpha \to \infty$ and $C \to \infty$ in such a way that $C/\alpha = \mathrm{const}$, then the correlation function (2.125) approaches zero for any $\tau \neq 0$, while $B(0) \to \infty$ in such a way that

$$\int_{-\infty}^{\infty} B(\tau)\,d\tau$$

remains constant. Thus, the spectral density (2.127) corresponds to a random process $\xi(t)$ whose values are uncorrelated at any two different instants of time (which may be arbitrarily close together). This is the so-called "purely random" or "absolutely random" process (see W1, Sec. 12). It is clear that we obtain the same spectral density (2.127) if we start from any other correlation function $B(\tau)$ which falls off at infinity, and then "compress" $B(\tau)$ [by going from $B(\tau)$ to $B(\alpha\tau)$, and

---

[13] A stationary process $\xi(t)$ with a constant spectral density is sometimes called "white noise," or one says that $\xi(t)$ has a "white" spectrum (by analogy with white light in optics).

then letting $\alpha \to \infty$], while simultaneously increasing $B(0)$ in such a way that

$$\int_{-\infty}^{\infty} B(\tau)\, d\tau = \text{const} \neq 0. \qquad (2.128)$$

In other words, we take $\alpha B(\alpha\tau)$ as the new correlation function, and then let $\alpha \to \infty$. The condition (2.128) is necessary for the limiting value

$$f(0) = \frac{1}{2\pi} \int_{-\infty}^{\infty} B(\tau)\, d\tau$$

to be finite and different from zero. As a matter of fact, the relation between the "spectral bandwidth" and the "correlation time" [discussed in connection with the correlation function (2.125)] is of a completely general character, and expresses a familiar result from the theory of Fourier transforms.

(b) When we say that only a restricted frequency range is important in a given problem, we thereby imply that the total power of the process, determined by the whole spectral distribution, does not play an important role in the problem. Under these circumstances, the fact that the spectral density (2.127) corresponds to infinite total power does not bother us. Thus, the value $f_0$ of the spectral density at $\lambda = 0$ is the basic spectral characteristic of a process for which the representation (2.127) is possible, and the determination of $f_0$ must be first and foremost in the theory of such processes.

In the case where the process $\xi(t)$ is the fluctuating force acting on a Brownian particle, statistical physics leads to the formula

$$f_0 = \frac{1}{\pi} \beta kT,$$

where $\beta$ is the proportionality constant between the force of friction and the velocity ($\beta = 6\pi a\mu$ for a spherical particle of radius $a$ immersed in a fluid of viscosity $\mu$), $T$ is the absolute temperature, and $k$ is the Boltzmann constant (see e.g., M2, Chap. 10, and W2). In the case where $\xi(t)$ is the fluctuating e.m.f. in a conductor, we need only replace the coefficient $\beta$ in the preceding formula by $R$, the resistance of the conductor, thereby obtaining the so-called *Nyquist formula*

$$f_0 = \frac{1}{\pi} RkT.$$

Finally, if $\xi(t)$ is the fluctuating current due to the shot effect, we have

$$f_0 = \frac{1}{2\pi} eI,$$

where $e$ is the charge of the electron, and $I$ is the mean value of the current flowing through the vacuum tube.[14, 15]

*We now indicate another interpretation of stationary random processes with constant spectral densities, which is of considerable interest. Let $\xi(t)$ be a stationary process whose spectral density can be regarded as constant; for simplicity, we assume that $\xi(t)$ is real. Consider the indefinite integral of $\xi(t)$:

$$\zeta(t) = \int_0^t \xi(s) \, ds.$$

It is clear that $\zeta(t)$ is also a random process, which, however, is no longer stationary. As is easily seen, the mathematical expectation of $[\zeta(t)]^2$ is given by

$$\mathbf{E}[\zeta(t)]^2 = \int_0^t \int_0^t \mathbf{E}[\xi(s)\xi(s')] \, ds \, ds' = \int_0^t \int_0^t B(s - s') \, ds \, ds'. \qquad (2.129)$$

According to the interpretation given in the first of the above remarks, the integral (2.129) has to be calculated as follows: We set

$$B(s) = Ce^{-\alpha|s|}$$

in (2.129), evaluate the integral, and then pass to the limit $\alpha \to \infty$, $C \to \infty$ in the resulting expression, assuming that

$$\frac{C}{\alpha} = \pi f_0 = \text{const.}$$

(Twice this constant will henceforth be denoted by $c$.) These completely elementary transformations lead to the simple result

$$\mathbf{E}[\zeta(t)]^2 = c|t|. \qquad (2.130)$$

By carrying out a similar calculation, we find that

$$\mathbf{E}[\zeta(t_2) - \zeta(t_1)][\zeta(t_4) - \zeta(t_3)] = 0 \qquad (2.131)$$

if $t_1 < t_2 \leqslant t_3 < t_4$.

It is also easy to derive (2.131) directly from (2.130) as follows: First we generalize (2.130) in the obvious way to the case of an integral from $t_0$ to $t$, obtaining

$$\mathbf{E}\left\{\int_{t_0}^t \xi(s) \, ds\right\}^2 = \mathbf{E}[\zeta(t) - \zeta(t_0)]^2 = c|t - t_0|. \qquad (2.130')$$

Then we let $[t_1, t_2]$ and $[t_2, t_3]$ be two closed intervals, sharing the point $t_2$. (The general case is easily reduced to this case.) To be explicit, we assume

---

[14] See e.g., M2, Chap. 11.
[15] Specifically, a temperature-limited diode.

that $0 < t_1 < t_2 < t_3$. According to (2.130'), we have

$$E[\zeta(t_2) - \zeta(t_1)]^2 = c(t_2 - t_1),$$
$$E[\zeta(t_3) - \zeta(t_2)]^2 = c(t_3 - t_2),$$
$$E[\zeta(t_3) - \zeta(t_1)]^2 = c(t_3 - t_1).$$

On the other hand,

$$\begin{aligned} E[\zeta(t_3) - \zeta(t_1)]^2 &= E\{[\zeta(t_3) - \zeta(t_2)] + [\zeta(t_2) - \zeta(t_1)]\}^2 \\ &= E[\zeta(t_3) - \zeta(t_2)]^2 + E[\zeta(t_2) - \zeta(t_1)]^2 \\ &\quad + 2E[\zeta(t_3) - \zeta(t_2)][\zeta(t_2) - \zeta(t_1)], \end{aligned}$$

and hence

$$E[\zeta(t_3) - \zeta(t_2)][\zeta(t_2) - \zeta(t_1)] = 0.$$

Conversely, it is an easy consequence of (2.131) that

$$E[\zeta(t)]^2 = c_1|t|,$$

where $c_1$ is a constant. To see this, first let

$$E[\zeta(t)]^2 = \varphi(t).$$

Then, using (2.131) and the fact that

$$E[\zeta(t + \tau) - \zeta(t)]^2 = E[\zeta(\tau)]^2,$$

we can easily show that the function $\varphi(t)$ has to satisfy the functional equation

$$\varphi(t + s) = \varphi(t) + \varphi(s)$$

for $t > 0$, $s > 0$ [cf. the derivation of (2.131) from (2.130)]. Since $\varphi(0) = 0$, by the definition of $\varphi(t)$, if we assume that $\varphi(t)$ is continuous, it follows that

$$\varphi(t) = c_1 t$$

for $t > 0$, where $c_1$ is a constant. Finally, by considering negative values of $t$ as well, we find that

$$\varphi(t) = c_1|t|.$$

Returning to formulas (2.130) and (2.131), we see that $\zeta(t)$ is a random process with uncorrelated increments (cf. Sec. 9), which has the property that the mathematical expectation of the square of the increment of $\zeta(t)$ during any time interval is proportional to the length of the time interval. Such processes will henceforth be called *homogeneous random processes with uncorrelated increments*. These processes are a subclass of the broad class of *random processes with stationary increments*, studied in terms of curves in Hilbert space (see Sec. 6) by Kolmogorov (K10, K11) and by von Neumann and Schoenberg (V2). (See also Sec. 18 below.)

We have seen that although the process $\xi(t)$ itself, which has a constant spectral density, cannot be defined in a mathematically rigorous manner,[16] and

---

[16] See, however, Appendix I, p. 207

although use of $\xi(t)$ is only a convenient description of a certain passage to the limit, the process $\zeta(t)$, which is the integral of $\xi(t)$ and has the properties (2.130) and (2.131), can be defined in the ordinary way. However, $\zeta(t)$ is nondifferentiable, and hence, before we can talk about its derivative $\zeta'(t) = \xi(t)$, we must first explain what is meant by $\xi(t)$. Nevertheless, the concept of a process $\xi(t)$ which is the derivative $\zeta'(t)$ of a homogeneous process $\zeta(t)$ with uncorrelated increments often turns out to be useful. In fact, when using $\zeta'(t)$, we can often manage to express the final result in terms of its integral $\zeta(t)$, and then the result automatically acquires a precise meaning. We observe that in the three examples cited above, leading to processes $\xi(t)$ with constant spectral densities, the integral

$$\zeta(t) = \int_0^t \xi(s)\, ds$$

is more physically meaningful than the process $\xi(t)$ itself.

Now that we have introduced the concept of a process $\zeta'(t)$ with a constant spectral density and of its integral $\zeta(t)$, which is a homogeneous process with uncorrelated increments, we can immediately extend the considerations in Secs. 7 and 13, pertaining to *moving averages*, to the case of stationary processes. In the case of stationary processes, the analog of a sequence of uncorrelated random variables $\eta(t)$ is obviously a process $\zeta'(t)$ with constant spectral density (cf. Example 1 of Sec. 13), but such a process has only a provisional meaning. However, if we form the analog of the moving average (2.4), we are led to consider integrals of the form

$$\int f(\tau)\zeta'(t - \tau)\, d\tau,$$

and such an integral can be given an exact meaning by making the change of variable $\tau = t - s$ and rewriting it in the form

$$\int f(t - s)\, d\zeta(s).$$

Here, the integral is interpreted in the same way as the integral in formula (2.40) on p. 38. Then, the stationary process

$$\xi(t) = \int_{-\infty}^t f(t - s)\, d\zeta(s) \tag{2.132}$$

is the analog of the stationary sequence (2.6), and the process

$$\xi(t) = \int_{-\infty}^{\infty} f(t - s)\, d\zeta(s) \tag{2.133}$$

is the analog of (2.8), where now, instead of (2.7) and the second of the equations (2.8), we must require that

$$\int_0^{\infty} |f(t)|^2\, dt < \infty \quad \text{or} \quad \int_{-\infty}^{\infty} |f(t)|^2\, dt < \infty. \tag{2.134}$$

By analogy with (2.110), it is clear that we can obtain a process whose correlation function is of the form (2.125) by setting

$$f(t) = Ae^{-\alpha t}$$

in (2.132). In fact, the correlation function of the process

$$\xi(t) = \int_{-\infty}^{t} Ae^{-\alpha(t-s)}\, d\zeta(s), \tag{2.135}$$

where $\zeta(s)$ satisfies (2.130) and (2.131), equals

$$B(\tau) = \mathbf{E}\xi(t+\tau)\overline{\xi(t)} = |A|^2 c \int_{-\infty}^{t} e^{-\alpha(\tau+2t-2s)}\, ds = \frac{|A|^2 c}{2\alpha} e^{-\alpha\tau} \tag{2.136}$$

for $\tau > 0$. [The proof is just like the derivation of (2.105) from (2.108).]

We conclude our discussion of Example 1 by showing how formula (2.125) for the correlation function can often be obtained from physical considerations. As a typical example, we consider the Brownian motion of a particle immersed in a fluid. Here we can assume that the surrounding medium acts on the particle in two ways. First, it offers resistance to the motion of the particle, which takes the form of a frictional force equal to $\beta v(t)$, where $v(t)$ is the velocity of the particle, and the coefficient $\beta$ equals $6\pi a\mu$ for a spherical particle of radius $a$ immersed in a fluid of viscosity $\mu$. Secondly, the fluctuations in the number of collisions of molecules of the fluid with the particle appear as an additional "purely random" force $\xi(t)$ [see p. 64]. Thus, the equation of motion for a particle of mass $m$ becomes

$$\frac{dv(t)}{dt} + \alpha v(t) = \zeta'(t), \tag{2.137}$$

where $\alpha = \beta/m > 0$, and $\zeta'(t) = \xi(t)/m$ is the derivative of a homogeneous random process with uncorrelated increments. In the theory of Brownian motion, (2.137) is known as *Langevin's equation*. However, it should be noted that it would be more legitimate to write (2.137) in the form

$$dv(t) + \alpha v(t) = d\zeta(t), \tag{2.137'}$$

since, strictly speaking, the process $\zeta(t)$ has no derivative.

We now assume that there exists a stationary random process $v(t)$ satisfying equation (2.137) [or (2.137')], and we denote the spectral density of the process $v(t)$ by $f(\lambda)$. The spectral density of $\alpha v(t)$ is $\alpha^2 f(\lambda)$, which, according to (2.84), means that the left-hand side of (2.137) is a process with spectral density $(\lambda^2 + \alpha^2)f(\lambda)$. But then, since the right-hand side of (2.137) is a process with constant spectral density, the function $f(\lambda)$ must have the form

$$f(\lambda) = \frac{A}{\lambda^2 + \alpha^2},$$

and we have arrived at formula (2.126). This implies that the correlation function of the process $v(t)$ must have the form (2.125).

We note that despite our use of the process $\zeta'(t)$ with constant spectral density, we obtain a formula for the correlation function of the process $v(t)$ which shows that $v(t)$ can be regarded as an ordinary stationary process. The special character of $\zeta'(t)$ reveals itself only in the fact that $v(t)$ does not satisfy the condition (2.85), so that $v(t)$ is a nondifferentiable process, which, strictly speaking, cannot satisfy an equation of the form (2.137). In other words, in the theory of Brownian motion which starts from equation (2.137), the velocity of the Brownian particle exists (more precisely, the velocity has a definite probability distribution), but the acceleration of the particle has no meaning; to study the acceleration, we have to use a more exact theory. It should be noted that in the original Einstein-Smoluchowski theory (E1, V3), the inertia of the particle was neglected, i.e., the term $dv(t)/dt$ was omitted in equation (2.137). With this approximation, the Brownian particle does not even have a velocity, since then

$$v(t) = \frac{1}{\alpha} \zeta'(t),$$

and $\zeta(t)$ is nondifferentiable. However, the integral of the velocity, i.e., the path traversed by the particle, has real meaning, and represents a homogeneous random process with uncorrelated increments. Bearing in mind that the spectral density of $\zeta'(t) = \xi(t)/m$ equals $\beta kT/\pi m^2$ (cf. p. 65), while $\alpha = \beta/m$, we find, using (2.130), that the mathematical expectation of the square of the path length $x(t)$ traversed by the particle in time $t$ is given by formula

$$E[x(t)]^2 = \frac{2kTt}{\beta}$$

to this approximation. This is the basic formula of the Einstein theory (cf. E1, L4). In particular, this formula immediately implies that the particle cannot have a finite velocity, for otherwise, the mean value of the square of the path traversed by the particle in a small time $\Delta t$ would have to be proportional to $(\Delta t)^2$ and not to $\Delta t$. [Cf. the related remark concerning $Z(\lambda)$ in Sec. 10.]

Finally, we indicate a method of proving the existence of a stationary solution of equation (2.137). It is clear that the homogeneous equation corresponding to (2.137) has the solution

$$v(t) = Ce^{-\alpha t}.$$

Thus, formally applying the method of variation of parameters, we obtain the general solution of (2.137) in the form

$$v(t) = v_0 e^{-\alpha t} + \int_0^t e^{-\alpha(t-s)} \zeta'(s)\, ds = v_0 e^{-\alpha t} + \int_0^t e^{-\alpha(t-s)}\, d\zeta(s). \qquad (2.138)$$

If in this formula we regard $v_0$ as a random variable and $\zeta(s)$ as a homogeneous random process with uncorrelated increments, then we obtain a random process $v(t)$. It is not hard to see that this random process is nondifferentiable. However, (2.138) will satisfy (2.137) if we suitably interpret the equation (2.137), e.g., if we define its solution as the limit of a sequence of solutions of equations of the same type whose right-hand sides are stationary processes and whose spectral densities approach a constant, or if we interpret (2.137) in the sense of equation (2.137′) [cf. D4]. Moreover, if we calculate the correlation function of the process (2.138), we can easily verify that $v(t)$ will be stationary if and only if we choose the "constant" random variable $v_0$ to be

$$v_0 = \int_{-\infty}^{0} e^{\alpha s} \, d\zeta(s), \tag{2.139}$$

which implies that

$$v(t) = \int_{-\infty}^{t} e^{-\alpha(t-s)} \, d\zeta(s). \tag{2.140}$$

In this way, we have arrived at an explicit formula for a stationary solution of equation (2.137). It is interesting to note that this formula has the same form as (2.135), so that it also immediately implies our previous result about the form of the correlation function of the solution of (2.137).

We obtain an equation of the same form as (2.137) if we consider the result of passing the current due to thermal noise through a circuit containing a resistance and an inductance (M2, p. 438). Moreover, a great variety of other physical and mechanical processes are described by similar equations. [For example, see L1, Sec. 118, where some general conditions are given under which fluctuations of physical quantities are described by an equation like (2.137).] Consequently, in all these cases, we are dealing with stationary processes which have the correlation function (2.125).*

**Example 2.** Now let the correlation function $B(\tau)$ represent a "damped oscillation," i.e., let

$$B(\tau) = Ce^{-\alpha|\tau|} \cos \beta\tau, \tag{2.141}$$

where $C > 0$, $\alpha > 0$, $\beta > 0$. Then we have

$$\begin{aligned}
f(\lambda) &= \frac{1}{2\pi} \int_{-\infty}^{\infty} Ce^{-\alpha|\tau| - i\lambda t} \cos \beta\tau \, d\tau \\
&= \frac{C}{4\pi} \left\{ \int_{-\infty}^{\infty} e^{-\alpha|\tau| - i(\lambda - \beta)\tau} \, d\tau + \int_{-\infty}^{\infty} e^{-\alpha|\tau| - i(\lambda + \beta)\tau} \, d\tau \right\} \\
&= \frac{C\alpha}{2\pi} \left\{ \frac{1}{\alpha^2 + (\lambda - \beta)^2} + \frac{1}{\alpha^2 + (\lambda + \beta)^2} \right\}
\end{aligned} \tag{2.142}$$

[cf. the derivation of (2.126)], or equivalently

$$f(\lambda) = \frac{C\alpha}{2\pi} \left\{ \frac{1}{\lambda^2 - 2\beta\lambda + (\alpha^2 + \beta^2)} + \frac{1}{\lambda^2 + 2\beta\lambda + (\alpha^2 + \beta^2)} \right\}$$

$$= \frac{A(\lambda^2 + b^2)}{\lambda^4 + 2a\lambda^2 + b^4},$$

(2.143)

where

$$A = \frac{C\alpha}{\pi}, \qquad b = \sqrt{\alpha^2 + \beta^2}, \qquad a = \alpha^2 - \beta^2.$$  (2.144)

It is clear that $f(\lambda) > 0$, and hence (2.141) is actually a correlation function.

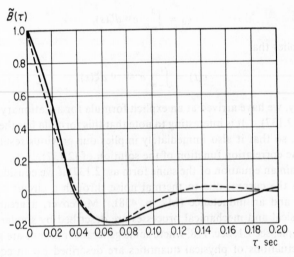

FIG. 8. Enlarged portion of the correlation function shown in Fig. 2.

The function (2.141) is very suitable for approximating certain empirical correlation functions, which change their signs several times (i.e., which alternate between positive and negative values). As an example, consider the correlation function shown in Figure 2 on p. 19. In Figure 8, the solid line indicates the initial portion of this curve on a larger scale; the dashed line indicates the curve

$$\tilde{B}(\tau) = e^{-24|\tau|} \cos 40\tau,$$  (2.145)

which has the form (2.141). We see that the curve (2.145) closely fits the experimental curve over the whole region where the experimental values can be considered reliable. Thus, for theoretical purposes, we can replace the correlation function shown in Figure 2 by the function (2.145). In fact, the spectral density in Figure 3 corresponding to the correlation function in Figure 2, was not actually calculated numerically by using formula (2.74);

instead, it was determined from (2.143) and (2.144), with $\beta = 40 \ \text{sec}^{-1}$ and $\alpha = 24 \ \text{sec}^{-1}$. [In this case, numerical integration of formula (2.74) would have given practically the same result.]

As can be seen from Figure 3, the spectral density takes its largest value when the frequency $\lambda$ is approximately equal to $\beta$, the frequency of the periodic factor in formula (2.145). In fact, an elementary calculation shows that if $\alpha < \sqrt{3}\beta$ (i.e., $a < b^2/2$), then the function (2.143) has its maximum at

$$\lambda = \lambda_0 = b \sqrt{\sqrt{2\left(1 - \frac{a}{b^2}\right)} - 1},$$

where

$$\lambda_0 \approx \beta \left[ 1 - \frac{1}{8} \left( \frac{\alpha}{\beta} \right)^4 \right] \approx \beta$$

if $\alpha$ is appreciably less than $\beta$. If $\alpha \geqslant \sqrt{3}\beta$ (i.e., $a \geqslant b^2/2$), then the spectral density (2.143) has its maximum at $\lambda = 0$ and decreases monotonically for positive $\lambda$. It is clear that the sharpness of the maximum of the function (2.143) is determined by the quantity $\alpha$, i.e., by the rate at which the function (2.141) falls off with $\tau$. For very large $\alpha$, the maximum is quite broad, so that in many cases we can even make the simple assumption that $f(\lambda) = f_{\max} = \text{const}$; on the other hand, for very small $\alpha$, the maximum occurs almost exactly at the point $\lambda = \beta$ and is quite sharp, so that $f(\lambda)$ takes values many times less than $f_{\max}$ even when $\lambda$ is very close to $\beta$. In this latter case, a realization of a random process with the spectral density (2.143) behaves just like a realization of the spectral component $\xi(\Delta\lambda; t)$ defined by formula (2.61) for small $\Delta\lambda$, and can be regarded as a sinusoidal oscillation of angular frequency $\beta$, whose amplitude and phase are very slowly varying with respect to $1/\beta$. Of course, the phrase "sinusoidal oscillation with slowly varying amplitude and phase" must be interpreted loosely, since, strictly speaking, any sinusoidal oscillation has fixed amplitude and phase. Such a realization can be represented analytically by the formula

$$\xi^{(1)}(t) = A(t) \sin [\beta t + \varphi(t)],$$

where

$$\left| \frac{A'(t)}{A(t)} \right| \ll \frac{1}{\beta} \quad \text{and} \quad \left| \frac{\varphi'(t)}{\varphi(t)} \right| \ll \frac{1}{\beta}.$$

Random functions of this type have been studied in the interesting investigations of Bunimovich (B12, B13).

*Example 3.* As our next example, we consider the correlation function corresponding to a spectral density of the form

$$f(\lambda) = \frac{A}{\lambda^4 + 2a\lambda^2 + b}, \tag{2.146}$$

where $A > 0$, $a$ and $b$ are real constants. In order for the spectral density (2.146) to be integrable and everywhere nonnegative, the numbers $a$ and $b$ have to satisfy the inequalities

$$b > 0, \quad \sqrt{b} + a > 0. \tag{2.147}$$

If these inequalities hold, we set

$$\sqrt{b} = \omega^2, \quad \sqrt{b} + a = 2\alpha^2,$$

and then (2.146) can be rewritten in the form

$$f(\lambda) = \frac{A}{(\lambda^2 - \omega^2)^2 + 4\alpha^2\lambda^2}, \tag{2.148}$$

from which it is immediately clear that $f(\lambda) > 0$.

According to (2.66), the correlation function corresponding to (2.148) is given by the integral

$$B(\tau) = \int_{-\infty}^{\infty} \frac{Ae^{i\lambda t}\, d\lambda}{(\lambda^2 - \omega^2)^2 + 4\alpha^2\lambda^2}. \tag{2.149}$$

This integral can easily be evaluated by using the theory of residues, and the form of the resulting expression depends on the sign of the difference

$$\omega^2 - \alpha^2 = \frac{\sqrt{b} - a}{2}.$$

If $\omega^2 - \alpha^2 = \beta^2 > 0$, we obtain

$$B(\tau) = \frac{\pi A}{2\alpha\omega^2} e^{-\alpha|\tau|}\left(\cos \beta\tau + \frac{\alpha}{\beta} \sin \beta|\tau|\right); \tag{2.150}$$

in particular, the correlation function corresponding to the spectral density

$$f(\lambda) = \frac{A}{\lambda^4 + 4\alpha^4} \tag{2.151}$$

is of this form (with $\alpha = \beta$). If $\omega^2 = \alpha^2$, the spectral density (2.148) becomes

$$f(\lambda) = \frac{A}{(\lambda^2 + \alpha^2)^2}, \tag{2.152}$$

and then

$$B(\tau) = \frac{\pi A}{2\alpha^3} e^{-\alpha|\tau|}(1 + \alpha|\tau|). \tag{2.153}$$

Finally, if $\omega^2 - \alpha^2 = -\beta_1^2 < 0$, the parameter $\beta$ in (2.150) is a pure imaginary, i.e., $\beta = i\beta_1$, and then (2.150) becomes

$$B(\tau) = \frac{\pi A}{4\alpha\omega^2\beta_1}\left[(\alpha + \beta_1)e^{-(\alpha-\beta_1)|\tau|} - (\alpha - \beta_1)e^{-(\alpha+\beta_1)|\tau|}\right]. \tag{2.154}$$

The graph of the spectral density (2.148) has the same character as the graph of the spectral density (2.143). For $\lambda \geqslant 0$, it is a "dome-shaped" curve, with its maximum at

$$\lambda = \sqrt{\omega^2 - 2\alpha^2}$$

if $\omega^2 > 2\alpha^2$, and it is a monotonically decreasing curve with its maximum at

$$\lambda = 0$$

if $\omega^2 \leqslant 2\alpha^2$. Formulas (2.148) and (2.150) to (2.154) are often used to approximate experimental spectral densities and correlation functions. Moreover, as we now show, these formulas can often be justified on theoretical grounds.

*The great importance of the spectral density $f(\lambda)$ and the correlation function $B(\tau)$ just analyzed is in large measure due to the fact that $f(\lambda)$ and $B(\tau)$ correspond to a stationary process $\xi(t)$ which is the "solution" of the very simple second order differential equation

$$\frac{d^2\xi}{dt^2} + 2\alpha \frac{d\xi}{dt} + \omega^2\xi = \zeta'(t), \tag{2.155}$$

where $\zeta'(t)$ is a process with constant spectral density $A$. [It would be more accurate to write (2.155) in the form

$$d\frac{d\xi}{dt} + 2\alpha d\xi + \omega^2\xi dt = d\zeta(t), \tag{2.155'}$$

analogous to (2.137').] For example, this differential equation may describe the Brownian motion of a harmonic oscillator.  In this case, $\xi(t)$ denotes the coordinate of the oscillator, the term $2\alpha \, (d\xi/dt)$ is due to the action of the friction force which causes "damping," and the right-hand side contains a "purely random" force due to fluctuations in the number of molecular collisions.[17]   The analogous electrical problem is obtained by passing thermal noise through a resonant circuit containing a resistance, a capacitance and an inductance.   Moreover, it is not hard to find various other physical problems leading to the same differential equation.

Now let $f(\lambda)$ be the spectral density of the stationary process $\xi(t)$ satisfying (2.155).  Then, using the formulas of Sec. 11, we see that the process

$$\frac{d^2\xi}{dt^2} + 2\alpha \frac{d\xi}{dt} + \omega^2\xi$$

has the spectral density

$$f_1(\lambda) = |-\lambda^2 + 2i\alpha\lambda + \omega^2|^2 f(\lambda).$$

---

[17] The term $\omega^2\xi$ is due to the action of the restoring force.

But since a process with constant spectral density appears in the right-hand side of (2.155), we obviously must have

$$f(\lambda) = \frac{A}{|-\lambda^2 + 2i\alpha\lambda + \omega^2|^2} = \frac{A}{(\lambda^2 - \omega^2)^2 + 4\alpha^2\lambda^2}.$$

Thus, we have arrived at formula (2.148). We note that the process $\xi(t)$ turns out to be differentiable, but it does not have a second derivative [recall (2.86)]; this is explained by the fact that the process $\zeta'(t)$ has only a provisional meaning.

So far, we have only shown that the stationary solution of equation (2.155) cannot have a spectral density different from (2.148), but we have in no way proved the existence of such a solution. To construct a solution of (2.155), we use the method of variation of parameters, just as in the case of the simpler equation (2.137) analyzed in Example 1. The general solution of the homogeneous equation corresponding to (2.155) has the form

$$\xi(t) = C_1 e^{-\alpha t} \sin \beta t + C_2 e^{-\alpha t} \cos \beta t,$$

where, to be explicit, we assume that $\omega^2 - \alpha^2 = \beta^2 > 0$ (the case of damped oscillations). By formally applying the method of variation of parameters, we find that one particular solution of the inhomogeneous equation (2.155) has the form

$$\xi(t) = \frac{1}{\beta} \int_{-\infty}^{t} e^{-\alpha(t-s)} \sin \beta(t - s) \, d\zeta(s), \tag{2.156}$$

which, as is easily seen, defines a stationary random process. It can be shown that this process does in fact satisfy (2.155), if we suitably interpret (2.155). [Some interpretation is required here, since the process (2.156) has no second derivative; a detailed investigation of this problem can be found, for example, in the paper D4.] The fact that the process (2.156) has a correlation function of the form (2.150) is easily verified directly, in complete analogy to what was done in the case of the process (2.135). Naturally, the cases $\omega^2 - \alpha^2 = 0$ and $\omega^2 - \alpha^2 < 0$ can be examined in just the same way.*

***Example 4.*** As our last example, we consider the formula representing the most general damped oscillation

$$B(\tau) = Ce^{-\alpha|\tau|} \cos (\beta|\tau| - \psi), \tag{2.157}$$

where $C > 0, \alpha > 0, \beta > 0$. It follows from property (c) of (1.33) that a necessary condition for the function (2.157) to be a correlation function is that $B'(0) \leqslant 0$, i.e.,

$$|\psi| \leqslant \arctan \frac{\alpha}{\beta}. \tag{2.158}$$

It is not hard to see that the Fourier transform of the function (2.157) is

$$f(\lambda) = \frac{c\lambda^2 + d}{(\lambda^2 + \alpha^2 - \beta^2)^2 + 4\alpha^2\beta^2}, \tag{2.159}$$

where

$$c = \frac{C}{\pi}(\alpha \cos \psi - \beta \sin \psi),$$
$$d = \frac{C(\alpha^2 + \beta^2)}{\pi}(\alpha \cos \psi + \beta \sin \psi). \tag{2.160}$$

Equations (2.159) and (2.160) imply that $f(\lambda)$ is nonnegative if and only if the condition (2.158) holds. Hence, this condition is actually necessary and sufficient for the function (2.157) to be a correlation function.

# 3

# FURTHER DEVELOPMENT
# OF THE CORRELATION THEORY
# OF RANDOM FUNCTIONS

In this final chapter of Part 1, we shall give without proof some results which generalize and strengthen the results obtained in the first two chapters. We shall devote most of our attention to generalizations of the spectral representations of stationary random functions and of their correlation functions (see Secs. 9, 10, and 12).

## 15. The Multidimensional Case

By analogy with the case of multidimensional random variables

$$\boldsymbol{\xi} = (\xi_1, \xi_2, \ldots, \xi_n)$$

(see the Introduction), we can also consider multidimensional random processes and sequences

$$\boldsymbol{\xi}(t) = (\xi_1(t), \xi_2(t), \ldots, \xi_n(t)).$$

Just as in the one-dimensional case, if we talk simultaneously about multidimensional random processes and multidimensional random sequences, we use the term *multidimensional random functions*. In studying such functions, we have to consider not only ordinary correlation functions,[1] but also the

---

[1] Sometimes called *autocorrelation functions*, to distinguish them from the *cross-correlation functions*.

so-called *cross-correlation functions* of the random functions $\xi_j(t)$ and $\xi_k(t)$, where $j \neq k$.  Thus, we must consider all the quantities[2]

$$B_{jk}(t, s) = \mathbf{E}\xi_j(t)\overline{\xi_k(s)} \qquad (j, k = 1, 2, \ldots, n). \qquad (3.1)$$

In the multidimensional case, the role of the correlation function $B(t, s)$ is played by the *correlation matrix*

$$B(t, s) = \|B_{jk}(t, s)\|. \qquad (3.2)$$

We shall only consider random functions whose mean values are constant. If the cross-correlation function $B_{\xi\eta}(t, s)$ of two such functions $\xi(t)$ and $\eta(t)$ depends only on the difference $t - s$, then $\xi(t)$ and $\eta(t)$ are said to be *stationarily correlated* (*with each other*).  A multidimensional random function $\boldsymbol{\xi}(t) = (\xi_1(t), \xi_2(t), \ldots, \xi_n(t))$ is said to be *stationary* if all its components $\xi_1(t), \xi_2(t), \ldots, \xi_n(t)$ are stationary and if they are stationarily correlated in pairs.  In other words, the random function $\boldsymbol{\xi}(t)$ with constant mean value is said to be stationary if all the elements of the correlation matrix $B(t, s)$ of $\boldsymbol{\xi}(t)$ depend only on the difference $t - s$, i.e., if

$$B(t, s) = B(t - s).$$

This definition is obviously analogous to the wide-sense definition of stationarity given in Sec. 3 for one-dimensional random functions.  It is also not hard to give a definition of stationarity for multidimensional random functions $\boldsymbol{\xi}(t)$ which resembles the strict-sense definition given in Sec. 2; we leave it to the reader to do this.  Similar remarks can be made in connection with all subsequent generalizations of the definition of stationarity given in this chapter.

It follows from (3.1) that the elements of the correlation matrix of a multi-dimensional random function satisfy the condition

$$B_{jk}(t, s) = \overline{B_{kj}(s, t)}. \qquad (3.3)$$

In the stationary case, (3.3) reduces to

$$B_{jk}(\tau) = \overline{B_{kj}(-\tau)}. \qquad (3.4)$$

The diagonal elements of the correlation matrix of a multidimensional stationary random function $\boldsymbol{\xi}(t)$ are the correlation functions of the separate components of $\boldsymbol{\xi}(t)$, and hence, according to (2.35) and (2.98), they can be represented in the form

$$B_{jj}(\tau) = \int e^{i\lambda\tau} \, dF_{jj}(\lambda) \qquad (j = 1, 2, \ldots, n), \qquad (3.5)$$

where the $F_{jj}(\lambda)$ are real nondecreasing functions, and the limits of integration

---

[2] If we wish to be even more explicit, we can write $B_{\xi_j\xi_k}(t, s)$ here, and $F_{\xi_j\xi_k}(\lambda)$ and $f_{\xi_j\xi_k}(\lambda)$ below.

in (3.5) are $-\infty$, $+\infty$ in the case of stationary processes, and $-\pi$, $+\pi$ in the case of stationary sequences. It turns out that it is also always possible to represent the nondiagonal elements of the correlation matrix [i.e., the cross-correlation functions of different components of $\xi(t)$] in the form of Fourier-Stieltjes integrals

$$B_{jk}(\tau) = \int e^{i\lambda\tau} \, dF_{jk}(\lambda) \qquad (j, k = 1, 2, \ldots, n; j \neq k), \qquad (3.6)$$

where the limits of integration are the same as in the case of (3.5), but where the functions $F_{jk}(\lambda)$ are now (in general) complex functions of bounded variation [see Cramér (C3) and Kolmogorov (K12)]. The function $F_{jk}(\lambda)$ will be called the *cross-spectral distribution function* of the stationary random functions $\xi_j(t)$ and $\xi_k(t)$, and the matrix

$$F(\lambda) = \|F_{jk}(\lambda)\| \qquad (3.7)$$

will be called the *spectral distribution matrix* of the multidimensional stationary random function $\xi(t)$.

It is an immediate consequence of (3.4) that

$$F_{jk}(\lambda) = \overline{F_{kj}(\lambda)}, \qquad (3.8)$$

so that the spectral distribution matrix is Hermitian. In the one-dimensional case, we have a single spectral distribution function $F(\lambda)$ instead of a spectral distribution matrix, and $F(\lambda)$ is a real nondecreasing function, i.e.,

$$\Delta F(\lambda) = F(\lambda + \Delta\lambda) - F(\lambda) \geqslant 0 \qquad (3.9)$$

for any $\lambda$ and $\Delta\lambda > 0$. The generalization of this property of the spectral distribution function is given by the following property of the spectral distribution matrix: *For any $\lambda$ and $\Delta\lambda > 0$, the matrix*

$$\|\Delta F_{jk}(\lambda)\| = \|F_{jk}(\lambda + \Delta\lambda) - F_{jk}(\lambda)\| \qquad (3.10)$$

*must be Hermitian and nonnegative definite*, i.e., the elements of (3.10) must satisfy the inequality

$$\sum_{j,k=1}^{n} \Delta F_{jk}(\lambda) x_j \bar{x}_k \geqslant 0 \qquad (3.11)$$

for any complex numbers $x_1, x_2, \ldots, x_n$ whatsoever.[3] (See C3, and for another proof, K12.) The converse assertion is also true: Given any matrix (3.7) satisfying the condition (3.11), we can find a multidimensional stationary random function $\xi(t)$ which has (3.7) as its spectral distribution matrix (C3, K12).

In the case where all the correlation functions $B_{jk}(\tau)$ fall off sufficiently

---

[3] Actually, (3.11) implies that $\|\Delta F_{jk}(\lambda)\|$ is Hermitian, and similarly for (3.13).

rapidly at infinity, i.e., satisfy the condition (2.65) or (2.99), the cross-spectral distribution functions $F_{jk}(\lambda)$ can be written as indefinite integrals

$$F_{jk}(\lambda) = \int^{\lambda} f_{jk}(\lambda)d\lambda \qquad (j, k = 1, 2, \ldots, n) \qquad (3.12)$$

of the *cross-spectral densities* $f_{jk}(\lambda)$.  In this case, the condition (3.11) can be replaced by the condition

$$\sum_{j,\,k=1}^{n} f_{jk}(\lambda)x_j\bar{x}_k \geq 0 \quad \text{for all} \quad \lambda, x_1, x_2, \ldots, x_n, \qquad (3.13)$$

i.e., *the spectral density matrix* $\|f_{jk}(\lambda)\|$ *must be Hermitian and nonnegative definite.*

The results just presented are intimately connected with the possibility of finding spectral representations for the multidimensional stationary random functions themselves.  According to (2.40) and (2.97), each component of a multidimensional stationary random function $\boldsymbol{\xi}(t)$ can be represented in the form

$$\xi_j(t) = \int e^{i\lambda t}\, dZ_j(\lambda) \qquad (j = 1, 2, \ldots, n), \qquad (3.14)$$

where the $Z_j(\lambda)$ are random functions with uncorrelated increments and zero mean values.  Using vector notation, we can rewrite (3.14) as

$$\boldsymbol{\xi}(t) = \int e^{i\lambda t}\, d\mathbf{Z}(\lambda), \qquad (3.15)$$

where

$$\mathbf{Z}(\lambda) = (Z_1(\lambda), Z_2(\lambda), \ldots, Z_n(\lambda))$$

is a vector random function.  It is not hard to show that the fact that the components of $\boldsymbol{\xi}(t)$ are stationarily correlated implies that

$$\mathbf{E}[Z_j(\lambda_2) - Z_j(\lambda_1)]\overline{[Z_k(\lambda_4) - Z_k(\lambda_3)]} = 0 \qquad (3.16)$$

if $\lambda_1 < \lambda_2 \leq \lambda_3 < \lambda_4$, for any $j$ and $k$ (see K12).  Then, formula (3.6) follows at once from (3.14), (3.16) and (3.1), where

$$F_{jk}(\lambda + \Delta\lambda) - F_{jk}(\lambda) = \mathbf{E}[Z_j(\lambda + \Delta\lambda) - Z_j(\lambda)]\overline{[Z_k(\lambda + \Delta\lambda) - Z_k(\lambda)]}. \qquad (3.17)$$

Using (3.17), we immediately obtain the inequality (3.11):

$$\sum_{j,\,k=1}^{n} \Delta F_{jk}(\lambda)x_j\bar{x}_k = \mathbf{E}\left| \sum_{j=1}^{n} [Z_j(\lambda + \Delta\lambda) - Z_j(\lambda)]x_j \right|^2 \geq 0.$$

## 16. Homogeneous Random Fields

Next, we generalize the concept of a stationary random process, by introducing the concept of a (statistically) homogeneous random field in an

$n$-dimensional affine space $R_n$. By a (*one-dimensional*) *random field* in the space $R_n$ (more precisely, *defined on* $R_n$), we mean a random function

$$\xi(P) = \xi(x_1, x_2, \ldots, x_n)$$

of the point

$$P = (x_1, x_2, \ldots, x_n)$$

in $R_n$. The field $\xi(P)$ is said to be *homogeneous* if its mean value $m = \mathbf{E}\xi(P)$ is a constant (which, for simplicity, we take to be zero), and if its correlation function

$$B(P, P') = \mathbf{E}\xi(P)\overline{\xi(P')} \tag{3.18}$$

depends only on the vector $\mathbf{r} = \overrightarrow{P'P}$, i.e.,

$$B(P, P') = B(\mathbf{r}). \tag{3.19}$$

It is clear that the concept of a homogeneous random field in $R_1$ coincides with the concept of a stationary random process, so that stationary processes are included as a special case in the class of homogeneous random fields.

Any homogeneous random field in an $n$-dimensional space $R_n$ has a spectral representation, which we write in a form similar to (2.37):

$$\xi(P) = \xi(x_1, x_2, \ldots, x_n) = \int_{-\infty}^{\infty} \int_{-\infty}^{\infty} \cdots \int_{-\infty}^{\infty} e^{i\lambda \cdot \mathbf{x}} Z(d\lambda). \tag{3.20}$$

Here,

$$\lambda \cdot \mathbf{x} = \lambda_1 x_1 + \lambda_2 x_2 + \cdots + \lambda_n x_n$$

is the scalar product of the two $n$-dimensional vectors $\lambda = (\lambda_1, \lambda_2, \ldots, \lambda_n)$ and $\mathbf{x} = (x_1, x_2, \ldots, x_n)$, $d\lambda$ is an element of volume of the $n$-dimensional space of vectors $\lambda$, and $Z(\Delta\lambda)$ is a random function of the $n$-dimensional interval $\Delta\lambda = (\Delta\lambda_1, \Delta\lambda_2, \ldots, \Delta\lambda_n)$, with the following properties:

(a) $$\mathbf{E}Z(\Delta\lambda) = 0$$

for all $\Delta\lambda$;

(b) $$Z(\Delta_1\lambda + \Delta_2\lambda) = Z(\Delta_1\lambda) + Z(\Delta_2\lambda)$$

if $\Delta_1\lambda$ and $\Delta_2\lambda$ are disjoint intervals;

(c) $$\mathbf{E}Z(\Delta_1\lambda)\overline{Z(\Delta_2\lambda)} = 0$$

if $\Delta_1\lambda$ and $\Delta_2\lambda$ are disjoint intervals.

The correlation function of a homogeneous random field can also be represented as a Fourier-Stieltjes integral

$$B(\mathbf{r}) = \int_{-\infty}^{\infty} \int_{-\infty}^{\infty} \cdots \int_{-\infty}^{\infty} e^{i\lambda \cdot \mathbf{r}} \, dF(\lambda), \tag{3.21}$$

where $F(\lambda) = F(\lambda_1, \lambda_2, \ldots, \lambda_n)$ has the property that the increment of the function $F(\lambda)$ on an arbitrary $n$-dimensional interval $\Delta\boldsymbol{\lambda} = (\Delta\lambda_1, \Delta\lambda_2, \ldots, \Delta\lambda_n)$, defined by the formula

$$\Delta F(\boldsymbol{\lambda}) = F(\lambda_1 + \Delta\lambda_1, \ldots, \lambda_n + \Delta\lambda_n)$$
$$- \sum_k F(\lambda_1 + \Delta\lambda_1, \ldots, \lambda_{k-1} + \Delta\lambda_{k-1}, \lambda_k, \lambda_{k+1} + \Delta\lambda_{k+1}, \ldots, \lambda_n + \Delta\lambda_n)$$
$$+ \sum_{i<j} F(\lambda_1 + \Delta\lambda_1, \ldots, \lambda_i, \ldots, \lambda_j, \ldots, \lambda_n + \Delta\lambda_n) \qquad (3.22)$$
$$- \cdots + (-1)^n F(\lambda_1, \lambda_2, \ldots, \lambda_n),$$

is nonnegative, and the relation between $F(\boldsymbol{\lambda})$ and $Z(\Delta\boldsymbol{\lambda})$ is given by

$$\Delta F(\boldsymbol{\lambda}) = \mathbf{E}|Z(\Delta\boldsymbol{\lambda})|^2. \qquad (3.23)$$

The function $F(\lambda)$ is called the *spectral distribution function* of the field $\xi(P)$. In the case where the correlation function $B(\mathbf{r})$ falls off sufficiently rapidly at infinity, the $n$-dimensional Fourier-Stieltjes integral (3.21) can be written as an ordinary $n$-dimensional Fourier integral of the *spectral density*

$$f(\boldsymbol{\lambda}) = f(\lambda_1, \lambda_2, \ldots, \lambda_n) = \frac{\partial^n F(\lambda_1, \lambda_2, \ldots, \lambda_n)}{\partial\lambda_1 \partial\lambda_2 \ldots \partial\lambda_n}, \qquad (3.24)$$

i.e., (3.21) becomes

$$B(\mathbf{r}) = \int_{-\infty}^{\infty} \int_{-\infty}^{\infty} \cdots \int_{-\infty}^{\infty} e^{i\boldsymbol{\lambda}\cdot\mathbf{r}} f(\boldsymbol{\lambda}) \, d\lambda_1 \, d\lambda_2 \ldots d\lambda_n. \qquad (3.21')$$

Moreover, the density $f(\boldsymbol{\lambda})$ can be found from the correlation function by using the inversion formula for $n$-dimensional Fourier integrals.

In addition to homogeneous fields in the $n$-dimensional space $R_n$, we can also consider homogeneous fields defined on the $n$-dimensional lattice consisting of all points in $R_n$ with integral coordinates; such fields are a generalization of the concept of a stationary random sequence. All the results concerning homogeneous fields defined on the whole space $R_n$ can be carried over to homogeneous fields defined on this lattice, provided we make a single change, i.e., the integration in formulas (3.20) and (3.21) must no longer be over the $n$-dimensional space of vectors $\boldsymbol{\lambda}$, but only over the cube in this space defined by the inequalities

$$|\lambda_j| \leqslant \pi \qquad (j = 1, 2, \ldots, n).$$

According to a general theorem of Karhunen (K2, Theorem 10), once we have proved the spectral representation (3.21) of the correlation function $B(\mathbf{r})$ of a homogeneous field $\xi(P)$, the spectral representation (3.20) of $\xi(P)$ itself follows at once. Moreover, by using an argument similar to that given in the paper K6, the proof of the spectral representation of $B(\mathbf{r})$ reduces to proving the multidimensional analog of the Bochner theorem (Sec. 10). The multidimensional Bochner theorem can be proved in a way completely analogous to the proof of the corresponding one-dimensional theorem (see

B9).   (See also B7, where a proof is given which is similar to the proof indicated at the end of Sec. 10.)

*It should also be noted that the generalization of the Bochner theorem needed here can be proved from the outset in a very general form, which includes as special cases the space $R_n$ and the lattice of points in $R_n$ with integral coordinates.   In fact, if we consider an arbitrary commutative topological group $G$ with a Haar measure, e.g., a locally bicompact group (see W3), then there is a theorem about nonnegative definite functions defined on $G$ which suitably generalizes the Bochner theorem with the complex exponential functions replaced by the characters of $G$, as one might expect (see R1, W3). This immediately implies that the theorems on the spectral representation of stationary random processes and their correlation functions can be generalized to the case of homogeneous random fields defined on any commutative topological group $G$ with a Haar measure.   This remark constitutes the content of the note K1.*

The concept of a homogeneous random field can be generalized further, leading to the concept of a multidimensional homogeneous random field (see e.g., B3, O1, T1, Y1, Y5).   In fact, an *n-dimensional random field* is an ordered set $\xi(P) = (\xi_1(P), \xi_2(P), \ldots, \xi_n(P))$ of $n$ one-dimensional random fields, called its *components* , and $\xi(P)$ is said to be *homogeneous* if all the elements of its correlation matrix

$$B(P, P') = \|\mathbf{E}\xi_j(P)\overline{\xi_\kappa(P')}\| = \|B_{jk}(P, P')\|$$

depend only on the vector $\mathbf{r} = \overrightarrow{P'P}$, i.e.,

$$B(P, P') = B(\mathbf{r}), \tag{3.25}$$

where $B(\mathbf{r})$ is now a matrix (cf. p. 82).

## 17. Homogeneous and Isotropic Random Fields

A homogeneous random field $\xi(P)$ in a Euclidean space $R_n$ is said to be *homogeneous and isotropic* if its correlation function $B(P, P') = B(\mathbf{r})$ depends only on the distance between the points $P$ and $P'$, i.e., only on the length $r$ of the vector $\mathbf{r}$ and not on its direction, so that

$$B(P, P') = B(r). \tag{3.26}$$

Such fields are encountered in the statistical theory of turbulence (see e.g., B3, O3, T1, Y1).

In the case of a homogeneous and isotropic random field, we can first transform to $n$-dimensional spherical coordinates in (3.21), and then explicitly perform the integration with respect to the angular variables, obtaining the formula

$$B(r) = \int_0^\infty \frac{J_{(n-2)/2}(\lambda r)}{(\lambda r)^{(n-2)/2}} \, dG(\lambda), \tag{3.27}$$

where $J_{(n-2)/2}$ is the Bessel function of the first kind of order $(n-2)/2$, and $G(\lambda)$ is a real nondecreasing bounded function of the variable $\lambda$. This formula gives the general form of the correlation function of a homogeneous and isotropic random field in an $n$-dimensional Euclidean space $R_n$. Alternatively, (3.27) can be characterized as the formula giving the general form of a function of nonnegative definite type defined on $R_n$ which depends only on the distance between its arguments.[4] From this point of view, (3.27) was first derived in Schoenberg's paper S1. [See also Krein (K25) and Fan (F1).]

The concept of a homogeneous and isotropic random field in $R_n$ is susceptible to two different multidimensional generalizations, the first of which is the following: The multidimensional homogeneous random field $\xi(P) = (\xi_1(P), \xi_2(P), \ldots, \xi_n(P))$ is said to be *homogeneous and isotropic* if its correlation matrix (3.25) depends only on the distance between the points $P$ and $P'$, i.e., only on the length of the vector $\mathbf{r}$ and not on its direction. This definition corresponds to a system of several homogeneous and isotropic scalar fields which are "homogeneously and isotropically correlated" (e.g., the turbulent pressure and temperature fields considered in Y1). However, in the applications one also encounters a second, more interesting generalization: Let $\xi(P) = (\xi_1(P), \xi_2(P), \ldots, \xi_n(P))$ be a multidimensional homogeneous random field whose $n$ components transform linearly among themselves under spatial rotations. For example, the components of $\xi(P)$ may form a vector, a second-order tensor, or a tensor of higher order. Then, it is natural to call $\xi(P)$ a *homogeneous and isotropic field of the quantity* $\xi$ (e.g., a vector or tensor field) if the correlation matrix $B(P, P')$ does not change when we replace the pair of points $P$, $P'$ by a new pair of points $P_1$, $P_1'$ obtained from $P$, $P'$ by making a rotation (so that $\overrightarrow{P'P}$ and $\overrightarrow{P_1'P_1}$ have the same length), *and simultaneously carry out the linear transformation of the components of* $\xi(P)$ *corresponding to the rotation.* The general theory of such fields is intimately connected with the theory of representations of the group of spatial rotations. For the case of a homogeneous and isotropic random vector field in $R_n$, the general form of the correlation matrix is given in the paper Y5.

*Finally, we indicate another natural generalization of the concept of a homogeneous (or a homogeneous and isotropic) random field. Instead of

---

[4] In $R_1$, a function $B(t, t')$ of two variables is said to be of *nonnegative definite type* if

$$\sum_{j, k=1}^{n} B(t_j, t_k)x_j \bar{x}_k \geqslant 0$$

for all values $t_1, t_2, \ldots, t_n$ of the arguments and arbitrary complex numbers $x_1, x_2, \ldots, x_n$. Such a function is also sometimes called a *positive definite kernel*.

It can be shown (see L8, p. 466) that the class of functions which are correlation functions of *nonstationary* random processes coincides with the class of functions of nonnegative definite type (cf. p. 47).

the $n$-dimensional space $R_n$, we can consider an arbitrary "homogeneous" space $R$, i.e., a space with a "group of motions" defined on it. Then, a random field defined on the space $R$ is called *homogeneous* if the mean value of the field is constant, and if the correlation function of the field is the same for all pairs of points which can be carried into each other by some motion of the group. It is clear that this definition contains the definitions of a homogeneous random field in $R_n$ and of a homogeneous and isotropic random field in $R_n$ as special cases: In the first case, the space $R$ corresponds to the space $R_n$ with the translation group as the group of motions, while in the second case, $R$ corresponds to the space $R_n$ with the ordinary Euclidean group (consisting of both translations and rotations) as the group of motions. The problem of finding the general form of the correlation function of a homogeneous random field in $R$ can be reduced to the problem of finding the form of an arbitrary function of nonnegative definite type defined on $R$ which is invariant under the group of motions in $R$. In the special case where $R$ is an $n$-dimensional sphere, this problem has been solved by Schoenberg (S2). The case of an $n$-dimensional space with constant negative curvature is analyzed in Krein's paper K25. For further results along these lines, see the papers by Gelfand (G1), Krein (K25), and the monograph by Fan (F1). In a recent paper (Y7), the author attempts to develop a general theory of homogeneous fields defined on the rather wide class of homogeneous spaces.*

## 18. Processes with Stationary Increments

Consider the problem of the one-dimensional Brownian motion of a particle (Brownian motion on the real line). Suppose that at the time $t = 0$ the particle is located at the origin of the real line, and let $\xi(t)$ be the coordinate of the particle at the time $t$. The random process $\xi(t)$ is obviously not stationary, since it is clear that $E[\xi(t)]^2$ increases without limit as $t$ increases. However, there is still something resembling stationarity involved here: For example, the set of "paths" traversed by the particle during consecutive and equal time intervals, i.e., the set of random variables

$$\Delta_k \xi = \xi((k + 1)\Delta) - \xi(k\Delta) \qquad (k = \cdots, -2, -1, 0, 1, 2, \ldots),$$

where $\Delta$ is arbitrary, form a stationary random sequence. This fact suggests that we try to generalize the definition of stationarity in such a way as to include the random process $\xi(t)$ and processes like it. Such a generalization actually exists, and will be given below.

In the example just considered, the increment $\xi(t + \tau) - \xi(t)$ of the process $\xi(t)$ is stationary, but the process $\xi(t)$ itself is not. Because of this, in studying processes like $\xi(t)$ it is appropriate to choose as the basic characteristic of $\xi(t)$ the function

$$D(t, s; u, v) = E[\xi(u) - \xi(t)]\overline{[\xi(v) - \xi(s)]}, \qquad (3.28)$$

and not the correlation function $B(t, s) = E\xi(t)\overline{\xi(s)}$. The function (3.28) is called the *structure function* of the process $\xi(t)$, and was first used as a basic characteristic of a process $\xi(t)$ in Kolmogorov's papers on statistical turbulence theory (K14, K15). This function was also studied in a somewhat different context by Kolmogorov (K10, K11) and by von Neumann and Schoenberg (V2).

The structure function obviously tells us less about the process $\xi(t)$ than the correlation function. In fact, if we know $B(t, s)$, we can determine $D(t, s; u, v)$ from the formula

$$D(t, s; u, v) = B(u, v) - B(u, s) - B(t, v) + B(t, s), \qquad (3.29)$$

but if we only know $D(t, s; u, v)$ we cannot uniquely determine $B(t, s)$. It follows at once from (3.28) that the structure function has the following properties:

$$D(t, s; u, v) = \overline{D(s, t; v, u)}, \qquad (3.30)$$

$$D(t, s; u, v) + D(u, s; w, v) = D(t, s; w, v), \qquad (3.31)$$

$$D(t, s; u, v) = -D(u, s; t, v). \qquad (3.32)$$

In particular, (3.31) and (3.32) imply that instead of the general structure function (3.28), we need only consider the values of (3.28) for $t = s$, i.e., the function

$$D(t; u, v) = \mathbf{E}[\xi(u) - \xi(t)]\overline{[\xi(v) - \xi(t)]} = D(t, t; u, v), \qquad (3.33)$$

which depends on three variables. In terms of $D(t; u, v)$, the function (3.28) is given by

$$D(t, s; u, v) = D(t; u, v) - D(t; u, s). \qquad (3.34)$$

In the real case, we can go even further in this direction by introducing the function

$$D(t, u) = \mathbf{E}|\xi(u) - \xi(t)|^2 = D(t; u, u) = D(t, t; u, u), \qquad (3.35)$$

which depends on only two variables. In fact, in the real case, it is easily seen that

$$D(t; u, v) = \tfrac{1}{2}\{D(t, u) + D(t, v) - D(u, v)\}. \qquad (3.36)$$

However, in the complex case, the relation (3.36) is no longer true, and from a knowledge of just the function (3.35) we cannot uniquely determine the more general function (3.33). Henceforth, we shall not use the general function (3.28), and we shall reserve the term "structure function of the random process $\xi(t)$" to designate the simpler functions (3.33), (3.35), or the functions (3.39), (3.40) introduced below.

*We* now make the following basic definition: *The random process $\xi(t)$ is called a process with stationary increments if the mathematical expectation of the increment of $\xi(t)$ during any time interval is proportional to the length of the interval, so that*

$$\mathbf{E}[\xi(s) - \xi(t)] = a(s - t), \qquad (3.37)$$

*where a is a constant, and if the structure function $D(t; u, v)$ of the process $\xi(t)$ depends only on the differences $u - t$ and $v - t$:*[5]

$$D(t; u, v) = D(u - t, v - t). \qquad (3.38)$$

[More generally, equation (3.37) can be replaced by the requirement that the function $\mathbf{E}[\xi(s) - \xi(t)]$ depend (continuously) only on the difference $s - t$.] Thus, a process with stationary increments is characterized by the constant $a$ [which in practice can be taken to be zero, since, if necessary, we can consider the fluctuation of $\xi(t)$ instead of $\xi(t)$ itself[6]] and by the function of two variables

$$D(\tau_1, \tau_2) = \mathbf{E}[\xi(t + \tau_1) - \xi(t)][\overline{\xi(t + \tau_2) - \xi(t)}]. \qquad (3.39)$$

In the real case, instead of (3.39), we can consider the simpler function of one variable

$$D(\tau) = \mathbf{E}|\xi(t + \tau) - \xi(t)|^2. \qquad (3.40)$$

In what follows, we shall always assume that the functions $D(\tau_1, \tau_2)$ and $D(\tau)$ are continuous.

It is clear that every stationary process is also a process with stationary increments.   For a stationary process, $a = 0$ and the functions $D(\tau_1, \tau_2)$ and $D(\tau)$ can be expressed in terms of the correlation function $B(\tau)$ as follows:

$$D(\tau_1, \tau_2) = B(\tau_1 - \tau_2) - B(\tau_1) - B(-\tau_2) + B(0), \qquad (3.41)$$

$$D(\tau) = 2B(0) - B(\tau) - B(-\tau). \qquad (3.42)$$

As an example of a process with stationary increments which is not a stationary process, we cite the process $\xi(t)$ considered above, which describes how the coordinate of a Brownian particle changes in time.

It is easy to see that the derivative $\xi'(t)$ of a process $\xi(t)$ with stationary increments is a stationary process [provided $\xi'(t)$ exists], and that conversely, the indefinite integral of a stationary process is a process with stationary increments.   It is an immediate consequence of the results of Sec. 11 that every differentiable process with stationary increments can be represented in the form

$$\xi(t) = \int_{-\infty}^{\infty} \frac{e^{it\lambda} - 1}{i\lambda} \, dZ(\lambda) + \xi_0, \qquad (3.43)$$

where $Z(\lambda)$ is a random function with uncorrelated increments such that

$$\lim_{\Lambda \to \infty} \mathbf{E}|Z(\Lambda) - Z(-\Lambda)|^2 < \infty, \qquad (3.44)$$

---

[5] The reader should note that the function $D(\cdot, \cdot)$ in (3.38), and subsequently, is not the same as in (3.35).

[6] By the *fluctuation* of $\xi(t)$ is meant the quantity $\xi(t) - \mathbf{E}\xi(t)$ [cf. p.22]. However, it should be noted that in the case of a process $\xi(t)$ with stationary increments, the mean value $\mathbf{E}\xi(t)$ need not exist in general.

while $\xi_0 = \xi(0)$ is a "constant" (i.e., time-independent) random variable. The structure function $D(\tau_1, \tau_2)$ of the process (3.43) obviously equals

$$D(\tau_1, \tau_2) = \int_{-\infty}^{\infty} \frac{(e^{i\tau_1\lambda} - 1)(e^{-i\tau_2\lambda} - 1)}{\lambda^2} dF(\lambda), \qquad (3.45)$$

where $F(\lambda)$ is a bounded real nondecreasing function such that

$$\mathbf{E}|Z(\lambda_2) - Z(\lambda_1)|^2 = F(\lambda_2) - F(\lambda_1) \qquad (3.46)$$

if $\lambda_1 < \lambda_2$. It follows from (3.45) that

$$D(\tau) = D(\tau, \tau) = 2\int_{-\infty}^{\infty} \frac{1 - \cos\tau\lambda}{\lambda^2} dF(\lambda) = \int_0^{\infty} \frac{\sin^2 \tau\lambda}{\lambda^2} dG(\lambda), \qquad (3.47)$$

where $G(\lambda)$ is another bounded real nondecreasing function. Conversely, every random process of the form (3.43), where $Z(\lambda)$ has uncorrelated increments and satisfies (3.44), is a differentiable process with stationary increments, and every function of the form (3.45) [or (3.47)], where $F(\lambda)$ and $G(\lambda)$ are nondecreasing and bounded, is the structure function $D(\tau_1, \tau_2)$ [or $D(\tau)$] of a differentiable process with stationary increments.

It turns out that any *nondifferentiable* random process $\xi(t)$ with stationary increments can also be represented in the form (3.43), and that the corresponding structure functions $D(\tau_1, \tau_2)$ and $D(\tau)$ can be represented in the forms (3.45) and (3.47), respectively. However, in this case, the condition (3.44) imposed on the random function $Z(\lambda)$ with uncorrelated increments appearing in (3.43) is not satisfied, and the nondecreasing functions $F(\lambda)$ and $G(\lambda)$ in (3.45) and (3.47) are unbounded. Thus, for example, if we take $\xi(t)$ to be a random process with uncorrelated increments such that $\mathbf{E}\xi(t) = 0$, $\mathbf{E}[\xi(t)]^2 = c_1|t|$, then the functions $D(\tau)$ and $D(\tau_1, \tau_2)$ are obviously

$$D(\tau) = c_1|t|, \quad D(\tau_1, \tau_2) = \begin{cases} c_1 \min\{|\tau_1|, |\tau_2|\} & \text{for } \tau_1\tau_2 > 0, \\ 0 & \text{for } \tau_1\tau_2 \leqslant 0, \end{cases} \qquad (3.48)$$

where $c_1$ is a positive constant. These functions can be represented in the form (3.47) and (3.45), respectively, if we set

$$G(\lambda) = \frac{2c_1}{\pi}\lambda, \quad F(\lambda) = \frac{c_1}{2\pi}\lambda. \qquad (3.49)$$

Moreover, by using the representation (3.45) of $D(\tau_1, \tau_2)$, it is not hard to prove that the random process $\xi(t)$ itself can be written as an integral of the form (3.43) if

$$Z(\lambda) = \frac{1}{\sqrt{2\pi}}\zeta^*(\lambda), \qquad (3.50)$$

where $\zeta^*(\lambda)$ is a random function of the variable $\lambda$ with uncorrelated increments, such that $E\xi^*(\lambda) = 0$ and

$$E|\zeta^*(\lambda_2) - \zeta^*(\lambda_1)|^2 = c_1|\lambda_2 - \lambda_1|.$$

(See e.g., K2, p. 53.) The functions (3.49) are obviously nondecreasing and unbounded, and (3.50) is a random function with uncorrelated increments which does not satisfy (3.44).

It can be shown that in the general case of an arbitrary process with stationary increments, we have to replace the condition that the functions $F(\lambda)$ and $G(\lambda)$ be bounded, by the weaker conditions

$$\int_\varepsilon^\infty \frac{dF(\lambda)}{\lambda^2} < \infty, \qquad \int_{-\infty}^{-\varepsilon} \frac{dF(\lambda)}{\lambda^2} < \infty \tag{3.51}$$

and

$$\int_\varepsilon^\infty \frac{dG(\lambda)}{\lambda^2} < \infty, \tag{3.51'}$$

for any $\varepsilon > 0$. Correspondingly, in the general case the condition (3.44) is replaced by the requirement that the function

$$F(\lambda) = E|Z(\lambda) - Z(0)|^2 \tag{3.52}$$

satisfy the condition (3.51).

The representation (3.43) of a random process $\xi(t)$ with stationary increments, and the representations (3.45) and (3.47) of the structure functions $D(\tau_1, \tau_2)$ and $D(\tau)$ of $\xi(t)$, where $Z(\lambda)$, $F(\lambda)$, and $G(\lambda)$ satisfy the conditions just indicated, are called the *spectral representations* of the process $\xi(t)$ and of the functions $D(\tau_1, \tau_2)$ and $D(\tau)$. The proof that such representations are possible (given in terms of the geometry of the Hilbert space $H$ introduced in Sec. 6) can be found in papers by Kolmogorov (K10, K11) and by von Neumann and Schoenberg (V2).[7]   (Only the real case is considered in V2.) The converse theorem is also true: If $Z(\lambda)$, $F(\lambda)$ and $G(\lambda)$ have the specified properties, then any random process $\xi(t)$ which has the representation (3.43) is a random process with stationary increments, and any functions $D(\tau_1, \tau_2)$ and $D(\tau)$ which have the representations (3.45) and (3.47) are the structure functions of such a process.

We now explain the meaning of the spectral representations (3.43) and (3.45) for the case of stationary processes, which are special cases of processes with stationary increments. According to (3.41), the spectral representations (2.40) and (2.35) of a stationary process and of its correlation function are equivalent to

$$\xi(t) = \int_{-\infty}^\infty (e^{it\lambda} - 1)\, dZ_1(\lambda) + \xi_0, \tag{3.53}$$

$$D(\tau_1, \tau_2) = \int_{-\infty}^\infty (e^{i\tau_1\lambda} - 1)(e^{-i\tau_2\lambda} - 1)\, dF_1(\lambda), \tag{3.54}$$

---

[7] See also D6, Chap. 11, Sec. 11.

where $\xi_0 = \xi(0)$, and we write $Z_1(\lambda)$ and $F_1(\lambda)$ rather than $Z(\lambda)$ and $F(\lambda)$, in order not to confuse these functions with the functions appearing in (3.43) and (3.45). On the other hand, we can write the general spectral representations (3.43) and (3.45) of a process with stationary increments and of its structure function in a form completely analogous to (3.53) and (3.54). To do so, we introduce the functions

$$
Z_1(\lambda) = \begin{cases} \displaystyle\int_{-\infty}^{\lambda} \frac{dZ(\mu)}{i\mu} & \text{for} \quad \lambda < 0, \\[3mm] -\displaystyle\int_{\lambda}^{\infty} \frac{dZ(\mu)}{i\mu} & \text{for} \quad \lambda > 0, \end{cases}
$$

$$
F_1(\lambda) = \begin{cases} \displaystyle\int_{-\infty}^{\lambda} \frac{dF(\mu)}{\mu^2} & \text{for} \quad \lambda < 0, \\[3mm] -\displaystyle\int_{\lambda}^{\infty} \frac{dF(\mu)}{\mu^2} & \text{for} \quad \lambda > 0. \end{cases} \tag{3.55}
$$

[These functions exist, because of (3.51) and (3.52).] Then, $F_1(\lambda)$ is obviously a nondecreasing function on the half-lines $(-\infty, 0)$ and $(0, \infty)$ such that

$$
\int_{-\infty}^{-\epsilon} dF_1(\lambda) + \int_{\epsilon}^{\infty} dF_1(\lambda) < \infty, \quad \int_{-\epsilon}^{0} \lambda^2 dF_1(\lambda) + \int_{0}^{\epsilon} \lambda^2 dF_1(\lambda) < \infty, \tag{3.56}
$$

for any $\epsilon > 0$ [$F_1(\lambda)$ is not defined at the point $\lambda = 0$]. Moreover, on the half-lines $(-\infty, 0)$ and $(0, \infty)$, $Z_1(\lambda)$ is a random function with uncorrelated increments, and

$$
\mathbf{E}|Z_1(\lambda_2) - Z_1(\lambda_1)|^2 = F_1(\lambda_2) - F_1(\lambda_1) \tag{3.57}
$$

if $\lambda_1 \leqslant \lambda_2 < 0$ or $0 < \lambda_1 \leqslant \lambda_2$.

Now, using the functions (3.55), we can write (3.43) and (3.45) as follows:

$$
\xi(t) = \int_{-\infty}^{\infty} (e^{it\lambda} - 1) \, dZ_1(\lambda) + \xi_0 + \xi_1 t, \tag{3.58}
$$

$$
D(\tau_1, \tau_2) = \int_{-\infty}^{\infty} (e^{i\tau_1\lambda} - 1)(e^{-i\tau_2\lambda} - 1) \, dF_1(\lambda) + c^2 \tau_1 \tau_2. \tag{3.59}
$$

Here, $\xi_1 = Z(+0) - Z(-0)$ is a "constant" random variable (like $\xi_0$),

$$
c^2 = F(+0) - F(-0) = \mathbf{E}|\xi_1|^2
$$

is a nonnegative constant, and the integral from $-\infty$ to $\infty$ in (3.58) and (3.59) is interpreted as the improper integral

$$
\int_{-\infty}^{\infty} = \lim_{\epsilon \to 0} \int_{-\infty}^{-\epsilon} + \lim_{\epsilon \to 0} \int_{\epsilon}^{\infty}.
$$

Thus, the spectral representation (3.43) of a process with stationary increments and the spectral representation (3.45) of its structure function are equivalent to (3.58) and (3.59), and constitute a generalization of the representations (2.40) and (2.35), which are valid for the stationary case. If $\xi(t)$ is stationary, the function $F_1(\lambda)$ gives the distribution of the (average) power of the process over the frequency spectrum, and hence $F_1(\lambda)$ must satisfy the relation

$$\int_{-\infty}^{\infty} dF_1(\lambda) = F_1(\infty) - F_1(-\infty) = \mathbf{E}|\xi(t)|^2 < \infty, \qquad (3.60)$$

whereas in the general case, the increment of the function $F_1(\lambda)$ over a small interval containing the point $\lambda = 0$, i.e., the quantity

$$\Delta_\varepsilon F_1(\lambda) = \int_{-\varepsilon}^{\varepsilon} dF_1(\lambda)$$

can be infinitely large, and then (3.60) is replaced by the condition (3.56). Moreover, in this general case, $\xi(t)$ can contain an extra purely linear term $\xi_1 t$, which is a process with stationary increments, but not a stationary process.

Above, we have given two different ways of writing the spectral representation of a process with stationary increments and of its structure function, namely (3.43), (3.45) and (3.58), (3.59). However, it is sometimes convenient to use a third way of writing $\xi(t)$ and $D(\tau_1, \tau_2)$, which is equivalent to the first two:

$$\xi(t) = \int_{-\infty}^{\infty} (e^{it\lambda} - 1) \frac{i\lambda + 1}{i\lambda} dZ_2(\lambda),$$

$$D(\tau_1, \tau_2) = \int_{-\infty}^{\infty} (e^{i\tau_1\lambda} - 1)(e^{-i\tau_2\lambda} - 1) \frac{\lambda^2 + 1}{\lambda^2} dF_2(\lambda).$$

Here, $F_2(\lambda)$ is a bounded nondecreasing function, and $Z_2(\lambda)$ is a random function with uncorrelated increments, satisfying (3.44). (See K22 and D6, p. 552.)

*The representation (3.58) allows us to construct examples of nonstationary processes with stationary increments by taking the limit of a sequence of stationary processes whose low frequency components have a spectral density which increases without limit. As an example which is both typical and physically intuitive, we consider how such a limit is taken in the case of the Brownian motion of a harmonic oscillator. As we know (see Example 3 of Sec. 14), for a special initial probability distribution, the coordinate $\xi(t)$ of such an oscillator is a stationary random process with spectral density

$$f(\lambda) = \frac{A}{(\lambda^2 - \omega^2)^2 + 4\alpha^2\lambda^2}.$$

In order to increase the spectral density at $\lambda = 0$ without limit, we let $\omega \to 0$. Physically, this is equivalent to letting the stiffness of the spring "binding" the oscillator go to zero. Then

$$\int_{-\varepsilon}^{\varepsilon} f(\lambda) \, d\lambda \to \infty$$

as $\omega \to 0$, with $A$ and $\alpha$ held constant, and hence it is clear that as $\omega \to 0$, $\xi(t)$ does not converge in the mean to any random variable, so that it is impossible to pass to the limit directly here. However, from the fact that $\lambda^2 f(\lambda)$ remains bounded as $\omega \to 0$ [cf. the condition (3.56)], it is easily deduced that the difference

$$\xi(t) - \xi(0) = \int_{-\infty}^{\infty} (e^{it\lambda} - 1) \, dZ(\lambda)$$

approaches a definite limit as $\omega \to 0$. By choosing $\xi(0)$ to be arbitrary, we obtain in the limit a process with stationary increments which describes the Brownian motion of a free particle, and hence is no longer stationary.

Random processes with stationary increments can be generalized further by introducing processes with stationary increments of the second or higher orders. The simplest special case of such processes are processes which have a second or higher-order derivative which is a stationary process. The general theory of processes with stationary increments of higher order can be developed by analogy with the theory of ordinary processes with stationary increments (see e.g., Y3).

In addition to random processes with stationary increments, we can also consider random sequences with stationary increments. However, the study of such sequences is of little theoretical interest, since it can always be reduced to the study of the stationary sequence which consists of the differences of neighboring elements in the original sequence. [It will be recalled that in the case of processes with stationary increments, it is in general not possible to consider the stationary process $\xi'(t)$ instead of the original process $\xi(t)$, since the spectral distribution function of $\xi'(t)$ may diverge at infinity.]

Random processes with stationary increments are susceptible to another generalization, resembling the generalization leading from a stationary random process to a homogeneous or a homogeneous and isotropic random field (see Secs. 16 and 17). The corresponding kinds of random functions are called *locally homogeneous* and *locally homogeneous and locally isotropic* random fields.[8] Such random fields were first considered by Kolmogorov in his papers on turbulence theory (K14, K15), and at present, they play an important role in most investigations devoted to the statistical theory of "developed" turbulence (see e.g., B3, O3, O4, T1).[9*]

---

[8] For a different generalization of stationarity bearing the name *local stationarity*, see the paper S4.

[9] A detailed treatment of the mathematical theory of this class of random fields can be found in the author's paper Y5. This paper uses the theory of *generalized random processes* [due to Gelfand (G2) and Itô (I1)], discussed in Appendix I, p. 207.

# LINEAR EXTRAPOLATION
## AND FILTERING OF
## STATIONARY RANDOM FUNCTIONS

Part 2

LINEAR EXTRAPOLATION
AND FILTERING OF
STATIONARY RANDOM FUNCTIONS

# 4

# LINEAR EXTRAPOLATION
# OF STATIONARY
# RANDOM SEQUENCES

## 19. Statement of the Problem

Suppose that we are dealing with a quantity whose values at the times $t = \ldots, -2, -1, 0, 1, 2, \ldots$ form a stationary random sequence $\xi(t)$. In the applications, it is often necessary to be able to predict the value of $\xi(t)$ at the future time $t + m$, given the observed values of $\xi(t)$ at the past times $t - 1$, $t - 2, \ldots, t - n$. In other words, using the values

$$\xi^{(1)}(t - 1), \xi^{(1)}(t - 2), \ldots, \xi^{(1)}(t - n)$$

belonging to a given realization of the stationary random sequence $\xi(t)$, we are often required to make a prediction of the value $\xi^{(1)}(t + m)$ belonging to the same realization. This problem will be called the *extrapolation problem* for a stationary random sequence.

In general, the predicted value $\bar{\xi}^{(1)}(t + m)$ of the quantity $\xi^{(1)}(t + m)$ will be a function of all the values $\xi^{(1)}(t - 1), \xi^{(1)}(t - 2), \ldots, \xi^{(1)}(t - n)$, i.e.,

$$\bar{\xi}^{(1)}(t + m) = g[\xi^{(1)}(t - 1), \xi^{(1)}(t - 2), \ldots, \xi^{(1)}(t - n)]. \qquad (4.1)$$

Expressions of the form (4.1) will be called *extrapolation formulas*, and we have to find the best such formula. From the very outset, the problem arises of which extrapolation formula should be regarded as the "best." It is clear that we cannot simply take the absolute value of the prediction error

$$\varepsilon_{m,n}^{(1)} = \xi^{(1)}(t + m) - g[\xi^{(1)}(t - 1), \xi^{(1)}(t - 2), \ldots, \xi^{(1)}(t - n)] \qquad (4.2)$$

as an index of the quality of the extrapolation formula, since $|\varepsilon_{m,n}^{(1)}|$ will be different for different realizations of our stationary random sequence. In a probability theory context, the quality of the formula (4.1) can only be evaluated statistically, by averaging it over all realizations. In other words, instead of (4.2), we ought to consider the *random variable*

$$\varepsilon_{m,n} = \xi(t + m) - g[\xi(t - 1), \xi(t - 2), \ldots, \xi(t - n)], \qquad (4.3)$$

and then use some statistical characteristic of this random variable as a measure of the quality of the formula (4.1). In this book, we shall be guided by the classical method of least squares, and we shall consider the best of the extrapolation formulas (4.1) to be the formula for which the quantity

$$\sigma_{m,n}^2 = \mathbf{E}\,|\xi(t + m) - g[\xi(t - 1), \xi(t - 2), \ldots, \xi(t - n)]|^2, \qquad (4.4)$$

called the *mean square extrapolation error*, takes its minimum value. With this interpretation, the problem of extrapolating stationary random sequences becomes a rigorous problem of probability theory.

A more complete solution of the problem of extrapolating the sequence $\xi(t)$ would consist in giving the conditional probability distribution for the random variable $\xi(t + m)$, given that the random variables $\xi(t - 1)$, $\xi(t - 2)$, $\ldots, \xi(t - n)$ take specified values $\xi^{(1)}(t - 1), \xi^{(1)}(t - 2), \ldots, \xi^{(1)}(t - n)$. However, finding such a conditional distribution is usually an extremely complicated problem, and moreover, the conditional distribution itself is too formidable for the majority of applications. Therefore, in practice we confine ourselves to finding the simplest numerical characteristics of the conditional distribution. (Cf. the remarks made at the beginning of Sec. 3.) It is not hard to show that the function (4.1) which corresponds to the minimum value of the quantity (4.4) is just the mean value (mathematical expectation) of our conditional distribution.[1] It follows that the quantity (4.4) itself is the variance of the conditional distribution.

In the case where the sequence $\xi(t)$ is normal, the conditional probability distribution of $\xi(t + m)$ is also normal,[2] and its mean value, corresponding to the maximum of the Gaussian curve, is also simultaneously the most probable value (the mode of the distribution), the median, etc. Thus, in this case, all reasonable criteria for choosing $\xi^{(1)}(t + m)$ lead to the same result. However, in the case where the distributions (1.3) for $\xi(t)$ are not normal, there are no theoretical grounds for always choosing the "conditional mathematical expectation" as an approximate value of the random variable $\xi(t + m)$, and this choice is justified only by the relative simplicity of the analytic evaluation of this quantity as compared with other suitable quantities.

---

[1] This is an immediate consequence of the formula

$$\mathbf{E}|\xi - a|^2 = \mathbf{E}|\xi - \mathbf{E}\xi|^2 + |\mathbf{E}\xi - a|^2,$$

where $\xi$ is a random variable and $a$ is a constant.

[2] See the last paragraph of this section.

We must now make more precise which characteristics of the stationary random sequence $\xi(t)$ are assumed to be available for solving our problem. Clearly, the most interesting solution from a practical point of view would take as known only the values $\xi^{(1)}(t - 1), \xi^{(1)}(t - 2), \ldots, \xi^{(1)}(t - n)$, and nothing else. However, to obtain such a solution we would require a much more developed theory of correlated time series[3] than has yet been achieved. Here, we shall solve the problem in the context of the correlation theory of random functions, i.e., we shall assume that the correlation function $B(\tau)$ of the random sequence $\xi(t)$ is known. If we want to apply the solution so obtained to cases where nothing is known but the quantities $\xi^{(1)}(t - 1)$, $\xi^{(1)}(t - 2), \ldots, \xi^{(1)}(t - n)$, it is only necessary to first use (1.22) to calculate an empirical correlation function. Of course, then, in addition to the error given by the extrapolation formula, there will be an extra error (which will not be considered here), due to the lack of precision in determining $B(\tau)$.

It is not hard to see that the general problem of finding the function

$$g[\xi(t - 1), \xi(t - 2), \ldots, \xi(t - n)]$$

of arbitrary form for which (4.4) takes its minimum value cannot be solved within the framework of the correlation theory. In fact, for any complicated function $g$, the quantity (4.4) will in turn be a very complicated statistical characteristic of the sequence $\xi(t)$, which cannot be expressed in terms of the correlation function $B(\tau)$. Therefore, we shall restrict ourselves here to choosing the best function $g$ from the comparatively narrow class of *linear* functions. In this case, we have

$$g[\xi(t - 1), \xi(t - 2), \ldots, \xi(t - n)] \\ = \alpha_1\xi(t - 1) + \alpha_2\xi(t - 2) + \cdots + \alpha_n\xi(t - n), \qquad (4.5)$$

and the mean square error can be expressed very simply in terms of the correlation function $B(\tau)$. Then, from a knowledge of $B(\tau)$, we can easily find the values of the coefficients $\alpha_1, \alpha_2, \ldots, \alpha_n$ for which the expression (4.4) takes its minimum value. In what follows, we shall be concerned only with such *linear extrapolation*.

It should be noted that when the stationary sequence $\xi(t)$ is normal, the assumption that $g$ is a linear function is not a restriction. In fact, in the next paragraph, we shall show that in this case, the minimum value of the mean square error (4.4) is attained when $g$ is linear, so that the best *linear* extrapolation formula is automatically the best *possible* extrapolation formula. In the general case, this is not so, and as a rule the extrapolation error can be decreased by using a nonlinear extrapolation formula. However, even in the general case, the best linear extrapolation formula is of considerable practical interest, since, aside from the fact that at present there are no general results concerning nonlinear extrapolation, linear extrapolation formulas are always

---

[3] The terms *time series* and *random function* are synonymous.

very convenient in the applications, because of the simplicity of the resulting calculations. Moreover, in many cases, the mean square error of the best linear extrapolation only slightly exceeds the mean square error of the best nonlinear extrapolation (cf. the examples given in Y8).

Finally, we supply a few details concerning normal random sequences. If $\xi(t)$ is a normal sequence, then the probability distribution for the $n + 1$ random variables $\xi(t + m)$, $\xi(t - 1), \ldots, \xi(t - n)$ is normal, and the corresponding probability density at the point $(x_0, x_1, \ldots, x_n)$ of an $(n + 1)$-dimensional space, where $x_0$ is the value of the random variable $\xi(t + m)$, $x_1$ the value of the random variable $\xi(t - 1)$, etc., is proportional to the quantity

$$\exp\left\{-\frac{1}{2}\sum_{k=0}^{n}\sum_{l=0}^{n}b_{kl}x_kx_l\right\}.$$

Here,

$$\sum_{k=0}^{n}\sum_{l=0}^{n}b_{kl}x_kx_l$$

is a positive definite quadratic form in $n + 1$ variables [for simplicity, we assume that the distribution is nonsingular (see C5, p. 311)]. Then, the conditional probability distribution for $\xi(t + m)$, given that

$$\xi(t - k) = \xi^{(1)}(t - k) \qquad (k = 1, 2, \ldots, n),$$

has a probability density proportional to

$$\exp\left\{-\frac{1}{2}\left[b_{00}x^2 + 2\sum_{k=1}^{n}b_{k0}x\xi^{(1)}(t - k)\right]\right\},$$

which is a one-dimensional normal distribution whose mean value depends linearly on $\xi^{(1)}(t - 1)$, $\xi^{(1)}(t - 2), \ldots, \xi^{(1)}(t - n)$. But, as already remarked, the best extrapolation formula is just the mean value of this conditional distribution, and hence in this case is a linear formula.

## 20. Extrapolation When the Values of a Finite Number of Elements of the Sequence are Known

Substituting the expression (4.5) for the function $g$ into the formula (4.4), we obtain

$$\sigma_{m,n}^2 = \mathbf{E}\left|\xi(t + m) - \sum_{k=1}^{n}\alpha_k\xi(t - k)\right|^2$$

$$= B(0) - 2\,\mathrm{Re}\sum_{k=1}^{n}\bar{\alpha}_kB(k + m) + \sum_{k=1}^{n}\sum_{l=1}^{n}\bar{\alpha}_k\alpha_lB(k - l). \tag{4.6}$$

We have to find the values

$$\alpha_1 = a_1, \quad \alpha_2 = a_2, \ldots, \quad \alpha_n = a_n$$

for which (4.6) takes its minimum value. These values are determined from the following system of equations:

$$\left.\frac{\partial \sigma_{m,n}^2}{\partial \bar{\alpha}_k}\right|_{\alpha_1 = a_1, \ldots, \alpha_n = a_n} = -B(m+k) + \sum_{l=1}^{n} a_l B(k-l) = 0 \tag{4.7}$$
$$(k = 1, 2, \ldots, n).$$

In the real case, (4.7) reduces to the familiar condition

$$\frac{\partial \sigma_{m,n}^2}{\partial \alpha_k} = 0 \qquad (k = 1, 2, \ldots, n)$$

for a function of several variables to have an extremum. However, in the complex case, we can write either

$$\frac{\partial \sigma_{m,n}^2}{\partial \alpha_k} = 0 \quad \text{or} \quad \frac{\partial \sigma_{m,n}^2}{\partial \bar{\alpha}_k} = 0 \qquad (k = 1, 2, \ldots, n),$$

as we prefer, since both conditions are equivalent, and can be transformed into each other by replacing all terms of the equations by their complex conjugates.

The system of equations (4.7) has a simple geometric meaning, which can be explained by regarding random variables as vectors in the Hilbert space $H$ introduced in Sec. 6. When $\alpha_1, \alpha_2, \ldots, \alpha_n$ take all possible values, the vectors of the form (4.5) range over a finite-dimensional (linear) subspace $H_n(t)$ of the space $H$, spanned by the $n$ vectors $\xi(t-1), \xi(t-2), \ldots, \xi(t-n)$, and the expression (4.6) is the square of the distance between the point $\xi(t+m)$ of the space $H$ and the point (4.5). Thus, the problem of finding the best linear extrapolation formula reduces to the problem of finding the point

$$L_{m,n}(t) = a_1 \xi(t-1) + a_2 \xi(t-2) + \cdots + a_n \xi(t-n) \tag{4.8}$$

of the subspace $H_n(t)$ for which the distance to the point $\xi(t+m)$ is a minimum. This problem is equivalent to the geometric problem of dropping a perpendicular from the point $\xi(t+m)$ onto the subspace $H_n(t)$, and, as is well known, the foot of the perpendicular is just the point with the desired minimal property. Hence, if $L_{m,n}(t)$ is the linear combination we are looking for, the difference $\xi(t+m) - L_{m,n}(t)$ is the perpendicular dropped from the point $\xi(t+m)$ onto $H_n(t)$, i.e., $\xi(t+m) - L_{m,n}(t)$ is orthogonal to all the vectors $\xi(t-1), \xi(t-2), \ldots, \xi(t-n)$, which constitute a basis for $H_n(t)$. The corresponding orthogonality conditions are

$$(\xi(t+m) - L_{m,n}(t), \xi(t-k)) = 0 \qquad (k = 1, 2, \ldots, n), \tag{4.9}$$

with the scalar product defined by (1.41). In view of (1.41), it is clear that (4.9) and (4.7) are the same.

The geometric derivation of the equations (4.7) gives more than the derivation which uses the conditions for the extremum of a function of several variables. In fact, it follows at once from the geometric derivation that the system (4.7) always has a solution (i.e., the perpendicular exists), where the solution is unique if the vectors $\xi(t-1), \xi(t-2), \ldots, \xi(t-n)$ are linearly independent[4] and where otherwise any two solutions of (4.7) differ only by a linear combination of the vectors $\xi(t-1), \xi(t-2), \ldots, \xi(t-n)$ which is identically zero (i.e., the perpendicular is unique). In other words, this means that the mean square error (the "square of the length of the perpendicular")

$$
\begin{aligned}
\sigma_{m,n}^2(a_1, a_2, \ldots, a_n) &= \left\| \xi(t+m) - \sum_{k=1}^{n} a_k \xi(t-k) \right\|^2 \\
&= B(0) - 2 \operatorname{Re} \sum_{k=1}^{n} \bar{a}_k B(m+k) + \sum_{k=1}^{n} \sum_{l=1}^{n} \bar{a}_k a_l B(k-l)
\end{aligned}
\tag{4.10}
$$

has the same value for all $a_1, a_2, \ldots, a_n$ which are roots of the system (4.7). We also note that because of (4.7),

$$
\sum_{k=1}^{n} \bar{a}_k B(m+k) = \sum_{k=1}^{n} \sum_{l=1}^{n} \bar{a}_k a_l B(k-l),
\tag{4.11}
$$

from which it is clear that

$$
\sum_{k=1}^{n} \bar{a}_k B(m+k)
$$

is real, and hence (4.10) can be rewritten in the form

$$
\begin{aligned}
\sigma_{m,n}^2(a_1, a_2, \ldots, a_n) &= B(0) - \sum_{k=1}^{n} \bar{a}_k B(m+k) \\
&= B(0) - \sum_{k=1}^{n} \sum_{l=1}^{n} \bar{a}_k a_l B(k-l).
\end{aligned}
\tag{4.12}
$$

Moreover, it is an immediate consequence of the geometric interpretation of the system of equations (4.7) that the solution of (4.7) actually corresponds to a minimum of the quadratic form (4.6), and not to a maximum, say.

Of course, all these facts can also be proved analytically, without recourse to the geometric interpretation of the system (4.7). Thus, for example, the fact that the solution $a_1, a_2, \ldots, a_n$ of (4.7) corresponds to an absolute minimum of (4.6) is a consequence of the following identity, which is easily deduced from (4.7):

$$
\begin{aligned}
\sigma_{m,n}^2(\alpha_1, \alpha_2, \ldots, \alpha_n) &- \sigma_{m,n}^2(a_1, a_2, \ldots, a_n) \\
&= \sum_{k=1}^{n} \sum_{l=1}^{n} (\bar{\alpha}_k - \bar{a}_k)(\alpha_l - a_l) B(k-l).
\end{aligned}
$$

---

[4] This is also clear from the fact that in this case the determinant of the system (4.7) is nonzero.

Here $\sigma_{m,n}^2(\alpha_1, \alpha_2, \ldots, \alpha_n)$ is the expression (4.6), and $\sigma_{m,n}^2(a_1, a_2, \ldots, a_n)$ is the expression (4.10). Since the correlation function $B(\tau)$ is nonnegative definite (see Secs. 5, 6), it follows at once that we always have

$$\sigma_{m,n}^2(\alpha_1, \alpha_2, \ldots, \alpha_n) \geqslant \sigma_{m,n}^2(a_1, a_2, \ldots, a_n).$$

Now that we have obtained the system of linear equations (4.7), it is a trivial matter to find the best linear extrapolation formula (4.5). In fact, we need only solve the system (4.7) by ordinary methods, e.g., by using determinants. If the determinant of the system is nonzero, the solution will be unique, and otherwise, we choose one of the solutions of (4.7), according to our preference. If $a_1, a_2, \ldots, a_n$ is such a solution, then the best linear extrapolation formula

$$\xi^{(1)}(t + m) = a_1\xi^{(1)}(t - 1) + a_2\xi^{(1)}(t - 2) + \cdots + a_n\xi^{(1)}(t - n), \quad (4.13)$$

and the corresponding mean square error can be calculated from (4.12). However, in the next section, we shall see that from the standpoint of making practical applications of the theory, the situation is by no means so simple.

## 21. Extrapolation When the Entire Past of the Sequence is Known

In Sec. 20, we reduced the problem of finding the best linear extrapolation formula to the solution of the system (4.7) of $n$ linear equations in $n$ unknowns, where $n$ is the number of observed values of the sequence $\xi(t)$. If the number $n$ is small ($< 10$, say), then our entire formulation of the problem is clearly quite unrealistic, since we would not be able to make a sufficiently reliable determination of the correlation function. In practice, the number $n$ is often large, of order $10^1$, $10^2$ and sometimes even $10^3$. However, the solution of systems of linear equations of high order is exceptionally tedious, and hence for large $n$, the method given in Sec. 20 for finding the best extrapolation formula is very unsuitable for practical purposes.[5] Thus, in cases where $n$ is large, it is often more convenient to assume that $n = \infty$, i.e., that we know the entire past behavior of the sequence $\xi(t)$. As we shall see below, in this limiting case it is very often possible to find comparatively simple extrapolation formulas, which are of considerable practical interest. The remainder of this chapter will be devoted to the problem of linear extrapolation of a random sequence by using its entire "past."

---

[5] In the special case of a stationary sequence with a correlation function of the form (2.3), the best linear extrapolation formulas have been found for all values of $m$ and $n$ by Kozulyaev in his dissertation (K19). (See also K18.) The basic difficulty here is that one has to calculate certain $n$th order determinants of a special form. The complexity of these calculations, and the formidable character of the resulting formulas, serve as a good illustration of the hopelessness of using this method in practice, when we have to deal with correlation functions which are not so simple.

It is easy to see that the problem of linear extrapolation of a stationary sequence $\xi(t)$ whose values are known at all past instants of time $t - 1, t - 2,$ $t - 3, \ldots$ can also be formulated as the problem of dropping a perpendicular in the space $H$. The only difference is that in the present case, instead of the finite-dimensional subspace $H_n(t)$ spanned by the vectors $\xi(t - 1), \xi(t - 2),$ $\ldots, \xi(t - n)$, we have to consider the subspace $H(t)$ spanned by the infinite set of vectors $\xi(t - 1), \xi(t - 2), \xi(t - 3), \ldots$. Of course, in general the subspace $H(t)$ will be infinite-dimensional. In fact, the subspace $H(t)$ consists of all possible linear combinations of the form

$$\sum_{k=1}^{N} \alpha_k \xi(t - k)$$

and of the limits of all such combinations in the sense of the metric of $H$, i.e., in the mean. (For a rigorous treatment of the geometric ideas used here, as applied to infinite-dimensional spaces, see e.g., A2, Chap. 1.)

The point $L_m(t)$ of $H(t)$ which is the foot of the perpendicular dropped from the point $\xi(t + m)$ onto $H(t)$ is the point of $H(t)$ which is closest to $\xi(t + m)$. Thus, for $L_m(t)$ the distance

$$\sigma_m^2 = \| \xi(t + m) - L_m(t) \|^2 \qquad (4.14)$$

is smaller than for all other points of $H(t)$. The vector $L_m(t)$ in $H(t)$ is obviously uniquely determined by the fact that the vector $\xi(t + m) -. L_m(t)$ is orthogonal to the subspace $H(t)$, i.e., orthogonal to all the vectors $\xi(t - 1),$ $\xi(t - 2), \xi(t - 3), \ldots$. In other words, $L_m(t)$ satisfies the infinite system of equations

$$(\xi(t + m) - L_m(t), \xi(t - k)) = 0 \qquad (k = 1, 2, 3, \ldots). \qquad (4.15)$$

In terms of the vectors $\xi(t - 1), \xi(t - 2), \xi(t - 3), \ldots$, the vector $L_m(t)$ belonging to $H(t)$ can often be expanded as an infinite series

$$L_m(t) = a_1 \xi(t - 1) + a_2 \xi(t - 2) + a_3 \xi(t - 3) + \cdots, \qquad (4.16)$$

where the series converges in the mean and may not be unique, since in general, the basis vectors chosen in $H(t)$ do not have to be linearly independent.[6] Finally, the best linear extrapolation formula (in the sense of the method of least squares) is given by

$$\xi^{(1)}(t + m) = a_1 \xi^{(1)}(t - 1) + a_2 \xi^{(1)}(t - 2) + a_3 \xi^{(1)}(t - 3) + \cdots \qquad (4.17)$$

---

[6] Generally speaking, the fact that $L_m(t)$ belongs to the subspace $H(t)$ means only that $L_m(t)$ is the limit of a sequence of finite linear combinations of the vectors $\xi(t - k), k > 0$. This still does not imply that the expansion (4.16) is possible, since the basis is not orthogonal. However, in the cases considered here, such an expansion will always exist, and in fact, in many cases, the series (4.16) will even converge with probability one (or as one says in such cases, *for almost all realizations*). This fact justifies the use of formula (4.17).

*Of course, in practice we never know the past values of all elements of the sequence $\xi(t)$. Therefore, in order for the solution of the extrapolation problem for $n = \infty$ to have any practical meaning, the solution must be close to the solution obtained when $n$ is finite but large. We now show that the solution corresponding to $n = \infty$ is in fact the limit as $n \to \infty$ of the solutions corresponding to finite values of $n$. Let $\sigma_{m,n}^2$ be the mean square error defined by (4.12) for extrapolating the sequence $\xi(t)$ ahead $m + 1$ steps when the values of $\xi(t)$ are known at the times $t - 1, t - 2, \ldots, t - n$, and let $L_{m,n}(t)$ be the linear combination of the random variables $\xi(t - 1), \xi(t - 2), \ldots, \xi(t - n)$ defined by formula (4.8), which gives the best approximation to $\xi(t + m)$. It is clear that as $n$ increases, the quantity $\sigma_{m,n}^2$ cannot increase, and hence the limit

$$\lim_{n \to \infty} \sigma_{m,n}^2 = \sigma_m^2 \qquad (4.18)$$

exists. Next, we prove that as $n \to \infty$, the sequence of random variables $L_{m,n}(t)$ also converges (with our interpretation of convergence for sequences of random variables, i.e., in the mean) to a limit

$$\lim_{n \to \infty} L_{m,n}(t) = L_m(t), \qquad (4.19)$$

where $L_m(t)$ belongs to $H(t)$, the difference $\xi(t + m) - L_m(t)$ is orthogonal to $\xi(t - k)$ for all $k > 0$, and

$$\|\xi(t + m) - L_m(t)\|^2 = \sigma_m^2.$$

This will prove that the limits (4.18) and (4.19) coincide with the corresponding quantities appearing in (4.14) to (4.16).

It is clear from the formula (4.8) and the orthogonality conditions (4.9) that $\xi(t + m) - L_{m,n}(t)$ is orthogonal to the vectors $L_{m,1}(t), L_{m,2}(t), \ldots, L_{m,n}(t)$. Then, it is easily deduced from the orthogonality of $\xi(t + m) - L_{m,n}(t)$ and $L_{m,n}(t)$ that

$$\|\xi(t + m) - L_{m,n}(t)\|^2 = \|\xi(t + m)\|^2 - \|L_{m,n}(t)\|^2,$$

and hence

$$\|\xi(t + m)\|^2 - \|L_{m,n}(t)\|^2 \geqslant 0 \qquad (4.20)$$

for any $n$. Moreover, it follows from the orthogonality of $\xi(t + m) - L_{m,n}(t)$ and $L_{m,\nu}(t)$, $1 \leqslant \nu \leqslant n$, that all the vectors

$$\delta_{m,n}(t) = L_{m,n+1}(t) - L_{m,n}(t) \qquad (n = 1, 2, 3, \ldots)$$

are orthogonal in pairs. From this fact and (4.20), we find that

$$\sum_{n=1}^{N} \|\delta_{m,n}(t)\|^2 = \left\| \sum_{n=1}^{N} \delta_{m,n}(t) \right\|^2$$

$$= \|L_{m,N+1}(t) - L_{m,1}(t)\|^2 \leqslant \|\xi(t + m)\|^2 + \|L_{m,1}(t)\|.$$

Consequently, the series

$$\sum_{n=1}^{\infty} \delta_{m,n}(t)$$

converges in the mean, and hence so does the sequence $L_{m,n}(t)$, $n = 1$, $2, 3, \ldots$, i.e., the limit (4.19) exists. It is clear that this limit belongs to the subspace $H(t)$, since all the $L_{m,n}(t)$ belong to $H(t)$. Moreover, the quantity $L_m(t)$ and the limit

$$\sigma_m^2 = \lim_{n \to \infty} \sigma_{m,n}^2 = \lim_{n \to \infty} \|\xi(t + m) - L_{m,n}(t)\|^2$$

are connected by the relation

$$\sigma_m^2 = \|\xi(t + m) - L_m(t)\|^2.$$

Finally, since $\xi(t + m) - L_{m,n}(t)$ is orthogonal to $\xi(t - k)$ for $1 \leqslant k \leqslant n$, it follows that $\xi(t + m) - L_m(t)$ is orthogonal to $\xi(t - k)$ for all $k > 0$, and our proof is complete.*

## 22. Spectral Formulation of the Problem of Linear Extrapolation

In what follows, we shall find it convenient to use the spectral expansion (2.101) of the correlation function $B(\tau)$, where for simplicity, we assume that the spectral density $f(\lambda)$ exists, although this restriction is not essential. We begin by considering the case of linear extrapolation for finite $n$ (see Sec. 20).

Because of (2.101), the basic system of linear equations (4.7) determining the best extrapolation formula can be rewritten as

$$\int_{-\pi}^{\pi} \left[ e^{i(m+k)\lambda} - \sum_{l=1}^{n} a_l e^{i(k-l)\lambda} \right] f(\lambda) \, d\lambda = 0 \qquad (k = 1, 2, \ldots, n), \quad (4.21)$$

or equivalently

$$\int_{-\pi}^{\pi} e^{ik\lambda} [e^{im\lambda} - \Phi_{m,n}(\lambda)] f(\lambda) \, d\lambda = 0 \qquad (k = 1, 2, \ldots, n), \quad (4.22)$$

where

$$\Phi_{m,n}(\lambda) = \sum_{l=1}^{n} a_l e^{-il\lambda}. \tag{4.23}$$

Thus, the problem reduces to finding a function $\Phi_{m,n}(\lambda)$ which is a linear combination of the functions $e^{-i\lambda}, e^{-2i\lambda}, \ldots, e^{-ni\lambda}$, and which satisfies the relations (4.22). The function $\Phi_{m,n}(\lambda)$ will be called the *spectral characteristic for extrapolation*. The extrapolation formula (4.13) is easily determined from $\Phi_{m,n}(\lambda)$, since the coefficients $a_k$ in (4.13) are the same as the coefficients in the linear combination (4.23). Moreover, it is an easy consequence of

(4.10), (4.12) and (2.101) that the mean square error $\sigma_{m,n}^2$ is given in terms of $\Phi_{m,n}(\lambda)$ by the formula

$$\sigma_{m,n}^2 = \int_{-\pi}^{\pi} |e^{im\lambda} - \Phi_{m,n}(\lambda)|^2 f(\lambda) \, d\lambda$$

$$= B(0) - \int_{-\pi}^{\pi} |\Phi_{m,n}(\lambda)|^2 f(\lambda) \, d\lambda. \tag{4.24}$$

If we now set $n = \infty$ in these formulas (in a purely formal way), we arrive at the problem of finding a function

$$\Phi_m(\lambda) \sim a_1 e^{-i\lambda} + a_2 e^{-2i\lambda} + a_3 e^{-3i\lambda} + \cdots \tag{4.25}$$

(for the time being, we do not specify the nature of the convergence of this infinite series) for which

$$\int_{-\pi}^{\pi} e^{ik\lambda}[e^{im\lambda} - \Phi_m(\lambda)]f(\lambda) \, d\lambda = 0 \qquad (k = 1, 2, 3, \ldots). \tag{4.26}$$

If we assume that the expansion (4.16) exists (in the sense of mean convergence), then the convergence in (4.25) must be regarded as mean convergence with respect to $f(\lambda) \, d\lambda$, i.e., with weighting $f(\lambda)d\lambda$. In other words, the sign $\sim$ in (4.25) has to be interpreted as meaning

$$\lim_{n \to \infty} \int_{-\pi}^{\pi} \left| \Phi_m(\lambda) - \sum_{k=1}^{n} a_k e^{-ik\lambda} \right|^2 f(\lambda) \, d\lambda = 0. \tag{4.25'}$$

Then, the coefficients $a_k$ in formulas (4.25) and (4.16) will coincide, so that the coefficients of the best extrapolation formula are simply the Fourier coefficients of the function $\Phi_m(\lambda)$. In the general case, however, the series (4.25) has a more complicated meaning. In fact, it is not hard to see that, strictly speaking, the condition (4.25) has to be interpreted as meaning that $\Phi_m(\lambda)$ is the limit in the mean, with respect to $f(\lambda) \, d\lambda$, of a sequence of finite linear combinations $\Phi_{m,n}(\lambda)$, $n = 1, 2, 3, \ldots$, of the functions $e^{-ik\lambda}, k > 0$; this is equivalent to the condition that $L_m(t)$ belong to $H(t)$. Then, a series like (4.25) which converges to $\Phi_m(\lambda)$ in the usual sense, or even in the mean with respect to $f(\lambda) \, d\lambda$, may not exist, but a series of the form (4.25) which converges to $\Phi_m(\lambda)$ in a certain generalized sense will always exist. However, these subtleties will not be needed here.

Thus, in all cases, if the series (4.25) is interpreted properly, the problem of finding a function $\Phi_m(\lambda)$ satisfying the conditions (4.25) and (4.26) is exactly equivalent to the problem of dropping a perpendicular from the point $\xi(t + m)$ onto the subspace $H(t)$. It follows immediately from the fact that such a perpendicular exists and is unique that a function $\Phi_m(\lambda)$ satisfying these conditions always exists, and moreover, $\Phi_m(\lambda)$ is unique in the case of a continuous spectral density $f(\lambda)$ which vanishes nowhere.[7] Finding the

---

[7] In the general case, the function $\Phi_m(\lambda)$ is determined only to within an additive term $\varphi(\lambda)$ which satisfies the condition

$$\int_{-\pi}^{\pi} |\varphi(\lambda)|^2 f(\lambda) \, d\lambda = 0.$$

function $\Phi_m(\lambda)$, which we shall call the *spectral characteristic for extrapolation* [just as in the case of $\Phi_{m,n}(\lambda)$], constitutes a complete solution of the extrapolation problem. Obviously, the expression for the mean square error $\sigma_m^2$ in terms of $\Phi_m(\lambda)$ is the same as the expression for $\sigma_{m,n}^2$ in terms of $\Phi_{m,n}(\lambda)$:

$$\sigma_m^2 = \int_{-\pi}^{\pi} |e^{im\lambda} - \Phi_m(\lambda)|^2 f(\lambda)\, d\lambda = B(0) - \int_{-\pi}^{\pi} |\Phi_m(\lambda)|^2 f(\lambda)\, d\lambda. \quad (4.27)$$

The problem of finding the function $\Phi_m(\lambda)$ in the general case is very complicated, because of the complexity of the interpretation of (4.25). A complete solution of the problem has been given by Kolmogorov (K12), but to present it here would require the use of delicate results from the theory of functions of a complex variable (cf. Sec. 36). Instead, we shall only consider the case where $L_m(t)$ can be expanded in a series (4.16) which converges in the mean, so that (4.25) is interpreted in the sense of (4.25'). In turn, the relation (4.25') will certainly be satisfied if the infinite series

$$\sum_{k=1}^{\infty} a_k e^{-ik\lambda}$$

converges to $\Phi_m(\lambda)$ *uniformly*, a condition which is met in the cases encountered most frequently in the applications. Then, the problem reduces to finding a function $\Phi_m(\lambda)$ which can be expanded (in the usual sense) in a one-sided Fourier series

$$\Phi_m(\lambda) = \sum_{k=1}^{\infty} a_k e^{-ik\lambda} \quad (4.28)$$

involving only negative powers of $e^{i\lambda}$, and is such that the Fourier series corresponding to the function

$$\Psi_m(\lambda) = [e^{im\lambda} - \Phi_m(\lambda)] f(\lambda) \quad (4.29)$$

contains only nonnegative powers of $e^{i\lambda}$:

$$\Psi_m(\lambda) \sim \sum_{k=0}^{\infty} c_k e^{ik\lambda}. \quad (4.30)$$

[Note that (4.26) means that the Fourier coefficients of the function $\Psi_m(\lambda)$ corresponding to negative powers of $e^{i\lambda}$ vanish.] If such a function can be found, then obviously it is just the spectral characteristic for extrapolation. Moreover, in what follows, we shall only consider examples where $f(\lambda)$ is continuous and everywhere nonzero, so that the spectral characteristic is unique.

Below, we shall examine in detail the special case (but one which is very important in the applications) where the spectral density $f(\lambda)$ is a rational function in $e^{i\lambda}$.   In this case, it is convenient to reformulate the problem of finding the function $\Phi_m(\lambda)$ satisfying the conditions (4.28) to (4.30), by introducing the complex variable $z$.   In fact, let $f^*(z)$ be the rational function of $z$ which reduces to $f(\lambda)$ on the unit circle $z = e^{i\lambda}$:

$$f^*(e^{i\lambda}) = f(\lambda).$$

If the series (4.28) converges uniformly, the function

$$\Phi_m^*(z) = \sum_{k=1}^{\infty} \frac{a_k}{z^k} \tag{4.31}$$

is an analytic function of $z$ outside and on the boundary of the unit circle, which satisfies the condition $\Phi_m^*(\infty) = 0$ and reduces to $\Phi_m(\lambda)$ on the unit circle:

$$\Phi_m^*(e^{i\lambda}) = \Phi_m(\lambda).$$

Moreover, the function

$$\Psi_m^*(z) = [z^m - \Phi_m^*(z)]f^*(z) \tag{4.32}$$

is analytic on the unit circle and

$$\Psi_m^*(e^{i\lambda}) = \Psi_m(\lambda),$$

where $\Psi_m(\lambda)$ is the function defined by formula (4.29).   If the function $\Psi_m^*(z)$ is analytic inside and on the boundary of the unit circle, then it can be represented as a power series

$$\Psi_m^*(z) = \sum_{k=0}^{\infty} c_k z^k \tag{4.33}$$

for $|z| \leqslant 1$.   If we set $z = e^{i\lambda}$ in (4.33), we can verify that $\Psi_m(\lambda)$ has a one-sided Fourier expansion of the form (4.30), as required.

Thus, in the case where $f^*(z)$ is rational, the following three conditions uniquely define $\Phi_m^*(z)$, and allow us to find $\Phi_m^*(z)$ without difficulty:

(a) $\Phi_m^*(z)$ is an analytic function outside and on the boundary of the unit circle, i.e., $\Phi_m^*(z)$ can only have singularities inside the unit circle;

(b) $\Phi_m^*(\infty) = 0$;

(c) $\Psi_m^*(z) = [z^m - \Phi_m^*(z)]f^*(z)$ is an analytic function inside and on the boundary of the unit circle, i.e., $\Psi_m^*(z)$ can only have singularities outside the unit circle.

In the next section, we shall give a variety of examples which show how one goes about determining $\Phi_m^*(z)$.

## 23. Examples of Linear Extrapolation of Stationary Sequences

**Example 1.** Let the sequence $\xi(t)$ have a correlation function of the form (2.106), i.e., let $B(\tau) = Ca^{|\tau|}$, where $C > 0$, $a$ is real and $|a| < 1$. Here we have

$$f(\lambda) = \frac{C}{2\pi} \frac{1 - a^2}{|e^{i\lambda} - a|^2} = \frac{C}{2\pi} \frac{1 - a^2}{(e^{i\lambda} - a)(e^{-i\lambda} - a)}$$

[cf. formula (2.107)], and hence $f(\lambda) = f^*(e^{i\lambda})$, where

$$f^*(z) = \frac{C(1 - a^2)}{2\pi} \frac{1}{(z - a)\left(\dfrac{1}{z} - a\right)}$$

$$= \frac{C_1 z}{(z - a)(1 - az)} \qquad \left[C_1 = \frac{C(1 - a^2)}{2\pi}\right]. \tag{4.34}$$

1. First, we consider the problem of extrapolating $\xi(t)$ one step ahead, i.e., we assume that $m = 0$. This problem reduces to finding a function $\Phi_0^*(z)$ with no singularities outside and on the boundary of the unit circle, which vanishes for $z = \infty$ and is such that the function

$$\Psi_0^*(z) = C_1 \frac{[1 - \Phi_0^*(z)]z}{(z - a)(1 - az)} \tag{4.35}$$

has no singularities inside and on the boundary of the unit circle. Since $|a| < 1$, this last condition can only be satisfied if $\Phi_0^*(a) = 1$. Moreover, it is clear from (4.35) that $\Phi_0^*(z)$ can have no singularities other than a simple pole at the point $z = 0$, since any other singularity of $\Phi_0^*(z)$ would also be a singularity of $\Psi_0^*(z)$, which is impossible because of conditions (a) and (c) of the preceding section. Thus, it follows that

$$\Phi_0^*(z) = \frac{\gamma_0(z)}{z},$$

where $\gamma_0(z)$ is an analytic function in the whole complex plane (i.e., an entire function) for which $\gamma_0(a) = a$ and

$$\lim_{z \to \infty} \frac{\gamma_0(z)}{z} = 0.$$

It is clear that the only function $\gamma_0(z)$ satisfying all these conditions is the constant function $\gamma_0(z) = a$, and hence

$$\Phi_0^*(z) = \frac{a}{z}, \qquad \Phi(\lambda) = ae^{-i\lambda}. \tag{4.36}$$

It should be recalled that the uniqueness of $\gamma_0(z)$ is actually an immediate consequence of the uniqueness of the perpendicular dropped from a point of the space $H$ onto a subspace of $H$. Therefore, we need not worry about uniqueness at all, provided we pick a function satisfying all the required conditions.

According to formula (4.27), the mean square extrapolation error equals

$$\sigma_0^2 = \int_{-\pi}^{\pi} |1 - \Phi_0(\lambda)|^2 f(\lambda)\, d\lambda$$
$$= \int_{-\pi}^{\pi} \frac{C(1 - a^2)}{2\pi} \frac{|1 - ae^{-i\lambda}|^2}{|e^{i\lambda} - a|^2}\, d\lambda = C(1 - a^2) = 2\pi C_1. \tag{4.37}$$

Moreover, recalling that the coefficients of the expansion (4.28) are the same as the coefficients of the best extrapolation formula (4.17), we find that here (4.17) is simply

$$\xi^{(1)}(t) = a\xi^{(1)}(t - 1). \tag{4.38}$$

Thus, in this example, we see that even when all the past values of the sequence $\xi(t)$ are known, the best linear prediction of the value of $\xi^{(1)}(t)$ uses only the last of these known values. In other words, if $\xi^{(1)}(t - 1)$ is known, further knowledge of values of the sequence at earlier times is of no use whatsoever in improving the prediction.

2. Next, we consider the general problem of extrapolating $\xi(t)$ ahead $m + 1$ steps, where $m > 0$. This problem reduces to finding a function $\Phi_m^*(z)$ with no singularities outside and on the boundary of the unit circle, which vanishes for $z = \infty$ and is such that the function

$$\Psi_m^*(z) = C_1 \frac{[z^m - \Phi_m^*(z)]z}{(z - a)(1 - az)} \tag{4.39}$$

has no singularities inside or on the boundary of the unit circle. By the same argument as before we find that

$$\Phi_m^*(z) = \frac{\gamma_m(z)}{z},$$

where $\gamma_m(z)$ is an analytic function in the whole complex plane which grows more slowly than the first power of $z$ (as $z$ approaches infinity) and is such that $\gamma_m(a) = a^{m+1}$. It follows that $\gamma_m(z)$ is a constant, i.e.,

$$\gamma_m(z) = a^{m+1}$$

and

$$\Phi_m^*(z) = \frac{a^{m+1}}{z}, \qquad \Phi_m(\lambda) = a^{m+1}e^{-i\lambda}. \tag{4.40}$$

According to formula (4.27), the mean square extrapolation error is now

$$\sigma_m^2 = B(0)[1 - a^{2(m+1)}] = C[1 - a^{2(m+1)}], \tag{4.41}$$

and the corresponding extrapolation formula is

$$\xi^{(1)}(t + m) = a^{m+1}\xi^{(1)}(t - 1). \tag{4.42}$$

Thus we see that in this case also, the extrapolation formula involves only the last known value of the sequence, and knowledge of earlier values is of no use whatsoever in improving the prediction.

The form of the extrapolation formula (4.42) suggests that $\xi(t)$ is a Markov sequence. As is familiar from probability theory, a *Markov sequence* (synonymously, a *Markov process with discrete time*) means a random sequence such that the conditional distribution of the random variable $\xi(t + m)$ if $\xi(t - k_1)$, $\xi(t - k_2), \ldots, \xi(t - k_n)$ take given values (where $m \geqslant 0$, and $1 \leqslant k_1 < k_2 < \cdots < k_n$ are arbitrary) depends only on the value of $\xi(t-k_1)$ and not on the values of the other random variables $\xi(t - k_2), \ldots,$ $\xi(t - k_n)$. This property is sometimes stated as follows (in a way which is not completely accurate): The future of a Markov sequence $\xi(t)$ is determined only by its present, and does not depend on the past behavior of the sequence.

If $\xi(t)$ is a stationary Markov sequence which is also normal, then, as remarked in Sec. 19, the best linear extrapolation formula for approximating $\xi(t + m)$ coincides with the mean value of the corresponding conditional probability distribution of $\xi(t + m)$, and hence has the form

$$L_m(t) = A(m)\xi(t - 1). \qquad (4.42')$$

However, it is not hard to show that if the entire past of the sequence is known, then the extrapolation formula has the form (4.42') only if the correlation function has the form (2.106) [where, however, it is possible that $|a| = 1$], and hence any normal stationary Markov sequence must have a correlation function of this form. Moreover, if we have a normal stationary sequence with a correlation function of the form (2.106), and if we write down the explicit form of the conditional probability distribution for $\xi(t + m)$ if $\xi(t - 1)$, $\xi(t - k_2), \ldots, \xi(t - k_n)$ take given values, where $2 \leqslant k_2 < \ldots < k_n$ (cf. the end of Sec. 19), then it is easily verified that any such sequence must be a Markov sequence. Thus, finally, for normal stationary sequences, the class of Markov sequences is precisely the class of sequences with correlation functions of the form (2.106).

The situation is more complicated in the case of stationary sequences $\xi(t)$ which are not normal. First of all, it is clear that in this case, the correlation function $B(\tau)$ no longer uniquely determines the conditional distribution of $\xi(t + m)$ when the past values of the sequence are known. Therefore, if we know only $B(\tau)$, we are in no position to say whether or not $\xi(t)$ is a Markov sequence. In particular, it is easy to construct examples of sequences with correlation functions of the form (2.106) which are not Markov sequences. On the other hand, there exist stationary Markov sequences with correlation functions different from (2.106).[8]   For such sequences, the best linear extra-

---

[8] For example, let $\xi(t)$ be a real normal stationary Markov sequence with correlation function (2.106), and let $\eta(t) = [\xi(t)]^3$. Then, $\eta(t)$ is obviously a (non-normal) stationary Markov sequence, and it can easily be verified that

$$B_{\eta\eta}(\tau) = \mathbf{E}\eta(t + \tau)\eta(t) = 3C^3[3a^{|\tau|} + 2a^{3|\tau|}].$$

See e.g., M2, formula (7.28).

polation formula will no longer have only one term; of course, the mean value of the corresponding conditional probability distribution for $\xi(t + m)$, i.e., the best extrapolation formula of *general* form (cf. Sec. 19), will still depend only on $\xi(t - 1)$ [since $\xi(t)$ is a Markov sequence], but this dependence will not be linear.    Thus, if we also consider stationary sequences which are not normal, then the correspondence between sequences with correlation functions of the form (2.106) and Markov sequences is completely lost.    However, in the context of correlation theory, there is no way at all of characterizing Markov sequences that are not normal, and the only analog of Markov sequences about which we can talk meaningfully are sequences such that the best linear extrapolation formula is of the type (4.42'); consequently, such sequences will be called *wide-sense Markov sequences* (cf. D6, pp. 90, 233) or *sequences of the Markov type*. These are the only "Markov" sequences which one can discuss without leaving the domain of correlation theory.

Finally, we point out the relation between the results just obtained and the representation of the sequence $\xi(t)$, with correlation function (2.106), as a moving average

$$\xi(t) = A \sum_{k=0}^{\infty} a^k \eta(t - k)$$

[see formula (2.110) of Sec. 13].    It will be recalled that here $\eta(t)$ is a sequence of uncorrelated random variables.    Therefore, if we know only the quantities $\xi(t - 1), \xi(t - 2), \xi(t - 3), \ldots$, we can obviously say nothing at all about the quantities $\eta(t + k)$, $k \geqslant 0$, which are uncorrelated with all the $\xi(t - n)$, $n \geqslant 1$.    From this it is clear that the best prediction of the value of $\xi(t + m)$ which can be made by using the past values of the sequence $\xi(t)$ is just the value of the sum

$$A \sum_{k=1}^{\infty} a^{k+m} \eta(t - k).$$

In fact, we observe that $\xi(t + m)$ differs from this sum only by the term

$$\xi_0 = A \sum_{k=0}^{m} a^{m-k} \eta(t + k),$$

and knowledge of the past values of $\xi(t)$ gives no information whatsoever about $\xi_0$, since $\xi_0$ is a random variable with mean value zero which is uncorrelated with all past values of the sequence $\xi(t)$.    However, it is easily seen that

$$A \sum_{k=1}^{\infty} a^{k+m} \eta(t - k) = a^{m+1} \xi(t - 1),$$

which explains why in the present case, the best extrapolation formula is given by (4.42).

***Example 2.*** Suppose now that the spectral density of the sequence $\xi(t)$ is of the form

$$f(\lambda) = \frac{C_1}{|e^{i\lambda} - a_1|^2 |e^{i\lambda} - a_2|^2}, \tag{4.43}$$

where $C_1 > 0$, $a_1$ and $a_2$ are real ($a_1 \neq a_2$), and $|a_1| < 1$, $|a_2| < 1$. (See Example 4, p. 61.)   Replacing $e^{i\lambda}$ by $z$, we find that

$$f^*(z) = \frac{C_1 z^2}{(z - a_1)(1 - a_1 z)(z - a_2)(1 - a_2 z)}. \tag{4.44}$$

1. First let $m = 0$.   Then, we have to find a function $\Phi_0^*(z)$ with no singularities outside or on the boundary of the unit circle, which vanishes for $z = \infty$ and is such that the function

$$\Psi_0^*(z) = C_1 \frac{[1 - \Phi_0^*(z)]z^2}{(z - a_1)(1 - a_1 z)(z - a_2)(1 - a_2 z)} \tag{4.45}$$

has no singularities inside or on the boundary of the unit circle.   By an argument like that given in Example 1, we conclude that

$$\Phi_0^*(a_1) = \Phi_0^*(a_2) = 1,$$

and that $\Phi_0^*(z)$ can have no singularities other than a pole of order two at the point $z = 0$.   In other words

$$\Phi_0^*(z) = \frac{\gamma_0(z)}{z^2},$$

where the function $\gamma_0(z)$ is regular (analytic) in the whole complex plane (i.e., is an entire function), grows more slowly than $z^2$ (as $z$ approaches infinity), and is such that

$$\gamma_0(a_1) = a_1^2, \qquad \gamma_0(a_2) = a_2^2.$$

In order for these conditions to be satisfied, we have to set $\gamma_0(z)$ equal to a linear function of $z$ which reduces to $a_1^2$ for $z = a_1$ and to $a_2^2$ for $z = a_2$, i.e.,

$$\gamma_0(z) = z^2 - (z - a_1)(z - a_2) = (a_1 + a_2)z - a_1 a_2. \tag{4.46}$$

Thus, we have

$$\Phi_0^*(z) = \frac{a_1 + a_2}{z} - \frac{a_1 a_2}{z^2}, \qquad \Phi_0(\lambda) = (a_1 + a_2)e^{-i\lambda} - a_1 a_2 e^{-2i\lambda} \tag{4.47}$$

and

$$\xi^{(1)}(t) = (a_1 + a_2)\xi^{(1)}(t - 1) - a_1 a_2 \xi^{(1)}(t - 2). \tag{4.48}$$

In this case, we see that the best linear prediction one step ahead uses only information about the last two values of the sequence $\xi(t)$, and knowledge of the earlier values of the sequence is of no use whatsoever in improving the prediction.    The mean square extrapolation error is easily calculated by using formula (4.27):

$$
\begin{aligned}
\sigma_0^2 &= \int_{-\pi}^{\pi} |1 - \Phi_0(\lambda)|^2 f(\lambda)\, d\lambda \\
&= C_1 \int_{-\pi}^{\pi} \frac{|1 - (a_1 + a_2)e^{-i\lambda} + a_1 a_2 e^{-2i\lambda}|^2}{|e^{i\lambda} - a_1|^2 |e^{i\lambda} - a_2|^2}\, d\lambda \\
&= C_1 \int_{-\pi}^{\pi} \frac{|1 - a_1 e^{-i\lambda}|^2 |1 - a_2 e^{-i\lambda}|^2}{|e^{i\lambda} - a_1|^2 |e^{i\lambda} - a_2|^2}\, d\lambda = 2\pi C_1.
\end{aligned}
\tag{4.49}
$$

2. Next let $m \neq 0$. Then, we have to find a function $\Phi_m^*(z)$ which is regular for $|z| \geqslant 1$ and vanishes for $z = \infty$, and which is such that the function

$$
\Psi_m^*(z) = C_1 \frac{[z^m - \Phi_m^*(z)]z^2}{(z - a_1)(1 - a_1 z)(z - a_2)(1 - a_2 z)}
\tag{4.50}
$$

is regular for $|z| \leqslant 1$.    Arguing as in the case $m = 0$, we find that

$$
\Phi_m^*(z) = \frac{\gamma_m(z)}{z^2},
$$

where $\gamma_m(z)$ is a linear function of $z$ which reduces to $a_1^{m+2}$ for $z = a_1$ and to $a_2^{m+2}$ for $z = a_2$.    Thus, we now have

$$
\begin{aligned}
\gamma_m(z) &= \frac{z - a_2}{a_1 - a_2} a_1^{m+2} + \frac{z - a_1}{a_2 - a_1} a_2^{m+2} \\
&= \frac{1}{a_1 - a_2} [(a_1^{m+2} - a_2^{m+2})z - a_1 a_2 (a_1^{m+1} - a_2^{m+1})],
\end{aligned}
\tag{4.51}
$$

so that

$$
\Phi_m^*(z) = \frac{1}{a_1 - a_2} \left[ \frac{a_1^{m+2} - a_2^{m+2}}{z} - \frac{a_1 a_2 (a_1^{m+1} - a_2^{m+1})}{z^2} \right],
\tag{4.52}
$$

$$
\Phi_m(\lambda) = \frac{1}{a_1 - a_2} [(a_1^{m+2} - a_2^{m+1})e^{-i\lambda} - a_1 a_2 (a_1^{m+1} - a_2^{m+1})e^{-2i\lambda}]
$$

and

$$
\xi^{(1)}(t + m) = \frac{a_1^{m+2} - a_2^{m+2}}{a_1 - a_2} \xi^{(1)}(t - 1) - a_1 a_2 \frac{a_1^{m+1} - a_2^{m+1}}{a_1 - a_2} \xi^{(1)}(t - 2).
\tag{4.53}
$$

Thus, the best linear formula for extrapolating $m + 1$ steps ahead also contains only the quantities $\xi^{(1)}(t - 1)$ and $\xi^{(1)}(t - 2)$.    By using this fact, it is easy to show that in the case of a normal stationary random process with the spectral density (4.43), the conditional probability distribution of the random variable $\xi(t + m)$ for known values of $\xi(t - 1), \xi(t - 2), \ldots, \xi(t - n)$, where $m$ and $n > 2$ are arbitrary, depends only on the values of $\xi(t - 1)$ and

$\xi(t - 2)$. Random sequences with this property are called *second-order* (or *compound*) *Markov sequences*. It can be shown (see D4) that with the exception of some degenerate cases, normal stationary second-order Markov sequences always have spectral densities of the form (4.43). In the general case (i.e., where normality is not assumed), stationary sequences with spectral densities of the form (4.43) can be called *wide-sense second-order Markov sequences* or *sequences of the second-order Markov type* (cf. p. 113).

**Example 3.** A generalization of the spectral densities of Examples 1 and 2 is the spectral density

$$f(\lambda) = \frac{1}{|a_0 e^{in\lambda} + a_1 e^{i(n-1)\lambda} + \cdots + a_n|^2}, \tag{4.54}$$

where $a_0 z^n + a_1 z^{n-1} + \cdots + a_n$ is a polynomial with real coefficients, all of whose zeros have absolute values less than unity.[9] By an argument just like that given in Examples 1 and 2, we can easily verify that in this case

$$\Phi_0^*(z) = \frac{\gamma_0(z)}{z^n}, \quad \text{where} \quad \gamma_0(z) = -\frac{a_1}{a_0} z^{n-1} - \frac{a_2}{a_0} z^{n-2} - \cdots - \frac{a_n}{a_0}, \tag{4.55}$$

so that the best linear formula for extrapolating one step ahead is

$$\xi^{(1)}(t) = -\frac{a_1}{a_0} \xi^{(1)}(t - 1) - \frac{a_2}{a_0} \xi^{(1)}(t - 2) - \cdots - \frac{a_n}{a_0} \xi^{(1)}(t - n). \tag{4.56}$$

We see that this formula involves only the quantities $\xi^{(1)}(t - 1)$, $\xi^{(1)}(t - 2)$, $\ldots, \xi^{(1)}(t - n)$, and knowledge of the more remote past cannot increase the accuracy of the prediction.

It is also not hard to show that in this case, the function $\Phi_m^*(z)$ has the form

$$\Phi_m^*(z) = \frac{\gamma_m(z)}{z^n}, \tag{4.57}$$

where $\gamma_m(z)$ is a polynomial of degree $n - 1$ which is easily determined in each specific case (cf. the general formula derived in the next section). From this it is clear that in extrapolating linearly $m + 1$ steps ahead, only the values of the sequence at the times $t - 1, t - 2, \ldots, t - n$ are of any use. Moreover, if it is known that the sequence $\xi(t)$ is normal, then the conditional probability distribution of $\xi(t + m)$ if the values of the sequence are known at the times $t - 1, t - 2, \ldots, t - N$ (where $N > n$) will depend only on the first $n$ of these values. It is natural to call random sequences with this property *n*th-order Markov sequences.

---

[9] It can be shown that the only essential restriction here is that all the zeros of the polynomial have absolute values different from 1, and the requirement that they all have absolute values *less* than 1 is not a further restriction (see Sec. 24 below).

*Example 4.* We now consider the extrapolation of a stationary sequence $\xi(t)$ with a spectral density of the form

$$f(\lambda) = C|e^{i\lambda} - b|^2, \tag{4.58}$$

where $b$ is real and $|b| < 1$. In this case, we have $f(\lambda) = f^*(e^{i\lambda})$, where

$$f^*(z) = C \frac{(z - b)(1 - bz)}{z}. \tag{4.58'}$$

1. First, we consider the problem of extrapolating one step ahead. We have to find a function $\Phi_0^*(z)$ which is regular for $|z| \geqslant 1$ and vanishes for $z = \infty$, and which is such that the function

$$\Psi_0^*(z) = C \frac{[1 - \Phi_0^*(z)](z - b)(1 - bz)}{z} \tag{4.59}$$

is regular for $|z| \leqslant 1$. The last condition shows that $\Phi_0^*(0) = 1$ and that $\Phi_0^*(z)$ can have no singularities other than a simple pole at the point $z = b$, since any other singularity of $\Phi_0^*(z)$ would also be a singularity of $\Psi_0^*(z)$, which is impossible. Therefore,

$$\Phi_0^*(z) = \frac{\gamma_0(z)}{z - b},$$

where $\gamma_0(z)$ is an entire function which grows more slowly than the first power of $z$ (as $z$ goes to infinity) and is such that $\gamma_0(0) = -b$. These conditions are satisfied by the constant function $\gamma_0(z) = -b$, and only by this function; hence, we have

$$\Phi_0^*(z) = - \frac{b}{z - b} = - \frac{b}{z} \frac{1}{1 - (b/z)} = - \sum_{k=1}^{\infty} \frac{b^k}{z^k} \tag{4.60}$$

and

$$\Phi_0(\lambda) = - \frac{b}{e^{i\lambda} - b} = - \sum_{k=1}^{\infty} b^k e^{-ik\lambda}. \tag{4.60'}$$

The corresponding extrapolation formula is

$$\begin{aligned}
\xi^{(1)}(t) &= - \sum_{k=1}^{\infty} b^k \xi^{(1)}(t - k) \\
&= -b\xi^{(1)}(t - 1) - b^2\xi^{(1)}(t - 2) - b^3\xi^{(1)}(t - 3) - \cdots
\end{aligned} \tag{4.61}$$

We see that in this example, to find the best approximation to the quantity $\xi^{(1)}(t)$, it is essential to use the values of the sequence at all past instants of time. The mean square error is easily calculated by using formula (4.27), and turns out to be $\sigma_0^2 = 2\pi C$.

2. Next, we consider the case $m = 1$, i.e., we predict the value of the quantity $\xi^{(1)}(t + 1)$. In this case, we have to find a function $\Phi_1^*(z)$ which

is regular for $|z| \geqslant 1$ and vanishes for $z = \infty$, and which is such that the function

$$\Psi_1^*(z) = C \frac{[z - \Phi_1^*(z)](z - b)(1 - bz)}{z} \tag{4.62}$$

is regular for $|z| \leqslant 1$. By the same argument as before we find that

$$\Phi_1^*(z) = \frac{\gamma_1(z)}{z - b},$$

where $\gamma_1(z)$ is an entire function which grows more slowly than $z$ and is such that $\gamma_1(0) = 0$. However, the only function satisfying these conditions is the function $\gamma_1(z) = 0$ which vanishes identically, and hence $\Phi_1^*(z) = 0$. Thus, the best linear prediction of the value $\xi^{(1)}(t + 1)$ which can be made is the prediction that this value will be zero. Recalling that $\mathbf{E}\xi(t + 1) = 0$, we see that this prediction uses no information at all about the value of the sequence in the past, and is just the prediction we would have made if we knew nothing at all about the random variable $\xi(t + 1)$ except its mean value.

In just the same way, it can be proved that $\Phi_m^*(z) = 0$ and $\Phi_m(\lambda) = 0$ for any $m \geqslant 1$. Thus, in the present case, knowledge of the behavior of the sequence in the past is of no use whatsoever in refining our information about the values of the sequence at the time $t + 1$ and at later times. At first glance, this result seems unexpected, but it has a very simple interpretation. In fact, the correlation function corresponding to the spectral density (4.58) is

$$B(\tau) = \int_{-\pi}^{\pi} e^{i\tau\lambda} f(\lambda) \, d\lambda = C \int_{-\pi}^{\pi} e^{i\tau\lambda}(e^{i\lambda} - b)(e^{-i\lambda} - b) \, d\lambda \tag{4.63}$$

$$= C \int_{-\pi}^{\pi} [(1 + b^2)e^{i\tau\lambda} - be^{i(\tau+1)\lambda} - be^{i(\tau-1)\lambda}] \, d\lambda.$$

Since

$$\int_{-\pi}^{\pi} e^{ik\lambda} \, d\lambda = 0 \quad \text{for} \quad k \neq 0,$$

it is clear that the only nonzero values of $B(\tau)$ are $B(0)$, $B(1)$ and $B(-1)$, i.e., $B(\tau) = 0$ for $|\tau| > 1$. In other words, only neighboring elements of our random sequence $\xi(t)$ are correlated. This implies that the random variables $\xi(t + 1)$, $\xi(t + 2), \ldots$ are completely uncorrelated with the past values of the sequence, and hence the values $\xi(t - 1)$, $\xi(t - 2), \ldots$ are of no use whatsoever in improving the prediction of the quantities $\xi(t + k), k \geqslant 1$. It should also be noted that despite the fact that only one of the quantities $\xi(t - 1)$, $\xi(t - 2), \ldots$ is correlated with $\xi(t)$, values of all the quantities $\xi(t - k), k > 0$ appear in the extrapolation formula (4.61) for $\xi(t)$.

A stationary sequence $\xi(t)$ for which only neighboring elements are correlated can be obtained from a sequence of uncorrelated random variables by using a moving average with only two terms.   In our case,

$$B(0) = 2\pi C(1 + b^2), \quad B(\pm 1) = -2\pi Cb, \quad B(\tau) = 0 \quad \text{for} \quad |\tau| \geqslant 2, \quad (4.64)$$

according to (4.63).   It is easily seen that a sequence with this correlation function is obtained by forming the sum

$$\xi(t) = \sqrt{2\pi C}\,[\eta(t) - b\eta(t - 1)], \tag{4.65}$$

where $\eta(t)$ is a sequence of uncorrelated random variables.   It follows from (4.65) that in extrapolating more than one step ahead, we can make no use whatsoever of known past values of the sequence.   However, in extrapolating one step ahead, i.e., in predicting the value of $\xi(t)$ by using the values of $\xi(t - 1), \xi(t - 2), \xi(t - 3), \ldots$, we ought to reproduce the value of the term containing $\eta(t - 1)$ in the sum (4.65).   This can be done by forming the sum

$$-\sum_{k=1}^{\infty} b^k \xi^{(1)}(t - k)$$

involving all past values of the sequence.   (The convergence of this sum is guaranteed by the fact that $|b| < 1$.)   These considerations make the meaning of (4.61) more intuitive.

Similarly, we can show that in the case of the spectral density

$$f(\lambda) = C|b_0 e^{in\lambda} + b_1 e^{i(n-1)\lambda} + \cdots + b_n|^2, \tag{4.66}$$

it is only possible to extrapolate ahead $n$ steps or less.   In fact, if $m > n$, the values of the sequence in the past are of no use whatsoever in predicting the quantity $\xi(t + m)$, and the best extrapolation formula is just

$$\xi^{(1)}(t + m) = 0.$$

This can be simply explained by the fact that when the spectral density is given by (4.66), the correlation function $B(\tau) = 0$ for $|\tau| > n$.

**Example 5.**   As a last example, we consider the extrapolation of a stationary random sequence with spectral density of the form

$$f(\lambda) = C_1 \frac{|e^{i\lambda} - b|^2}{|e^{i\lambda} - a|^2}, \tag{4.67}$$

where $C_1 > 0$, $a$ and $b$ are real, and $|a| < 1$, $|b| < 1$.   For example, we obtain such a sequence if we form the sum of two uncorrelated sequences, one of which is a sequence of Markov type and the other a sequence of uncorrelated random variables (see Example 3 of Sec. 13).   In this case, it is clear that

$$f^*(z) = C_1 \frac{(z - b)(1 - bz)}{(z - a)(1 - az)}. \tag{4.68}$$

1. Again, we begin by extrapolating one step ahead. We have to find a function $\Phi_0^*(z)$ which is regular for $|z| \geqslant 1$ and vanishes for $z = \infty$, and which is such that the function

$$\Psi_0^*(z) = C_1 \frac{[1 - \Phi_0^*(z)](z - b)(1 - bz)}{(z - a)(1 - az)} \tag{4.69}$$

is regular for $|z| \leqslant 1$. Since $|a| < 1$, we must have $\Phi_0^*(a) = 1$. Moreover, it is clear that $\Phi_0^*(z)$ can have no singularities other than a simple pole at the point $z = b$. It follows that

$$\Phi_0^*(z) = \frac{\gamma_0(z)}{z - b},$$

where $\gamma_0(z)$ is an entire function such that $\gamma_0(a) = a - b$, and

$$\lim_{z \to \infty} \frac{\gamma_0(z)}{z - b} = 0.$$

These conditions are satisfied only by the function

$$\gamma_0(z) = a - b,$$

and hence

$$\Phi_0^*(z) = \frac{a - b}{z - b} = \frac{a - b}{z} \frac{1}{1 - (b/z)} = \frac{a - b}{z} \sum_{k=0}^{\infty} \frac{b^k}{z^k}, \tag{4.70}$$

$$\Phi_0(\lambda) = \frac{a - b}{e^{i\lambda} - b} = (a - b) \sum_{k=1}^{\infty} b^{k-1} e^{-ik\lambda}, \tag{4.70'}$$

and

$$\xi^{(1)}(t) = (a - b)\xi^{(1)}(t - 1) + (a - b)b\xi^{(1)}(t - 2) \\ + (a - b)b^2\xi^{(1)}(t - 3) + \cdots \tag{4.71}$$

Using formula (4.27), the reader can easily verify that the mean square error of the extrapolation formula (4.71) equals $\sigma_0^2 = 2\pi C_1$.

2. Next, we consider extrapolation $m + 1$ steps ahead. In this case, we have to find a function $\Phi_m^*(z)$ satisfying the same conditions as $\Phi_0^*(z)$, except that now

$$\Psi_m^*(z) = C_1 \frac{[z^m - \Phi_m^*(z)](z - b)(1 - bz)}{(z - a)(1 - az)} \tag{4.72}$$

plays the role of the function (4.69). Here we obviously have $\Phi_m^*(a) = a^m$, which implies that

$$\Phi_m^*(z) = \frac{(a - b)a^m}{z - b} = \frac{(a - b)a^m}{z} \sum_{k=0}^{\infty} \frac{b^k}{z^k}, \tag{4.73}$$

$$\Phi_m(\lambda) = \frac{(a - b)a^m}{e^{i\lambda} - b} = (a - b)a^m \sum_{k=1}^{\infty} b^{k-1} e^{-ik\lambda}, \tag{4.73'}$$

and hence

$$\xi^{(1)}(t + m) = a^m(a - b)[\xi^{(1)}(t - 1) + b\xi^{(1)}(t - 2) \\ + b^2\xi^{(1)}(t - 3) + \cdots]. \quad (4.74)$$

(Concerning the meaning of this formula, see pp. 137–138.) By using (4.27), it is easily verified that the mean square error of the extrapolation formula (4.74) equals

$$\sigma_m^2 = 2\pi C_1 \left[ 1 + \frac{(a - b)^2(1 - a^{2m})}{1 - a^2} \right]. \quad (4.75)$$

## *24. The Case of a General Rational Spectral Density in $e^{i\lambda}$

All the examples given above involve extrapolation of stationary sequences whose spectral densities are rational functions of $e^{i\lambda}$. Hence, it is natural to consider the extrapolation of a sequence whose spectral density is a general rational function in $e^{i\lambda}$. By using the properties of the spectral density, it can be shown that a spectral density $f(\lambda)$ which is rational in $e^{i\lambda}$ can always be represented in the form

$$f(\lambda) = \frac{|B(e^{i\lambda})|^2}{|A(e^{i\lambda})|^2} = \frac{|B_0 e^{iM\lambda} + B_1 e^{i(M-1)\lambda} + \cdots + B_M|^2}{|A_0 e^{iN\lambda} + A_1 e^{i(N-1)\lambda} + \cdots + A_N|^2}, \quad (4.76)$$

where the coefficients $A_0$, $B_0$, $A_N$ and $B_M$ are nonzero (see D6, p. 501). Here, all the zeros of the polynomial $A(z)$ have absolute values less than 1, while all the zeros of the polynomial $B(z)$ have absolute values which do not exceed 1. Moreover, in the real case, all the coefficients of the polynomials $A(z)$ and $B(z)$ can be taken to be real.

The proof of these facts goes as follows : The function $f(\lambda)$ gives the boundary values of a rational function $f^*(z)$ on the unit circle, i.e.,

$$f(\lambda) = f^*(e^{i\lambda}).$$

Since $f(\lambda)$ is real, it follows from the Schwarz reflection principle that for every zero (or pole) of the function $f^*(z)$ at a point $\alpha$ where $|\alpha| \neq 1$, there is another zero (or pole) of the same multiplicity at the point $1/\bar{\alpha}$. Thus, either $|\alpha|$ or $1/|\alpha|$ is less than 1. Moreover, since $f^*(e^{i\lambda})$ is nonnegative and integrable, $f^*(z)$ cannot have poles on the circle $|z| = 1$, and all its zeros on the circle $|z| = 1$ must be of even order. This implies that $f(\lambda)$ can be represented in the form (4.76), where $A(z)$ and $B(z)$ satisfy the indicated conditions. In the case where the sequence $\xi(t)$ is real, we must have $f(-\lambda) = f(\lambda)$, and it is easy to see that then all the coefficients of $A(z)$ and $B(z)$ must have the same arguments, so that we can simply take these coefficients to be real.

We shall not consider the most general spectral density of the form (4.76), but instead we confine ourselves to the case where all the zeros of $B(z)$ have absolute values which are strictly less than 1, i.e., we assume that

$$
\begin{aligned}
A(z) &= A_0 z^N + A_1 z^{N-1} + \cdots + A_N \\
&= A_0(z - a_1)(z - a_2)\cdots(z - a_N),
\end{aligned}
\tag{4.77}
$$

$$
\begin{aligned}
B(z) &= B_0 z^M + B_1 z^{M-1} + \cdots + B_M \\
&= B_0(z - b_1)(z - b_2)\cdots(z - b_M),
\end{aligned}
\tag{4.78}
$$

where

$$
0 < |a_1| < 1, \ldots, 0 < |a_N| < 1;
$$
$$
0 < |b_1| < 1, \ldots, 0 < |b_M| < 1.
$$

The case where some of the zeros of $B(z)$ have absolute value 1 can be obtained from our case by passing to a limit, but then there may no longer exist a best extrapolation formula of the form (4.17) [see K18, K19]. Thus, the class of spectral densities considered here consists of functions $f(\lambda)$ which are rational in $e^{i\lambda}$ and which do not vanish for any value of $\lambda$. This class is sufficiently large to satisfy almost all practical requirements. This is because in the applications it almost always turns out that when a spectral density can be determined theoretically, it is a rational function belonging to our class,[10] and moreover, in cases where $f(\lambda)$ can only be determined empirically, it is usually not hard to choose a function from our class which is a sufficiently good approximation to the empirical spectral density.

1. We begin by considering the case of extrapolation one step ahead, i.e., we look for an approximation for $\xi^{(1)}(t)$ which depends linearly on the past values of the sequence. Thus, we have to find a function $\Phi_0^*(z)$ which satisfies the conditions (a), (b) and (c) given on p. 109. Because of (4.76), (4.77), and (4.78), we have

$$
\begin{aligned}
f^*(z) &= \frac{|B_0|^2}{|A_0|^2} \\
&\times \frac{(z - b_1)(z - b_2)\cdots(z - b_M)(1 - \bar{b}_1 z)(1 - \bar{b}_2 z)\cdots(1 - \bar{b}_M z)}{(z - a_1)(z - a_2)\cdots(z - a_N)(1 - \bar{a}_1 z)(1 - \bar{a}_2 z)\cdots(1 - \bar{a}_N z)} z^{N-M}.
\end{aligned}
\tag{4.79}
$$

It will simplify matters somewhat to look for the function

$$
\tilde{\Phi}_0(z) = 1 - \Phi_0^*(z),
$$

instead of $\Phi_0^*(z)$. Obviously, the function $\tilde{\Phi}_0(z)$ is regular for $|z| \geqslant 1$, equals 1 for $z = \infty$, and must be such that the product $\tilde{\Phi}_0(z) f^*(z)$ has no

---

[10] An explanation of this fact is provided by Theorem 3.9 of D4, according to which the spectral densities of a large class of stationary sequences of practical interest are always rational in $e^{i\lambda}$.

singularities for $|z| \leq 1$. It is easy to see that these conditions are satisfied by the function

$$\tilde{\Phi}_0(z) = \frac{(z - a_1)(z - a_2)\cdots(z - a_N)}{(z - b_1)(z - b_2)\cdots(z - b_M)} z^{M-N}, \qquad (4.80)$$

and only by this function. It follows that

$$\tilde{\Phi}_0^*(z) = 1 - \frac{(z - a_1)(z - a_2)\cdots(z - a_N)}{(z - b_1)(z - b_2)\cdots(z - b_M)} z^{M-N} \qquad (4.81)$$

and

$$\Phi_0(\lambda) = 1 - \frac{(e^{i\lambda} - a_1)(e^{i\lambda} - a_2)\cdots(e^{i\lambda} - a_N)}{(e^{i\lambda} - b_1)(e^{i\lambda} - b_2)\cdots(e^{i\lambda} - b_M)} e^{i(M-N)\lambda}. \qquad (4.82)$$

To obtain the coefficients of the extrapolation formula, we must now expand the function $\Phi_0^*(z)$, which is regular for $|z| \geq 1$, as a power series in $z^{-1}$. As a matter of technique, this is most simply done by first expanding $\Phi_0^*(z)$ in partial fractions. Using formula (4.27), we obtain the following remarkably simple expression for the mean square extrapolation error:

$$\sigma_0^2 = \int_{-\pi}^{\pi} |1 - \Phi_0(\lambda)|^2 f(\lambda)\, d\lambda = \int_{-\pi}^{\pi} \left|\frac{B_0}{A_0}\right|^2 d\lambda = 2\pi \left|\frac{B_0}{A_0}\right|^2. \qquad (4.83)$$

It is easily verified that all the results obtained above pertaining to the extrapolation one step ahead of various special stationary sequences are all contained in the general formulas obtained here. In particular, formula (4.81) immediately implies the particularly simple expression (4.56) for the extrapolation formula in the case where $B(z) \equiv 1$, as in Example 3.

2. Next, we consider the problem of extrapolating $m + 1$ steps ahead. In this case, we have to find a function $\tilde{\Phi}_m(z) = z^m - \Phi_m^*(z)$ which is regular for $|z| \geq 1$ and is such that the product $\tilde{\Phi}_m(z)f^*(z)$ is regular for $|z| \leq 1$; moreover, the difference $z^m - \tilde{\Phi}_m(z) = \Phi_m^*(z)$ has to vanish for $z = \infty$. It follows immediately from (4.79) that $\tilde{\Phi}_m(z)$ must have zeros at the points $a_1, a_2, \ldots, a_N$ (and also a zero of order $M - N$ at the point $z = 0$, if $M > N$), and that $\tilde{\Phi}_m(z)$ can have poles only at the points $b_1, b_2, \ldots, b_M$ (and also a pole of order $N - M$ at the point $z = 0$, if $N > M$). Thus

$$\tilde{\Phi}_m(z) = z^m - \Phi_m^*(z) = \gamma_m(z) \frac{(z - a_1)(z - a_2)\cdots(z - a_N)}{(z - b_1)(z - b_2)\cdots(z - b_M)} z^{M-N}, \qquad (4.84)$$

where $\gamma_m(z)$ is an analytic function in the whole complex plane (i.e., an entire function). All that we have to do now is satisfy the condition $\Phi_m^*(\infty) = 0$. But by (4.84)

$$\Phi_m^*(z) = z^m - \tilde{\Phi}_m(z) = \frac{z^m \tilde{\Psi}_0(z) - \gamma_m(z)}{\tilde{\Psi}_0(z)}, \qquad (4.85)$$

where

$$\tilde{\Psi}_0(z) = \frac{1}{\tilde{\Phi}_0(z)} = \frac{(z - b_1)(z - b_2)\cdots(z - b_M)}{(z - a_1)(z - a_2)\cdots(z - a_N)} z^{N-M} \qquad (4.86)$$

is an analytic function outside and on the boundary of the unit circle, which is such that $\tilde{\Psi}_0(\infty) = 1$. This implies that for $|z| \geqslant 1$, we can expand $\tilde{\Psi}_0(z)$ as a power series in $z^{-1}$, beginning with the constant term 1:

$$\tilde{\Psi}_0(z) = 1 + \frac{d_1}{z} + \frac{d_1}{z^2} + \cdots \qquad (4.87)$$

Substituting this expansion for $\tilde{\Psi}_0(z)$ in the numerator of (4.85) and bearing in mind that $\tilde{\Psi}_0(\infty) = 1$, we find that all the conditions imposed on $\Phi_m^*(z)$ will be satisfied if we set

$$\gamma_m(z) = z^m + d_1 z^{m-1} + d_2 z^{m-2} + \cdots + d_m. \qquad (4.88)$$

Thus, we finally find the formula

$$\Phi_m(\lambda) = e^{im\lambda} - \frac{e^{im\lambda} + d_1 e^{i(m-1)\lambda} + \cdots + d_m}{\tilde{\Psi}_0(e^{i\lambda})} \qquad (4.89)$$

for the spectral characteristic for extrapolating $m + 1$ steps ahead, where the function $\tilde{\Psi}_0(z)$ and the coefficients $d_1, d_2, \ldots, d_m$ are uniquely determined by (4.86) and (4.87). For practical calculations of the coefficients $d_k$ and the coefficients of the extrapolation function, it is convenient to first make partial fraction expansions of the rational functions involved.

In the special case where $B(z) = 1$, the function $z^N/\tilde{\Psi}_0(z)$ is a polynomial of degree $N$, and then it follows at once from (4.89) that $\Phi_m(\lambda)$ contains only powers of $e^{-i\lambda}$ which do not exceed $N$, so that the best extrapolation formula contains no more than $N$ terms. Another special case is where $A(z) = 1$. In this case, since

$$\tilde{\Psi}_0(z) = \frac{B(z)}{B_0 z^M},$$

we have

$$\tilde{\Psi}_0(z) = 1 + \frac{B_1}{B_0}\frac{1}{z} + \frac{B_2}{B_0}\frac{1}{z^2} + \cdots + \frac{B_M}{B_0}\frac{1}{z^M}$$

and

$$d_k = \frac{B_k}{B_0} \quad \text{for} \quad k = 1, 2, \ldots, M,$$
$$d_k = 0 \quad \text{for} \quad k > M.$$

Then, it is clear from (4.87) and (4.89) that

$$\Phi_m(\lambda) = 0 \quad \text{for} \quad m \geqslant M.$$

Both of these special cases have already been noted, in Examples 3 and 4 of Sec. 23.

We now calculate the mean square extrapolation error corresponding to formula (4.89). From (4.76), (4.77), (4.78) and (4.86), we obtain

$$f(\lambda) = \frac{|B_0|^2}{|A_0|^2} |\tilde{\Psi}_0(e^{i\lambda})|^2. \qquad (4.90)$$

Therefore

$$\sigma_m^2 = \int_{-\pi}^{\pi} |e^{im\lambda} - \Phi_m(\lambda)|^2 f(\lambda)\, d\lambda$$

$$= \frac{|B_0|^2}{|A_0|^2} \int_{-\pi}^{\pi} |e^{im\lambda} + d_1 e^{i(m-1)\lambda} + \cdots + d_m|^2\, d\lambda,$$

and since the integral of $e^{ik\lambda}$ is nonzero only for $k = 0$, it follows at once that

$$\sigma_m^2 = 2\pi \frac{|B_0|^2}{|A_0|^2} [1 + |d_1|^2 + |d_2|^2 + \cdots + |d_m|^2], \qquad (4.91)$$

which is the required expression for $\sigma_m^2$.

We note that the coefficients $1, d_1, d_2, \ldots$ are the Fourier coefficients of the function $\overset{*}{\Psi}_0(e^{i\lambda})$, since

$$\overset{*}{\Psi}_0(e^{i\lambda}) = 1 + d_1 e^{-i\lambda} + d_2 e^{-2i\lambda} + \cdots, \qquad (4.92)$$

according to (4.87).   If we now apply Parseval's theorem to the series (4.92), we find that

$$1 + \sum_{k=1}^{\infty} |d_k|^2 = \frac{1}{2\pi} \int_{-\pi}^{\pi} |\overset{*}{\Psi}_0(e^{i\lambda})|^2\, d\lambda. \qquad (4.93)$$

Using (4.91), we can write (4.93) in the form

$$\lim_{m \to \infty} \sigma_m^2 = \int_{-\pi}^{\pi} f(\lambda)\, d\lambda = B(0). \qquad (4.94)$$

Since $B(0) = \mathbf{E}|\xi(t + m)|^2$, we see that as $m \to \infty$, the mean square prediction error approaches the mean square of the predicted quantity.   Clearly, this is related to the fact that as the prediction time is increased, the "predictable part" (4.17) of $\xi(t + m)$ becomes smaller and smaller, and the best linear prediction of the distant future is close to zero, i.e., close to the mathematical expectation of $\xi(t + m)$.   In other words, in making predictions very far ahead, the past values of the stationary sequence $\xi(t)$ can only be used in a trivial way to find the mathematical expectation of the elements of $\xi(t)$.   This result can be explained by the fact that in the case being considered, the function $B(\tau)$ falls off very rapidly (in fact, exponentially) as $\tau \to \infty$, so that for very large $m$, the quantity $\xi(t + m)$ is practically independent of the past values of the sequence.   Therefore, it is not surprising that for large values of $m$, knowledge of the past values of $\xi(t)$ gives very meager information about the value of $\xi(t + m)$.   However, later we shall see that in general, explanations of this kind cannot be taken too seriously (see Example 2.2 on p.190, concerning the extrapolation of a particular stationary process).

# 5

---

# LINEAR FILTERING
## OF STATIONARY
## RANDOM SEQUENCES

---

## 25. Filtering When the Values of a Finite Number of Elements of the Sequence are Known

In the preceding chapter, we were concerned with "predicting" the value $\xi^{(1)}(t + m)$ of a stationary random sequence $\xi(t)$ at the future time $t + m$, by using the observed past values of $\xi(t)$. However, in practice, the measurement of any value of the quantity $\xi(t)$ always entails a certain error $\eta(t)$, depending on the measuring device and on the observer making the measurement. Therefore, one actually observes the past values of a new sequence

$$\zeta(t) = \xi(t) + \eta(t).$$

The results of Chap. 4 pertain to the case where the measurement errors $\eta(t)$ are so small that they can be neglected; however, it can hardly be said that neglecting $\eta(t)$ is always justified. In cases where the measurement errors have to be taken into consideration, we are no longer dealing with the problem of linear extrapolation. Instead, we must then consider a new problem, the so-called *filtering problem*.

The simplest version of the problem of filtering a stationary random sequence can be formulated as follows: Given the known values $\zeta^{(1)}(t - 1)$, $\zeta^{(1)}(t - 2), \ldots, \zeta^{(1)}(t - n)$ of the stationary sequence $\zeta(t) = \xi(t) + \eta(t)$ at the times $t - 1, t - 2, \ldots, t - n$, it is required to approximate as closely as possible the value $\xi^{(1)}(t + m)$ taken by the sequence $\xi(t)$ at the time $t + m$. In other words, the problem consists in finding the function

$$\xi^{(1)}(t + m) = g[\zeta^{(1)}(t - 1), \zeta^{(1)}(t - 2), \ldots, \zeta^{(1)}(t - n)] \qquad (5.1)$$

which gives the best approximation to $\xi^{(1)}(t + m)$. Of course, it is natural to assume that both sequences $\xi(t)$ and $\eta(t)$ are stationary random sequences, and we take the mean values of $\xi(t)$ and $\eta(t)$ to be zero, as always. Moreover, we also assume that the sequences $\xi(t)$ and $\eta(t)$ are not correlated with each other, or more precisely that

$$\mathbf{E}\xi(t)\overline{\eta(t)} = 0 \qquad (5.2)$$

for any $t$ and $s$. The condition (5.2) can be replaced by a more general condition,[1] but we shall not strive for the greatest generality. Since in practice the measurement errors can most often be regarded as independent of the values of the measured quantity, with a high degree of accuracy, the case where the condition (5.2) is satisfied is of greatest practical interest.

It will also be assumed that we know the correlation functions $B_{\zeta\zeta}(\tau)$ and $B_{\eta\eta}(\tau)$ of the sequences $\zeta(t)$ and $\eta(t)$. In practice, the correlation function $B_{\zeta\zeta}(\tau)$ is usually calculated from the observed values of the sequence $\zeta(t)$ by using formula (1.22), while the function $B_{\eta\eta}(\tau)$ can be determined by using the same measuring device and the same observer as involved in measuring $\zeta(t)$ to make a series of measurements of any quantity whose value is known precisely, e.g., which equals zero because of the conditions of the experiment. Knowing $B_{\zeta\zeta}(\tau)$ and $B_{\eta\eta}(\tau)$, we can also find the function $B_{\xi\xi}(\tau)$, since, according to (5.2),

$$\mathbf{E}\zeta(t + \tau)\overline{\zeta(t)} = \mathbf{E}[\xi(t + \tau) + \eta(t + \tau)]\overline{[\xi(t) + \eta(t)]}$$
$$= \mathbf{E}\xi(t + \tau)\overline{\xi(t)} + \mathbf{E}\eta(t + \tau)\overline{\eta(t)},$$

so that

$$B_{\xi\xi}(\tau) = B_{\zeta\zeta}(\tau) - B_{\eta\eta}(\tau). \qquad (5.3)$$

As in Chap. 4, we restrict ourselves to the case of a linear approximation to $\xi^{(1)}(t + m)$, i.e., we choose a *filtering formula* (5.1) of the form

$$\xi^{(1)}(t + m) = \alpha_1\zeta^{(1)}(t - 1) + \alpha_2\zeta^{(1)}(t - 2) + \cdots + \alpha_n\zeta^{(1)}(t - n). \qquad (5.4)$$

(This is the case of "linear filtering.") As before, we characterize the quality of the formula (5.4) by the mean square (over all the realizations) of the error

$$\varepsilon_{m,n}^{(1)} = |\xi^{(1)}(t + m) - \overset{\underset{\smile}{\xi}(1)}{\xi}^{(1)}(t + m)|.$$

---

[1] In fact, it is easy to see that all the theory developed here can be extended with no difficulty to the case where it is assumed only that the sequences $\xi(t)$ and $\eta(t)$ are stationarily correlated (p. 79). (Cf. the end of Sec. 28.)

In other words, we look for the values $a_1, a_2, \ldots, a_n$ of the coefficients $\alpha_1, \alpha_2, \ldots, \alpha_n$ for which the *mean square filtering error*

$$\sigma_{m,n}^2 = \mathbf{E}\left|\xi(t+m) - \sum_{k=1}^{n} \alpha_k \zeta(t-k)\right|^2 \tag{5.5}$$

takes its minimum value.   It is easy to see that knowledge of the correlation functions $B_{\zeta\zeta}(\tau)$, $B_{\eta\eta}(\tau)$ and $B_{\xi\xi}(\tau)$ is sufficient to find these values $a_1, a_2, \ldots, a_n$.

Clearly, the problem just formulated, i.e., linear filtering of the sequence $\zeta(t)$, has a simple geometric interpretation.   In fact, consider the set of all vectors of the form

$$\alpha_1 \zeta(t-1) + \alpha_2 \zeta(t-2) + \cdots + \alpha_n \zeta(t-n) \tag{5.6}$$

in the Hilbert space $H$ of random variables (introduced in Sec. 6).   This set of vectors represents a finite-dimensional (linear) subspace $H_n^\zeta(t)$ of the space $H$.   Our problem is equivalent to finding the vector

$$L_{m,n}(t) = a_1 \zeta(t-1) + a_2 \zeta(t-2) + \cdots + a_n \zeta(t-n) \tag{5.7}$$

of the subspace $H_n^\zeta(t)$ for which the quantity

$$\sigma_{m,n}^2 = \|\xi(t+m) - L_{m,n}(t)\|^2, \tag{5.8}$$

the square of the distance from $\xi(t+m)$ to $L_{m,n}(t)$, takes its minimum value. But this is just the problem of dropping a perpendicular from the point $\xi(t+m)$ onto the subspace $H_n^\zeta(t)$, and the vector $\xi(t+m) - L_{m,n}(t)$ is just this perpendicular.

It is now clear that the required values $a_1, a_2, \ldots, a_n$ are determined by the equations

$$(\xi(t+m) - L_{m,n}(t), \zeta(t-k)) = 0 \qquad (k = 1, 2, \ldots, n). \tag{5.9}$$

Substituting (5.7) into (5.9), and taking account of (5.2) and the fact that

$$\zeta(t-k) = \xi(t-k) + \eta(t-k),$$

we finally reduce the solution of the linear filtering problem to the solution of the following system of linear algebraic equations:

$$B_{\xi\xi}(m+k) - \sum_{l=1}^{n} a_l B_{\zeta\zeta}(k-l) = 0 \qquad (k = 1, 2, \ldots, n). \tag{5.10}$$

The mean square filtering error for known $a_1, a_2, \ldots, a_n$ is given by the formula

$$\sigma_{m,n}^2 = \left\|\xi(t+m) - \sum_{k=1}^{n} a_k \zeta(t-k)\right\|^2$$

$$= B_{\xi\xi}(0) - 2\,\mathrm{Re}\sum_{k=1}^{n} \bar{a}_k B_{\xi\xi}(m+k) + \sum_{k=1}^{n}\sum_{l=1}^{n} \bar{a}_k a_l B_{\zeta\zeta}(k-l). \tag{5.11}$$

Since (5.10) implies that

$$\sum_{k=1}^{n} \bar{a}_k B_{\xi\xi}(m + k) = \sum_{k=1}^{n} \sum_{l=1}^{n} \bar{a}_k a_l B_{\zeta\zeta}(k - l),$$

we can write (5.11) in the form

$$\sigma_{m,n}^2 = B_{\xi\xi}(0) - \sum_{k=1}^{n} \sum_{l=1}^{n} \bar{a}_k a_l B_{\zeta\zeta}(k - l). \tag{5.12}$$

## 26. Filtering When the Entire Past of the Sequence is Known

Unfortunately, the very simple method of solving the linear filtering problem given in the preceding section turns out to have little practical value when $n$ is large, because of the difficulty of solving systems of linear equations of high order. Thus, it is of great interest to find simple and convenient limiting formulas, corresponding to the case $n \to \infty$. Such formulas can be obtained by solving the problem of linear filtering for the case where the entire past of the sequence $\zeta(t)$ is known, i.e., by finding the random variable $L_m(t)$ which can be represented in the form

$$L_m(t) = a_1\zeta(t - 1) + a_2\zeta(t - 2) + a_3\zeta(t - 3) + \cdots, \tag{5.13}$$

and for which the quantity

$$\sigma_m^2 = \mathbf{E}|\xi(t + m) - L_m(t)|^2 \tag{5.14}$$

takes its minimum value. Obviously, the problem of determining $L_m(t)$ is equivalent to the problem of dropping a perpendicular from the point $\xi(t + m)$ of the space $H$ onto the subspace $H^\zeta(t)$ spanned by the random variables $\zeta(t - 1), \zeta(t - 2), \zeta(t - 3), \ldots$. [In general, $H^\zeta(t)$ is infinite-dimensional.] In the rest of this chapter, we shall consider only this problem.[2]

As in the case of the extrapolation problem, it is convenient to first go over to spectral densities, before passing from the case of finite $n$ to the limiting case $n = \infty$. We shall assume that the sequences $\xi(t)$ and $\eta(t)$ have spectral densities $f_{\xi\xi}(\lambda)$ and $f_{\eta\eta}(\lambda)$, respectively. Then, it is obvious that $\zeta(t)$ also has a spectral density, equal to

$$f_{\zeta\zeta}(\lambda) = f_{\xi\xi}(\lambda) + f_{\eta\eta}(\lambda). \tag{5.15}$$

---

[2] As in the case of linear extrapolation, equation (5.13) has to be replaced by the more precise condition that $L_m(t)$ belongs to $H^\zeta(t)$. [Cf. footnote 5, p. 104, and the explanation of the meaning of equation (4.25), p. 107.] However, here we shall only consider the case where $L_m(t)$ can be expanded in a convergent series of the form (5.13). Of course, this immediately implies that $L_m(t)$ belongs to $H^\zeta(t)$.

Because of (2.101), the system of equations (5.10) can be rewritten in the form

$$\int_{-\pi}^{\pi} \left[ e^{i(m+k)\lambda} f_{\xi\xi}(\lambda) - \sum_{l=1}^{n} a_l e^{i(k-l)\lambda} f_{\zeta\zeta}(\lambda) \right] d\lambda = 0 \qquad (k = 1, 2, \ldots, n), \quad (5.16)$$

or alternatively,

$$\int_{-\pi}^{\pi} e^{ik\lambda} [e^{im\lambda} f_{\xi\xi}(\lambda) - \Phi_{m,n}(\lambda) f_{\zeta\zeta}(\lambda)] \, d\lambda = 0 \qquad (k = 1, 2, \ldots, n), \quad (5.17)$$

where

$$\Phi_{m,n}(\lambda) = \sum_{l=1}^{n} a_l e^{-il\lambda}. \tag{5.18}$$

The function $\Phi_{m,n}(\lambda)$ will be called the *spectral characteristic for filtering*. It is obvious that if we know the expansion (5.18) of this function, we can immediately determine the coefficients $a_l$ in formula (5.7). It is also easy to see that the mean square filtering error (5.12) can be expressed as follows, in terms of $\Phi_{m,n}(\lambda)$:

$$\begin{aligned}
\sigma_{m,n}^2 &= \int_{-\pi}^{\pi} f_{\xi\xi}(\lambda) \, d\lambda - \int_{-\pi}^{\pi} |\Phi_{m,n}(\lambda)|^2 f_{\zeta\zeta}(\lambda) \, d\lambda \\
&= B_{\xi\xi}(0) - \int_{-\pi}^{\pi} |\Phi_{m,n}(\lambda)|^2 f_{\zeta\zeta}(\lambda) \, d\lambda.
\end{aligned} \tag{5.19}$$

If in (5.18) and (5.17) we set $n = \infty$ (in a purely formal way), we arrive at the problem of finding a function $\Phi_m(\lambda)$ which can be represented as a series

$$\Phi_m(\lambda) \sim a_1 e^{-i\lambda} + a_2 e^{-2i\lambda} + a_3 e^{-3i\lambda} + \cdots \tag{5.20}$$

and is such that

$$\int_{-\pi}^{\pi} e^{ik\lambda} [e^{im\lambda} f_{\xi\xi}(\lambda) - \Phi_m(\lambda) f_{\zeta\zeta}(\lambda)] \, d\lambda = 0 \qquad (k = 1, 2, 3, \ldots). \tag{5.21}$$

We see from (5.21) that the Fourier series of the function $e^{im\lambda} f_{\xi\xi}(\lambda) - \Phi_m(\lambda) f_{\zeta\zeta}(\lambda)$ contains only nonnegative powers of $e^{i\lambda}$:

$$e^{im\lambda} f_{\xi\xi}(\lambda) - \Phi_m(\lambda) f_{\zeta\zeta}(\lambda) \sim c_0 + c_1 e^{i\lambda} + c_2 e^{2i\lambda} + \cdots \tag{5.22}$$

With an appropriate interpretation of the sign $\sim$ in (5.20), the problem of determining a function $\Phi_m(\lambda)$ satisfying (5.20) and (5.21) is completely equivalent to the problem of dropping a perpendicular from the point $\xi(t + m)$ onto the subspace $H^\zeta(t)$.

We shall not make a detailed investigation here of the meaning of the relation (5.20). Instead, we merely note that the conditions on $\Phi_m(\lambda)$

expressed by this relation are certainly satisfied if $\Phi_m(\lambda)$ can be expanded as an infinite series in $e^{-ik\lambda}$, $k = 1, 2, 3, \ldots$, which converges in the mean with respect to $f_{\zeta\zeta}(\lambda)\, d\lambda$, i.e. if

$$\lim_{N \to \infty} \int_{-\pi}^{\pi} \left| \Phi_m(\lambda) - \sum_{k=1}^{N} a_k e^{-ik\lambda} \right|^2 f_{\zeta\zeta}(\lambda)\, d\lambda, \qquad (5.20')$$

or (*a fortiori*) if $\Phi_m(\lambda)$ can be expanded in a uniformly convergent series

$$\Phi_m(\lambda) = \sum_{k=1}^{\infty} a_k e^{-ik\lambda}. \qquad (5.20'')$$

If (5.20') or the stronger condition (5.20'') holds, then the series (5.13) will converge, and the coefficients $a_k$ in (5.13) and (5.20) will be the same. As in the case of the function $\Phi_{m,n}(\lambda)$, the function $\Phi_m(\lambda)$ which satisfies the conditions (5.20) and (5.21) will be called the *spectral characteristic for filtering*. It is obvious that finding this function is equivalent to solving the filtering problem for the case where the entire past of the sequence $\zeta(t)$ is known. In particular, we have the following expression, analogous to (5.19), which expresses the mean square filtering error $\sigma_m^2$ in terms of $\Phi_m(\lambda)$:

$$\sigma_m^2 = \int_{-\pi}^{\pi} f_{\xi\xi}(\lambda)\, d\lambda - \int_{-\pi}^{\pi} |\Phi_m(\lambda)|^2 f_{\zeta\zeta}(\lambda)\, d\lambda. \qquad (5.23)$$

Finally, we make the following important observation: Unlike the extrapolation problem, in the filtering problem the number $m$ can be positive, negative, or zero. (In the case of extrapolation, the problem makes sense only if $m \geqslant 0$.) For $m \geqslant 0$, we have filtering combined with extrapolation ("prediction based on corrupted data"), while for $m < -1$, we have filtering with *lag* (i.e., delay). Naturally, by introducing lag, we improve the quality of the filtering, and therefore in many cases it may be advisable to do so. In particular, the most exact restoration of the values of the sequence $\xi(t)$ based on the values of $\zeta(t) = \xi(t) + \eta(t)$ is obtained if we make the lag very large, by choosing $m$ to be a very large negative number. In this case, we have at our disposal not only the values of the sequence $\zeta(t)$ at all times prior to $t + m$, but also all values of $\zeta(t)$ during an interval of length $|m|$ (measured in time units) subsequent to $t + m$. Therefore, it is natural to suppose that the limit of the mean square filtering error $\sigma_m^2$ as $m \to -\infty$ can be obtained by considering the problem of filtering the sequence $\zeta(t)$ when the values of $\zeta(t)$ are known *at all instants of time*, and in fact, we shall begin our study of filtering theory by considering just this simplest situation. The results so obtained are of great practical interest, since they show the extent to which the quality of the filtering can be improved by introducing lag.

## 27. Examples of Linear Filtering of Stationary Sequences

**Example 1.** Suppose we know the values $\zeta^{(1)}(t)$ of the sequence $\zeta(t) = \xi(t) + \eta(t)$ at all times $t = \ldots, -2, -1, 0, 1, 2, \ldots$. Our problem is to find a random variable $L(t)$ which can be written as a convergent series

$$L(t) = \sum_{k=-\infty}^{\infty} a_k \zeta(t - k) \tag{5.24}$$

of random variables $a_k \zeta(t - k)$, and which is such that the mathematical expectation

$$\sigma^2 = \mathbf{E}|\xi(t) - L(t)|^2 \tag{5.25}$$

takes its minimum value. (We can set the number $m$ equal to zero, since it is clear that nothing depends on time here.[3]) This is equivalent to the problem of dropping a perpendicular from the point $\xi(t)$ of the space $H$ onto the subspace $H^\zeta$ spanned by the vectors $\zeta(t)$, $t = \ldots, -2, -1, 0, 1, 2, \ldots$, and hence the problem has a unique solution. By considerations analogous to those which led to formulas (5.20) and (5.21), we can show that to solve our problem it is only necessary to find a function

$$\Phi(\lambda) = \sum_{k=-\infty}^{\infty} a_k e^{-ik\lambda} \tag{5.26}$$

[where the convergence is interpreted as in (5.20)] for which

$$\int_{-\pi}^{\pi} e^{ik\lambda}[f_{\xi\xi}(\lambda) - \Phi(\lambda)f_{\zeta\zeta}(\lambda)]\, d\lambda = 0 \qquad (k = -2, -1, 0, 1, 2, \ldots). \tag{5.27}$$

However, (5.27) shows that all the Fourier coefficients of the function $f_{\xi\xi}(\lambda) - \Phi(\lambda)f_{\zeta\zeta}(\lambda)$ vanish, and it follows at once that (5.27) can only be satisfied if

$$f_{\xi\xi}(\lambda) - \Phi(\lambda)f_{\zeta\zeta}(\lambda) = 0. \tag{5.28}$$

Thus, the spectral characteristic $\Phi(\lambda)$ for filtering is uniquely determined by equation (5.27) alone, and is equal to

$$\Phi(\lambda) = \frac{f_{\xi\xi}(\lambda)}{f_{\zeta\zeta}(\lambda)}. \tag{5.29}$$

Equation (5.26) shows that the coefficients $a_k$ are just the Fourier coefficients of the function $\Phi(\lambda)$, so that

$$a_k = \frac{1}{2\pi} \int_{-\pi}^{\pi} e^{ik\lambda} \frac{f_{\xi\xi}(\lambda)}{f_{\zeta\zeta}(\lambda)}\, d\lambda \tag{5.30}$$

in the case where $f_{\xi\xi}(\lambda)/f_{\zeta\zeta}(\lambda)$ has a Fourier expansion in the usual sense.

---

[3] More precisely, the mean square filtering error $\sigma_m^2$ does not depend on $m$, but the formula for $\xi^{(1)}(t + m)$ depends on $m$ in a trivial way, i.e., to get the approximate expression for $\xi^{(1)}(t + m)$, we need only replace every $\zeta^{(1)}(t - k)$ by $\zeta^{(1)}(t + m - k)$ in the approximate expression for $\xi^{(1)}(t)$.

Next, we calculate the mean square filtering error

$$\sigma^2 = \int_{-\pi}^{\pi} f_{\xi\xi}(\lambda)\, d\lambda - \int_{-\pi}^{\pi} |\Phi(\lambda)|^2 f_{\zeta\zeta}(\lambda)\, d\lambda$$

$$= \int_{-\pi}^{\pi} \frac{f_{\xi\xi}(\lambda)[f_{\zeta\zeta}(\lambda) - f_{\xi\xi}(\lambda)]}{f_{\zeta\zeta}(\lambda)}\, d\lambda$$

or

$$\sigma^2 = \int_{-\pi}^{\pi} \frac{f_{\xi\xi}(\lambda) f_{\eta\eta}(\lambda)}{f_{\xi\xi}(\lambda) + f_{\eta\eta}(\lambda)}\, d\lambda, \tag{5.31}$$

where we have used the fact that $f_{\zeta\zeta}(\lambda) = f_{\xi\xi}(\lambda) + f_{\eta\eta}(\lambda)$. Formula (5.31) shows that even with infinite lag ($m = -\infty$), complete suppression of the error, i.e., complete restoration of the values of $\xi(t)$, by using the values of $\zeta(t) = \xi(t) + \eta(t)$, is possible only when the basic sequence $\xi(t)$ [the "signal"] and the error sequence $\eta(t)$ [the "noise"] have nonoverlapping spectra, so that $f_{\xi\xi}(\lambda) f_{\eta\eta}(\lambda) \equiv 0$. Obviously, the possibility of completely suppressing the error in this case is intimately related to the possibility of constructing a filter with a specified pass band, and in this regard, it should be noted that ideal band-pass filters require infinite lag times [cf. formulas (20) and (21) of K16]. However, in the general case where the spectra of $\xi(t)$ and $\eta(t)$ overlap, the mean square filtering error remains finite even when the lag is infinite. In fact, the limiting value as $m \to -\infty$ of the mean square error, given by formula (5.31), when suitably normalized, namely when divided by

$$\frac{1}{4} B_{\zeta\zeta}(0) = \frac{1}{4} \int_{-\pi}^{\pi} f_{\zeta\zeta}(\lambda)\, d\lambda,$$

can serve as a convenient measure of the "degree of overlap" of the spectra of the sequences $\xi(t)$ and $\eta(t)$.

In the next example, we consider the problem of filtering for finite values of $m$. This problem is much more complicated than that just considered, and a relatively elementary solution is possible only in the case where the spectral densities $f_{\xi\xi}(\lambda)$ and $f_{\zeta\zeta}(\lambda)$ are rational in $e^{i\lambda}$, i.e., are the boundary values on the circle $z = e^{i\lambda}$ of certain rational functions $f_{\xi\xi}^*(z)$ and $f_{\zeta\zeta}^*(z)$ of the complex variable $z$.

***Example 2.*** Suppose that the basic sequence $\xi(t)$, whose values we are interested in, is a stationary sequence of the "Markov type" (see p. 113), while the measurement errors $\eta(t)$ are uncorrelated at different instants of time. Then

$$f_{\xi\xi}(\lambda) = \frac{C_1}{2\pi} \frac{1 - a^2}{|e^{i\lambda} - a|^2} = \frac{A}{|e^{i\lambda} - a|^2} \quad \left(A = \frac{C_1(1 - a^2)}{2\pi}\right), \tag{5.32}$$

where $C_1 > 0$, $a$ is real, $|a| < 1$, and

$$f_{\eta\eta}(\lambda) = \frac{C_2}{2\pi} = \text{const.} \tag{5.33}$$

Moreover, the spectral density

$$f_{\zeta\zeta}(\lambda) = f_{\xi\xi}(\lambda) + f_{\eta\eta}(\lambda) \tag{5.34}$$

can be written in the form

$$f_{\zeta\zeta}(\lambda) = \frac{C}{2\pi} \frac{|e^{i\lambda} - b|^2}{|e^{i\lambda} - a|^2} = B \frac{|e^{i\lambda} - b|^2}{|e^{i\lambda} - a|^2} \quad \left(B = \frac{C}{2\pi}\right), \tag{5.35}$$

where $C$ and $b$ are the real numbers ($C > 0$, $|b| < 1$) determined by the system of equations (2.120), given on p. 60.

It is obvious that the problem of finding a function $\Phi_m(\lambda)$ satisfying the conditions (5.20) and (5.22) will be solved if we find a rational function $\Phi_m^*(z)$ which has all its singularities inside the unit circle, vanishes for $z = \infty$, and is such that the function

$$
\begin{aligned}
\Psi_m^*(z) &= z^m f_{\xi\xi}^*(z) - \Phi_m^*(z) f_{\zeta\zeta}^*(z) \\
&= z^m \frac{Az}{(z-a)(1-az)} - \Phi_m^*(z) \frac{B(z-b)(1-bz)}{(z-a)(1-az)}
\end{aligned} \tag{5.36}
$$

has no singularities for $|z| \leqslant 1$. Then, $\Phi_m(\lambda) = \Phi_m^*(e^{i\lambda})$, and the series (5.20) will even converge uniformly. We write the function (5.36) in the form

$$\Psi_m^*(z) = \frac{Az^{m+1} - B(z-b)(1-bz)\Phi_m^*(z)}{(z-a)(1-az)}, \tag{5.37}$$

and we consider separately the cases $m \geqslant -1$, corresponding to filtering without lag (in particular, if $m \geqslant 0$, this is filtering combined with prediction), and $m < -1$, corresponding to filtering with lag.

1. When $m \geqslant -1$, $Az^{m+1}$ is regular in the whole complex plane, and the numerator of the fraction (5.37) has no singularities at finite points other than the singularities of the function $(z-b)(1-bz)\Phi_m^*(z)$. On the one hand, these singularities cannot lie inside (or on the boundary of) the unit circle, since the function (5.37) cannot have singularities for $|z| \leqslant 1$, while on the other hand, they cannot lie outside the unit circle, since all three functions $z - b$, $1 - bz$ and $\Phi_m^*(z)$ are regular outside the unit circle. Consequently, the function $(z-b)(1-bz)\Phi_m^*(z)$ can have a singularity only at the point $z = \infty$. It follows that $\Phi_m^*(z)$ can have no singularities other than a simple pole at the point $z = b$, so that

$$\Phi_m^*(z) = \frac{\gamma_m(z)}{z - b}, \tag{5.38}$$

where $\gamma_m(z)$ is an analytic function in the entire complex plane (an entire function).

Substituting (5.38) into (5.37), we find that the function

$$\frac{Az^{m+1} - B(1-bz)\gamma_m(z)}{(z-a)(1-az)} \tag{5.39}$$

must be regular for $|z| \leqslant 1$, which is possible only if the numerator vanishes for $z = a$, i.e., if

$$\gamma_m(a) = \frac{A}{B} \frac{a^{m+1}}{1 - ab}. \qquad (5.40)$$

According to (5.38), to satisfy the condition $\Phi_m^*(\infty) = 0$ as well, we have to set $\gamma_m(z) = \gamma_m(a) = \text{const}$, so that for $m \geqslant -1$

$$\Phi_m^*(z) = \frac{A}{B} \frac{a^{m+1}}{1 - ab} \frac{1}{z - b} = \frac{A}{B} \frac{a^{m+1}}{1 - ab} \sum_{k=1}^{\infty} \frac{b^{k-1}}{z^k}, \qquad (5.41)$$

$$\Phi_m(\lambda) = \frac{A}{B} \frac{a^{m+1}}{1 - ab} \frac{1}{e^{i\lambda} - b} = \frac{A}{B} \frac{a^{m+1}}{1 - ab} \sum_{k=1}^{\infty} b^{k-1} e^{-ik\lambda}, \qquad (5.42)$$

and the best linear approximation to $\xi^{(1)}(t + m)$ has the form

$$\xi^{(1)}(t + m) = \frac{A}{B} \frac{a^{m+1}}{1 - ab} \sum_{k=1}^{\infty} b^{k-1} \zeta^{(1)}(t - k). \qquad (5.43)$$

The mean square error of the approximation (5.43), calculated by using (5.23), is equal to

$$\sigma_m^2 = \frac{2\pi A}{1 - a^2} \left[ 1 - \frac{A}{B} \frac{a^{2(m+1)}}{(1 - ab)^2} \right], \qquad (5.44)$$

and as $m \to \infty$, $\sigma_m^2$ gets arbitrarily close to

$$\frac{2\pi A}{1 - a^2} = B_{\xi\xi}(0) = \mathbf{E}|\xi(t)|^2.$$

This is completely natural, since as $m \to \infty$, the mean square prediction error approaches $\mathbf{E}|\xi(t)|^2$ even when the measured data is error-free.

We can modify (5.42), (5.43) and (5.44) somewhat, by replacing $A$ and $B$ by the expressions

$$A = \frac{C_1(1 - a^2)}{2\pi}, \qquad B = \frac{C}{2\pi}$$

[cf. (5.32) and (5.35)]. Then, using the second of the equations (2.120), we find that

$$\frac{C_1(1 - a^2)}{C} = \frac{(a - b)(1 - ab)}{a},$$

so that

$$\frac{A}{B(1 - ab)} = \frac{a - b}{a}$$

and

$$\Phi_m(\lambda) = \frac{(a-b)a^m}{e^{i\lambda} - b} = (a-b)a^m \sum_{k=1}^{\infty} b^{k-1} e^{-ik\lambda}, \qquad (5.42')$$

$$\xi^{(1)}(t+m) = (a-b)a^m \sum_{k=1}^{\infty} b^{k-1} \zeta^{(1)}(t-k) \qquad (5.43')$$

$$\sigma_m^2 = C_1 \left[ 1 - \frac{a-b}{1-ab} a^{2m+1} \right]. \qquad (5.44')$$

We note that for $m \geqslant 0$, (5.42') is exactly the same as (4.73').[4]   (See Example 5 of Sec. 23.)   Thus, if $m \geqslant 0$, the best linear approximation to $\xi(t+m)$ which can be constructed by using the past of the sequence $\zeta(t)$ coincides with the best linear approximation to $\zeta(t+m)$, i.e., the solution of the problem of *filtering* $\zeta(t)$ is identical with the solution of the problem of *extrapolating* $\zeta(t)$.   Of course, this fact could have been anticipated from the beginning, since in our example the "error"

$$\eta(t+m) = \zeta(t+m) - \xi(t+m)$$

at the time $t+m$, where $m \geqslant 0$, is uncorrelated with all the values of $\zeta(t-k)$, $k > 0$, so that within the context of correlation theory, knowledge of the past of the sequence $\zeta(t)$ gives no information whatsoever about $\eta(t+m)$.   Analytically, this is expressed by the fact that if $B_{\eta\eta}(\tau) = 0$ for $\tau \neq 0$, then for $m \geqslant 0$, the term $B_{\xi\xi}(m+k)$ in the system of equations (5.10) equals $B_{\zeta\zeta}(m+k)$, and the equations for filtering the sequence $\zeta(t)$ coincide with the equations for extrapolating $\zeta(t)$.   It follows that we did not really have to solve the filtering problem separately for $m \geqslant 0$; instead, we might have used the results of Example 5 of Sec. 23.   However, we preferred to give a new solution of the problem here, since this solution will be useful when we go over to the case $m \leqslant -1$ and to the general case of filtering, where the sequence $\eta(t)$ is no longer a sequence of uncorrelated random variables.

The fact that in the present example, the equations defining the solution of two different problems (extrapolation and filtering) are the same, is also apparent from the spectral formulation of the two problems.   In the filtering problem, we have to find an analytic function $\Phi_m^*(z)$ which has no singularities outside the unit circle and is such that the function

$$\Psi_m^*(z) = z^m f_{\xi\xi}^*(z) - \Phi_m^*(z) f_{\zeta\zeta}^*(z)$$

has no singularities for $|z| \leqslant 1$.   But if

$$f_{\eta\eta}(\lambda) = \text{const},$$

---

[4] However, note that (5.42') is applicable for $m \geqslant -1$.

i.e., if

$$f^*_{\zeta\zeta}(z) = f^*_{\xi\xi}(z) + \text{const},$$

then

$$z^m f^*_{\xi\xi}(z) - \Phi^*_m(z) f^*_{\zeta\zeta}(z) = [z^m - \Phi^*_m(z)] f^*_{\zeta\zeta}(z) - \text{const } z^m,$$

from which it is clear that if $m \geqslant 0$, the function $[z_m - \Phi^*_m(z)] f^*_{\zeta\zeta}(z)$, formed from the same $\Phi^*_m(z)$, will be regular for $|z| \leqslant 1$. Thus, $\Phi^*_m(z)$ is also the spectral characteristic for extrapolation of the sequence $\zeta(t)$. This is no longer the case if $m \leqslant -1$, since then the extrapolation problem has the trivial solution $\Phi^*_m(z) = z^m$, which is not the solution of the filtering problem.

Of course, the mean square filtering error (5.44') will not equal the mean square extrapolation error (4.75). In the first case, the mean square error equals

$$\mathbf{E}|\xi(t + m) - L_m(t)|^2,$$

where

$$L_m(t) = (a - b)a^m \sum_{k=1}^{\infty} b^{k-1}\zeta(t - k),$$

while in the second case, it equals

$$\mathbf{E}|\zeta(t + m) - L_m(t)|^2.$$

Bearing in mind that

$$\zeta(t + m) = \xi(t + m) + \eta(t + m),$$

where $\eta(t + m)$ is a random variable which is uncorrelated with $\xi(t + m)$ and all the random variables $\zeta(t - k)$, $k \geqslant 1$ [and hence with $L_m(t)$], we find that

$$\mathbf{E}|\zeta(t + m) - L_m(t)|^2 = \mathbf{E}|\xi(t + m) - L_m(t) + \eta(t + m)|^2$$
$$= \mathbf{E}|\xi(t + m) - L_m(t)|^2 + \mathbf{E}|\eta(t + m)|^2.$$

Therefore, in the present example, where $m \geqslant 0$, the mean square filtering error is exactly $\mathbf{E}|\eta(t)|^2 = C_2$ more than the mean square extrapolation error given by formula (4.75) of Example 5 of Sec. 23. This fact is easily verified directly by using the first of the equations (2.120).[5]

It is interesting to compare formula (5.43), or (5.43'), with formula (4.42), which gives the best approximation to the value of a stationary sequence $\xi(t)$ of "Markov type" at the time $t + m$ in terms of the "unperturbed" past values of $\xi(t)$. With this in mind, we rewrite (5.43') in the form

$$\xi^{(1)}(t + m) = \frac{a - b}{a(1 - b)} a^{m+1}\zeta^{(1)}(t - 1), \qquad (5.43'')$$

_____
[5] First replace the constant $C_1$ in (4.75) by $B$.

where

$$\hat{\zeta}^{(1)}(t - 1) = (1 - b) \sum_{k=0}^{\infty} b^k \zeta^{(1)}(t - k - 1). \qquad (5.45)$$

According to (5.45), $\hat{\zeta}^{(1)}(t - 1)$ is the "smoothed" value obtained when we add all the quantities $\zeta^{(1)}(t - 1)$, $\zeta^{(1)}(t - 2)$, $\zeta^{(1)}(t - 3), \ldots$, after multiplying them by the decreasing set of "weights" $1 - b$, $b(1 - b)$, $b^2(1 - b), \ldots$, whose sum is unity. [It was just to make the sum of these weights equal to unity that we introduced the factor $1 - b$ in formula (5.45).] Thus, in the case where the known past values of the sequence are unreliable, i.e., are perturbed by measurement errors, we have to replace $\xi^{(1)}(t - 1)$ in formula (4.42) by the smoothed value $\hat{\zeta}^{(1)}(t - 1)$, and then multiply the formula by

$$\frac{a - b}{a(1 - b)} = \frac{a - b}{a - ab}.$$

It is an easy consequence of the equations (2.120) which define $b$ and $C$, that as the "signal-to-noise ratio"

$$\rho = \sqrt{\frac{\mathbf{E}|\xi(t)|^2}{\mathbf{E}|\eta(t)|^2}} = \sqrt{\frac{C_1}{C_2}}, \qquad (5.46)$$

(characterizing the relative level of the signal and noise powers) approaches infinity, the quantity $b$ approaches zero, so that all the "smoothing weights" $(1 - b)b^k$, $k = 1, 2, 3, \ldots$ go to zero, except the first weight $1 - b$, which approaches unity, while the factor $(a - b)/(a - ab)$ also approaches unity. From this it is clear that in the limit as $\rho \to \infty$, formula (5.43) goes into formula (4.42). On the other hand, as $\rho$ decreases, the number $b$ increases monotonically and approaches $a$ as $\rho \to 0$, while the factor $(a - b)/(a - ab)$ decreases monotonically to 0 as $\rho \to 0$ (and hence is less than 1 for any $\rho$). This is quite understandable, for obviously, when the signal-to-noise ratio is very small, the accuracy of the prediction of future values of $\xi(t)$ must be very small, so that as $\rho \to 0$, the predicted value of $\xi(t + m)$ must go to zero for any $t$. Similar conclusions can be drawn if we analyze formula (5.44), or (5.44'), for the mean square error $\sigma_m^2$.

2. Next, we turn to the case $m < -1$, corresponding to filtering with lag. Setting $-m - 1 = \mu > 0$, we can write (5.37) in the form

$$\Psi_{-\mu-1}^*(z) = \frac{A - Bz^\mu(z - b)(1 - bz)\Phi_{-\mu-1}^*(z)}{z^\mu(z - a)(1 - az)}. \qquad (5.47)$$

Since the function (5.47) must not have singularities for $|z| \leqslant 1$, and since the function $\Phi_{-\mu-1}^*(z)$ must not have singularities for $|z| \geqslant 1$, it follows that the function $z^\mu(z - b)(1 - bz)\Phi_{-\mu-1}^*(z)$ can have no singularities other than the point $z = \infty$. Consequently, $\Phi^*_{-\mu-1}(z)$ can have at most a simple pole at the point $z = b$ and a pole of order $\mu$ at the origin, i.e.,

$$\Phi^*_{-\mu-1}(z) = \frac{\gamma_{-\mu-1}(z)}{z^\mu(z - b)}, \tag{5.48}$$

where $\gamma_{-\mu-1}(z)$ is an entire function. Then, (5.47) becomes

$$\Psi^*_{-\mu-1}(z) = \frac{A - B(1 - bz)\gamma_{-\mu-1}(z)}{z^\mu(z - a)(1 - az)}. \tag{5.49}$$

In order for the function (5.49) to have no singularities for $|z| \leqslant 1$, the entire function $A - B(1 - bz)\gamma_{-\mu-1}(z)$ must have a simple zero at the point $z = a$ and a zero of order no less than $\mu$ at the point $z = 0$. Therefore, we must have

$$A - (1 - Bz)\gamma_{-\mu-1}(z) = z^\mu(z - a)\tilde{\gamma}(z),$$
$$\gamma_{-\mu-1}(z) = \frac{A - z^\mu(z - a)\tilde{\gamma}(z)}{B(1 - bz)}, \tag{5.50}$$

where $\tilde{\gamma}(z)$ is also an entire function of $z$. But the function $\gamma_{-\mu-1}(z)$ has no singularities at all at finite points, and hence, in particular, $\gamma_{-\mu-1}(z)$ cannot have a singularity at the point $z = 1/b$, from which it follows that the difference $A - z^\mu(z - a)\tilde{\gamma}(z)$ must vanish at the point $z = 1/b$, i.e.,

$$\tilde{\gamma}(1/b) = \frac{A}{b^{-\mu}(b^{-1} - a)} = \frac{Ab^{\mu+1}}{1 - ab}. \tag{5.51}$$

Equations (5.51), (5.50) and (5.48) guarantee that $\Phi^*_{-\mu-1}(z)$ has no singularities for $|z| \geqslant 1$ and that $\Psi^*_{-\mu-1}(z)$ has no singularities for $|z| \leqslant 1$. Moreover, in order for the relation $\Phi^*_{-\mu-1}(\infty) = 0$ to hold, the function $\gamma_{-\mu-1}(z)$ must grow more slowly than $z^{\mu+1}$ (as $z$ goes to infinity), so that the function $\tilde{\gamma}(z)$ in (5.50) must be a constant. Thus, finally, all the conditions imposed on $\Phi^*_{-\mu-1}(z)$ will be satisfied if we set

$$\gamma_{-\mu-1}(z) = \frac{A - [z^\mu(z - a)Ab^{\mu+1}/(1 - ab)]}{B(1 - bz)}$$
$$= \frac{A}{B(1 - ab)} \frac{(1 - ab) - b^{\mu+1}z^\mu(z - a)}{1 - bz}. \tag{5.52}$$

(It is easy to see that this function is entire, since the denominator $1 - bz$ compensates for the zero of the numerator at $z = 1/b$.) Formula (5.52) implies that

$$\Phi^*_{-\mu-1}(z) = \frac{A}{B(1 - ab)} \frac{(1 - ab) - b^{\mu+1}z^\mu(z - a)}{z^\mu(z - b)(1 - bz)}. \tag{5.53}$$

If now we use (2.120) to replace $A/B(1 - ab)$ by $(a - b)/a$ [cf. the transition from (5.42) to (5.42')], we obtain

$$\Phi^*_{-\mu-1}(z) = \frac{a - b}{a} \frac{(1 - ab) - b^{\mu+1}z^\mu(z - a)}{z^\mu(z - b)(1 - bz)}, \tag{5.53'}$$

from which it follows that

$$\Phi_{-\mu-1}(\lambda) = \frac{a - b}{a} \frac{(1 - ab)e^{-i(\mu+1)\lambda} - b^{\mu+1}(1 - ae^{-i\lambda})}{|e^{i\lambda} - b|^2}. \tag{5.54}$$

In particular,

$$\begin{aligned}
\Phi^*_{-2}(z) &= \frac{a - b}{a} \frac{1 - ab - b^2z^2 + ab^2z}{z(z - b)(1 - bz)} = \frac{a - b}{a} \frac{1 - ab + bz}{z(z - b)} \\
&= \frac{a - b}{a} \left\{ \frac{1 - ab}{b} \left( \frac{1}{z - b} - \frac{1}{z} \right) + \frac{b}{z - b} \right\} \\
&= \frac{a - b}{a} \left\{ \frac{b}{z} + \frac{1 - ab + b^2}{b} \sum_{k=2}^\infty \frac{b^{k-1}}{z^k} \right\},
\end{aligned} \tag{5.55}$$

which makes it clear that for $m = -2$, the best linear filtering formula is

$$\xi^{(1)}(t - 2) = \frac{a - b}{a} \left\{ b\zeta^{(1)}(t - 1) + (1 - ab + b^2) \sum_{k=2}^\infty b^{k-2}\zeta^{(1)}(t - k) \right\}. \tag{5.55'}$$

In order to obtain the best filtering formula for arbitrary $m < -1$, we have to expand the general formula (5.53') as a power series in $1/z$. In doing this, it is convenient to first write (5.53') in the form

$$\Phi^*_{-\mu-1}(z) = \frac{a - b}{a} \frac{b^\mu}{z - b} \left[ 1 + (1 - ab) \sum_{k=1}^\mu \frac{1}{(bz)^k} \right], \tag{5.53''}$$

and only afterwards make the partial fraction expansion. The mean square filtering error $\sigma_m^2$ for $m < -1$ can be obtained from (5.54) by using formula (5.23). The resulting expression is

$$\begin{aligned}
\sigma_{-\mu-1}^2 &= C_1 \left\{ 1 - \frac{a - b}{a(1 - ab)} \left[ 1 + b^2(1 - b^{2\mu}) \frac{1 - a^2}{1 - b^2} \right] \right\} \\
&= C_1 \frac{b(1 - a^2)}{a(1 - ab)} \left\{ 1 - b \frac{a - b}{1 - b^2} (1 - b^{2\mu}) \right\}.
\end{aligned} \tag{5.56}$$

As $\mu \to \infty$, i.e., as $m \to -\infty$, formula (5.56) approaches the limit

$$\sigma_{-\infty}^2 = C_1 \frac{b(1 - a^2)}{a(1 - b^2)}. \tag{5.57}$$

This result is the same as that obtained by substituting (5.32) and (5.33) into (5.31), as must be the case.

## *28. The Case of a General Rational Spectral Density in $e^{i\lambda}$

By analogy with the solution of the filtering problem given in Example 2 of the preceding section, we can solve the filtering problem for sequences whose spectral densities $f_{\xi\xi}(\lambda)$ and $f_{\zeta\zeta}(\lambda)$ are general rational functions in $e^{i\lambda}$, provided only that $f_{\zeta\zeta}(\lambda)$ vanishes nowhere. Thus, let

$$
\begin{aligned}
f_{\zeta\zeta}(\lambda) &= \frac{|B(e^{i\lambda})|^2}{|A(e^{i\lambda})|^2} = \frac{|B_0 e^{iM\lambda} + B_1 e^{i(M-1)\lambda} + \cdots + B_M|^2}{|A_0 e^{iN\lambda} + A_1 e^{i(N-1)\lambda} + \cdots + A_N|^2} \\
&= \frac{|B_0(e^{i\lambda} - b_1)(e^{i\lambda} - b_2)\ldots(e^{i\lambda} - b_M)|^2}{|A_0(e^{i\lambda} - a_1)(e^{i\lambda} - a_2)\ldots(e^{i\lambda} - a_N)|^2},
\end{aligned} \tag{5.58}
$$

$$
\begin{aligned}
f_{\xi\xi}(\lambda) &= \frac{|D(e^{i\lambda})|^2}{|C(e^{i\lambda})|^2} = \frac{|D_0 e^{iK\lambda} + D_1 e^{i(K-1)\lambda} + \cdots + D_K|^2}{|C_0 e^{iL\lambda} + C_1 e^{i(L-1)\lambda} + \cdots + C_L|^2} \\
&= \frac{|D_0(e^{i\lambda} - d_1)(e^{i\lambda} - d_2)\ldots(e^{i\lambda} - d_K)|^2}{|C_0(e^{i\lambda} - c_1)(e^{i\lambda} - c_2)\ldots(e^{i\lambda} - c_L)|^2},
\end{aligned} \tag{5.59}
$$

where the coefficients $A_0$, $B_0$, $C_0$, $D_0$, $A_N$, $B_M$, $C_L$ and $D_K$ are different from zero, and all the zeros of the polynomials $A(z)$, $B(z)$, and $C(z)$ are strictly less than unity, while none of the zeros of the polynomial $D(z)$ has absolute value greater than unity:

$$
\begin{aligned}
0 < |a_1| < 1,\ldots,0 < |a_N| < 1; \quad 0 < |b_1| < 1,\ldots,0 < |b_M| < 1; \\
0 < |c_1| < 1,\ldots,0 < |c_L| < 1; \quad 0 < |d_1| \leqslant 1,\ldots,0 < |d_K| \leqslant 1.
\end{aligned} \tag{5.60}
$$

(For a discussion of this representation of the spectral densities, see Sec. 24.)

In this case, the linear filtering problem reduces to the problem of finding a function $\Phi_m^*(z)$ with no singularities for $|z| \geqslant 1$, which vanishes for $z = \infty$ and is such that the function

$$
\begin{aligned}
\Psi_m^*(z) &= z^m f_{\xi\xi}^*(z) - \Phi_m^*(z) f_{\zeta\zeta}^*(z) \\
&= \frac{|D_0|^2}{|C_0|^2} \frac{(z - d_1)\ldots(z - d_K)(1 - \bar{d}_1 z)\ldots(1 - \bar{d}_K z)}{(z - c_1)\ldots(z - c_L)(1 - \bar{c}_1 z)\ldots(1 - \bar{c}_L z)} z^{m+L-K} \\
&\quad - \Phi_m^*(z) \frac{|B_0|^2}{|A_0|^2} \frac{(z - b_1)\ldots(z - b_M)(1 - \bar{b}_1 z)\ldots(1 - \bar{b}_M z)}{(z - a_1)\ldots(z - a_N)(1 - \bar{a}_1 z)\ldots(1 - \bar{a}_N z)} z^{N-M'}
\end{aligned} \tag{5.61}
$$

has no singularities for $|z| \leqslant 1$. We shall assume that $\Phi_m^*(z)$ has the form

$$
\Phi_m^*(z) = \frac{(z - a_1)(z - a_2)\ldots(z - a_N)}{(z - b_1)(z - b_2)\ldots(z - b_M)} z^{M-N} \gamma_m(z), \tag{5.62}
$$

where $\gamma_m(z)$ is a new function. Then it is clear that the condition $\Phi_m^*(\infty) = 0$ and the condition concerning the location of the singularities of $\Phi_m^*(z)$ will be satisfied, provided that $\gamma_m(z)$ vanishes for $z = \infty$ and has no singularities outside or on the boundary of the unit circle. It is natural to write $\Phi^*(z)$ in the form (5.62), since, according to (5.61), $\Phi_m^*(z)$ must vanish at the points $a_1, a_2, \ldots, a_N$ if the function $\Psi_m^*(z)$ is to have no singularities at these points, and since $\Phi_m^*(z)$ can have poles at the points $b_1, b_2, \ldots, b_M$ without these poles leading to singularities of $\Psi_m^*(z)$.

Substituting (5.62) into (5.61), we obtain

$$\Psi_m^*(z) = \left\{ \frac{|D_0 A_0|^2}{|C_0 B_0|^2} \frac{(z - d_1)\ldots(z - d_K)(1 - \bar{d}_1 z)\ldots(1 - \bar{d}_K z)}{(z - c_1)\ldots(z - c_L)(1 - \bar{c}_1 z)\ldots(1 - \bar{c}_L z)} \right.$$
$$\times \frac{(1 - \bar{a}_1 z)\ldots(1 - \bar{a}_N z)}{(1 - \bar{b}_1 z)\ldots(1 - \bar{b}_M z)} z^{m+L-K} - \left. \gamma_m(z) \right\} \quad (5.63)$$
$$\times \frac{(1 - \bar{b}_1 z)\ldots(1 - \bar{b}_M z)}{(1 - \bar{a}_1 z)\ldots(1 - \bar{a}_N z)} \frac{|B_0|^2}{|A_0|^2}.$$

Since the function

$$\frac{(1 - \bar{b}_1 z)\ldots(1 - \bar{b}_M z)}{(1 - \bar{a}_1 z)\ldots(1 - \bar{a}_N z)}$$

is obviously regular for $|z| \leqslant 1$, we only have to see to it that the difference between the two functions appearing in braces is regular for $|z| \leqslant 1$. Thus, we consider the function

$$\chi(z) = \frac{|D_0 A_0|^2}{|C_0 B_0|^2}$$
$$\times \frac{(z - d_1)\ldots(z - d_K)(1 - \bar{d}_1 z)\ldots(1 - \bar{d}_K z)(1 - \bar{a}_1 z)\ldots(1 - \bar{a}_N z)}{(z - c_1)\ldots(z - c_L)(1 - \bar{c}_1 z)\ldots(1 - \bar{c}_L z)(1 - \bar{b}_1 z)\ldots(1 - \bar{b}_M z)} z^{L-K},$$
$$(5.64)$$

which is the first term of this difference, except for the factor $z^m$. The function $\chi(z)$ is rational and has no poles on the unit circle. Therefore, $\chi(z)$ is regular in some annulus containing the unit circle, and hence can be expanded in this annulus as a Laurent series

$$\chi(z) = \sum_{k=-\infty}^{\infty} \frac{\varkappa_k}{z^k}. \quad (5.65)$$

[In practice, the most convenient way to obtain this expansion is to first expand $\chi(z)$ in partial fractions.]

Using (5.65), we find that

$$\Psi_m^*(z) = \left\{ \sum_{k=-\infty}^{\infty} \varkappa_k z^{m-k} - \gamma_m(z) \right\} \frac{(1 - \bar{b}_1 z)\ldots(1 - \bar{b}_M z)}{(1 - \bar{a}_1 z)\ldots(1 - \bar{a}_N z)} \frac{|B_0|^2}{|A_0|^2}. \quad (5.66)$$

It is clear that (5.66) will be regular for $|z| \leqslant 1$, provided we set

$$\gamma_m(z) = \sum_{k=m+1}^{\infty} \varkappa_k z^{m-k}; \quad (5.67)$$

then, $\gamma_m(z)$ will be regular outside the unit circle (since it can be represented as a power series in $1/z$), and $\gamma_m(\infty) = 0$. It follows that all the conditions imposed on $\Phi_m^*(z)$ will be satisfied if we set

$$\Phi_m^*(z) = \frac{(z - a_1)(z - a_2)\ldots(z - a_N)}{(z - b_1)(z - b_2)\ldots(z - b_M)} z^{M-N} \sum_{k=m+1}^{\infty} \varkappa_k z^{m-k}, \qquad (5.68)$$

where the coefficients $\varkappa_k$ are determined from the expansion (5.67). It is not hard to see that the function $\Phi_m^*(z)$ which is defined in this way will always be rational [To see this, it is again convenient to imagine that $\chi(z)$ is expanded in partial fractions.] Moreover, if we expand $\Phi_m^*(z)$ in powers of $1/z$, we immediately obtain the coefficients $a_k$ of the best filtering formula.

The mean square filtering error can easily be calculated from $\Phi_m(\lambda) = \Phi_m^*(e^{i\lambda})$ by using formula (5.23); the result is

$$\sigma_m^2 = B_{\xi\xi}(0) - 2\pi \frac{|B_0|^2}{|A_0|^2} \sum_{k=m+1}^{\infty} |\varkappa_k|^2. \qquad (5.69)$$

As $m \to \infty$, we have $\sigma_m^2 \to B_{\xi\xi}(0)$, as is to be expected. In the other limiting case $m \to -\infty$, according to Parseval's theorem [cf. formula (4.93)], we have

$$\begin{aligned}
\lim_{m \to -\infty} \sigma_m^2 &= B_{\xi\xi}(0) - \frac{|B_0|^2}{|A_0|^2} \int_{-\pi}^{\pi} |\chi(e^{i\lambda})|^2 \, d\lambda \\
&= \int_{-\pi}^{\pi} f_{\xi\xi}(\lambda) \, d\lambda - \int_{-\pi}^{\pi} \frac{f_{\xi\xi}^2(\lambda)}{f_{\zeta\zeta}^2(\lambda)} \, d\lambda.
\end{aligned} \qquad (5.70)$$

Bearing in mind that $f_{\zeta\zeta}(\lambda) = f_{\xi\xi}(\lambda) + f_{\eta\eta}(\lambda)$, we can rewrite (5.70) in the form

$$\lim_{m \to -\infty} \sigma_m^2 = \int_{-\pi}^{\pi} \frac{f_{\xi\xi}(\lambda)f_{\eta\eta}(\lambda)}{f_{\xi\xi}(\lambda) + f_{\eta\eta}(\lambda)} \, d\lambda,$$

which agrees with formula (5.31), as one would expect.

It is not hard to verify that formulas (5.41), (5.44), (5.53), and (5.56), obtained in Sec. 27, are actually special cases of formulas (5.68) and (5.69). It is also clear that in the absence of measurement errors, i.e., for $\eta(t) \equiv 0$, $f_{\zeta\zeta}(\lambda) = f_{\xi\xi}(\lambda)$, formulas (5.68) and (5.69) go over into formulas (4.89) and (4.91), which were found earlier. (Of course, in this case, we have to assume that $m \geqslant 0$.)

Finally, we again point out that the solution of the filtering problem given here can be extended practically without change to the case where the measurement errors $\eta(t)$ are correlated with the measured quantities $\xi(t)$, provided that the sequences $\xi(t)$ and $\eta(t)$ are stationarily correlated (see Sec. 15) and that there exists a cross-spectral density $f_{\xi\eta}(\lambda)$ for $\xi(t)$ and $\eta(t)$ [equal to the Fourier transform of their cross-correlation function] which is rational in $e^{i\lambda}$. In this case, it is only necessary to replace $f_{\xi\xi}(\lambda)$ by $f_{\xi\zeta}(\lambda) = f_{\xi\xi}(\lambda) + f_{\xi\eta}(\lambda)$ in all the considerations of Chap. 5.

# 6

---

# LINEAR EXTRAPOLATION
# OF STATIONARY
# RANDOM PROCESSES

---

## 29. Statement of the Problem

In the two preceding chapters, we have considered the problem of extrapolation and filtering of stationary random sequences. However, in recent years, the demands of technology (especially, of radio engineering) have focussed attention on similar problems involving stationary random processes. The present chapter is devoted to the simpler of these two problems, namely, the problem of linear extrapolation of stationary processes.

Thus, suppose we know the values $\xi^{(1)}(t')$ of the stationary random process $\xi(t)$ [with correlation function $B(\tau)$] during the "past," i.e., for $t' \leqslant t$. It is required to use this data to make a prediction of the value of $\xi^{(1)}(t + \tau)$ of the process $\xi(t)$ at the time $t + \tau$, where $\tau > 0$. As in Chap. 4, we shall use the mean square error of the prediction as an index of its quality. Moreover, we shall assume that the predicted value $\overset{\approx}{\xi}{}^{(1)}(t + \tau)$ of the quantity $\xi^{(1)}(t + \tau)$ depends linearly on the given values $\xi^{(1)}(t')$; this last condition is necessary if in solving the problem, we do not want to use more complicated characteristics of the process than $B(\tau)$.[1] Finally, we shall assume as usual that the function

---

[1] However, we note that all other considerations notwithstanding, there are compelling practical grounds for restricting ourselves to linear predictions here. The point is that in practice, the value $\overset{\approx}{\xi}{}^{(1)}(t + \tau)$ is ordinarily obtained from the data $\xi^{(1)}(t')$, $t' \leqslant t$, by using special devices, and the construction and implementation of such devices is always much simpler when they are linear rather than nonlinear.

$B(\tau)$ falls off sufficiently rapidly at infinity to have a Fourier transform $f(\lambda)$ [the spectral density of the process].

We begin by considering the simplest case, where we know only a finite number of the past values of the process $\xi(t)$. For example, suppose that we know the values $\xi^{(1)}(t - s_1), \xi^{(1)}(t - s_2), \ldots, \xi^{(1)}(t - s_n)$ corresponding to the times $t - s_1, t - s_2, \ldots, t - s_n$, where $s_k \geqslant 0$, $k = 1, 2, \ldots, n$. Then, finding the best linear extrapolation formula is equivalent to determining the values of the coefficients $\alpha_1, \alpha_2, \ldots, \alpha_n$ in the formula

$$\alpha_1 \xi(t - s_1) + \alpha_2 \xi(t - s_2) + \cdots + \alpha_n \xi(t - s_n) \tag{6.1}$$

for which the mathematical expectation

$$\mathbf{E} \left| \xi(t + \tau) - \sum_{k=1}^{n} \alpha_k \xi(t - s_k) \right|^2 \tag{6.2}$$

takes its minimum value. Geometrically, this problem can be formulated as the problem of dropping a perpendicular from the point $\xi(t + \tau)$ of the space $H$ onto the subspace of vectors of the form (6.1). It follows at once that the desired values $a_1, a_2, \ldots, a_n$ of the coefficients $\alpha_1, \alpha_2, \ldots, \alpha_n$ can be found from the system of equations

$$\left( \xi(t + \tau) - \sum_{l=1}^{n} a_l \xi(t - s_l), \xi(t - s_k) \right) = 0 \qquad (k = 1, 2, \ldots, n), \tag{6.3}$$

or equivalently, from the system

$$B(\tau + s_k) - \sum_{l=1}^{n} a_l B(s_k - s_l) = 0 \qquad (k = 1, 2, \ldots, n) \tag{6.4}$$

[cf. formula (4.9)]. Then, the *mean square extrapolation error* can be written in a form completely analogous to (4.12):

$$\sigma_{\tau,n}^2 = \mathbf{E} \left| \xi(t + \tau) - \sum_{k=1}^{n} a_k \xi(t - s_k) \right|^2 = B(0) - \sum_{k=1}^{n} \bar{a}_k B(\tau + s_k)$$
$$= B(0) - \sum_{k=1}^{n} \sum_{l=1}^{n} \bar{a}_k a_l B(s_k - s_l). \tag{6.5}$$

Using (2.66), we can write the system (6.4) in the form

$$\int_{-\infty}^{\infty} e^{i s_k \lambda} [e^{i\tau\lambda} - \Phi_{\tau,n}(\lambda)] f(\lambda) \, d\lambda = 0 \qquad (k = 1, 2, \ldots, n), \tag{6.6}$$

where

$$\Phi_{\tau,n}(\lambda) = \sum_{l=1}^{n} a_l e^{-i s_l \lambda}. \tag{6.7}$$

Moreover, formula (6.5) can be written as

$$\sigma_{\tau,n}^2 = \int_{-\infty}^{\infty} |e^{i\tau\lambda} - \Phi_{\tau,n}(\lambda)|^2 f(\lambda)\, d\lambda$$
$$= \int_{-\infty}^{\infty} f(\lambda)\, d\lambda - \int_{-\infty}^{\infty} |\Phi_{\tau,n}(\lambda)|^2 f(\lambda)\, d\lambda \tag{6.8}$$

[cf. formulas (4.22) to (4.24)].

As in the case of random sequences, we shall be interested primarily in the problem of extrapolation when *all* past values of the process are known. This problem also reduces to the problem of dropping a perpendicular from the point $\xi(t + \tau)$ onto a certain subspace of $H$, namely, the subspace $H(t)$ spanned by the set of all vectors $\xi(t')$, $t' \leqslant t$, or equivalently, all vectors $\xi(t - s)$, $s \geqslant 0$. The subspace $H(t)$ consists of all possible linear combinations of the form (6.1), where $n$ is any positive integer, and $s_1, s_2, \ldots, s_n$ are arbitrary nonnegative numbers, and of all limits of sequences of such linear combinations, in the sense of the geometry of the space $H$, i.e., in the mean (square). Naturally, $H(t)$ will in general be infinite-dimensional. Thus, the problem of linear extrapolation of a stationary process $\xi(t)$ will be solved if we can find a vector (i.e., random variable) $L_\tau(t)$ in the space $H$ which is the limit of a sequence

$$\sum_{k=1}^{n_1} \alpha_k^{(1)} \xi(t - s_k^{(1)}), \ \sum_{k=1}^{n_2} \alpha_k^{(2)} \xi(t - s_k^{(2)}), \ldots, \ \sum_{k=1}^{n_m} \alpha_k^{(m)} \xi(t - s_k^{(m)}), \ldots \tag{6.9}$$

of linear combinations of the form (6.1), and which is such that

$$(\xi(t + \tau) - L_\tau(t), \xi(t - s)) = 0 \quad \text{for any} \quad s \geqslant 0. \tag{6.10}$$

These conditions uniquely define $L_\tau(t)$ [since perpendiculars in $H$ are unique!] and guarantee that the mean square extrapolation error

$$\sigma_\tau^2 = \mathbf{E}|\xi(t + \tau) - L_\tau(t)|^2 \tag{6.11}$$

takes its minimum value.

We now try to reduce the problem of finding $L_\tau(t)$ to a problem of function theory, just as we did in Chap. 4. In doing this, it is simplest to start from the spectral representation (2.40) of the stationary process $\xi(t)$ itself. Using (2.40), we can write the elements of the sequence (6.9) as integrals

$$\sum_{k=1}^{n_m} \alpha_k^{(m)} \xi(t - s_k^{(m)}) = \int_{-\infty}^{\infty} e^{it\lambda} \Phi_{\tau,m}(\lambda)\, dZ(\lambda), \tag{6.12}$$

where

$$\Phi_{\tau,m}(\lambda) = \sum_{k=1}^{n_m} \alpha_k^{(m)} e^{-is_k^{(m)}\lambda}, \tag{6.13}$$

and $Z(\lambda)$ is a random function with uncorrelated increments, satisfying the relation

$$\mathbf{E}|Z(\lambda + \Delta\lambda) - Z(\lambda)|^2 = \int_{\lambda}^{\lambda + \Delta\lambda} f(\lambda) \, d\lambda \qquad (6.14)$$

[cf. equation (2.62)].  Thus,

$$\sum_{k=1}^{n_m} \alpha_k^{(m)} \xi(t - s_k^{(m)}) - \sum_{k=1}^{n_{m+p}} \alpha_k^{(m+p)} \xi(t - s_k^{(m+p)})$$
$$= \int_{-\infty}^{\infty} e^{it\lambda} [\Phi_{\tau, m}(\lambda) - \Phi_{\tau, m+p}(\lambda)] \, dZ(\lambda), \qquad (6.15)$$

which means that

$$\mathbf{E}\left| \sum_{k=1}^{n_m} \alpha_k^{(m)} \xi(t - s_k^{(m)}) - \sum_{k=1}^{n_{m+p}} \alpha_k^{(m+p)} \xi(t - s_k^{(m+p)}) \right|^2$$
$$= \int_{-\infty}^{\infty} |\Phi_{\tau, m}(\lambda) - \Phi_{\tau, m+p}(\lambda)|^2 f(\lambda) \, d\lambda \qquad (6.16)$$

[cf. equation (2.63)].  From this it is clear that the existence of a limit of the sequence (6.9) is equivalent to the validity of the relation

$$\lim_{m \to \infty} \int_{-\infty}^{\infty} |\Phi_{\tau, m}(\lambda) - \Phi_{\tau, m+p}(\lambda)|^2 f(\lambda) \, d\lambda = 0 \qquad (6.17)$$

for any $p > 0$.  But (6.17) implies the existence of a function $\Phi_\tau(\lambda)$ such that

$$\lim_{m \to \infty} \int_{-\infty}^{\infty} |\Phi_\tau(\lambda) - \Phi_{\tau, m}(\lambda)|^2 f(\lambda) \, d\lambda = 0. \qquad (6.18)$$

In other words, $\Phi_\tau(\lambda)$ is the limit in the mean (square),[2] with respect to $f(\lambda) \, d\lambda$, of the sequence of functions $\Phi_{\tau, m}(\lambda)$, $m = 1, 2, 3, \ldots$.  Equations (6.12) and (6.18) show that

$$\lim_{m \to \infty} \mathbf{E}\left| \sum_{k=1}^{n_m} \alpha_k^{(m)} \xi(t - s_k^{(m)}) - \int_{-\infty}^{\infty} e^{it\lambda} \Phi_\tau(\lambda) \, dZ(\lambda) \right|^2 = 0, \qquad (6.19)$$

and consequently, we can write the limit $L_\tau(t)$ of the sequence (6.9) in the form

$$L_\tau(t) = \int_{-\infty}^{\infty} e^{it\lambda} \Phi_\tau(\lambda) \, dZ(\lambda). \qquad (6.20)$$

Thus, we see that the random variable $L_\tau(t)$, which is the best approximation to $\xi(t + \tau)$ depending linearly on $\xi(t - s)$, $s \geq 0$, is uniquely determined if we specify the function $\Phi_\tau(\lambda)$ satisfying (6.18).  The function $\Phi_\tau(\lambda)$ will be called the *spectral characteristic for extrapolation* of the process $\xi(t)$.

---

[2] Here, as in similar situations throughout the book, one can safely omit the word "square," since we never deal with mean convergence involving exponents other than 2.

The condition (6.10) imposes certain restrictions on the spectral characteristic, and in fact, it implies that

$$\int_{-\infty}^{\infty} e^{is\lambda}[e^{i\tau\lambda} - \Phi_\tau(\lambda)]f(\lambda)\,d\lambda = 0 \quad \text{for any} \quad s \geqslant 0, \qquad (6.21)$$

where we have used (6.20) and (2.40). Moreover, we note that since obviously

$$\mathbf{E}\left|\sum_{k=1}^{n_m} \alpha_k^{(m)}\xi(t - s_k^{(m)})\right|^2 = \int_{-\infty}^{\infty} |\Phi_{\tau,m}(\lambda)|^2 f(\lambda)\,d\lambda < \infty, \qquad (6.22)$$

the function $\Phi_\tau(\lambda)$ is also square integrable with respect to $f(\lambda)\,d\lambda$, i.e.,

$$\int_{-\infty}^{\infty} |\Phi_\tau(\lambda)|^2 f(\lambda)\,d\lambda < \infty, \qquad (6.23)$$

and moreover, the integral (6.23) equals $\mathbf{E}|L_\tau(t)|^2$. Bearing this in mind, we can rewrite (6.11) in the form

$$\sigma_\tau^2 = \int_{-\infty}^{\infty} |e^{i\tau\lambda} - \Phi_\tau(\lambda)|^2 f(\lambda)\,d\lambda, \qquad (6.24)$$

so that the mean square extrapolation error is also uniquely determined by the function $\Phi_\tau(\lambda)$. In fact, since the vector $\xi(t + \tau) - L_\tau(t)$ is orthogonal to the subspace $H(t)$ and the vector $L_\tau(t)$ belongs to $H(t)$, we must have

$$(\xi(t + \tau) - L_\tau(t), L_\tau(t)) = 0,$$

which leads to the relation [3]

$$\int_{-\infty}^{\infty} [e^{i\tau\lambda} - \Phi_\tau(\lambda)]\overline{\Phi_\tau(\lambda)}f(\lambda)\,d\lambda = 0.$$

This implies at once that the expression (6.24) for the mean square extrapolation error can also be written in the following two forms [cf. (4.12) and (6.5)]:

$$\begin{aligned}
\sigma_\tau^2 &= \int_{-\infty}^{\infty} e^{-i\tau\lambda}[e^{i\tau\lambda} - \Phi_\tau(\lambda)]f(\lambda)\,d\lambda \\
&= \int_{-\infty}^{\infty} f(\lambda)\,d\lambda - \int_{-\infty}^{\infty} |\Phi_\tau(\lambda)|^2 f(\lambda)\,d\lambda.
\end{aligned} \qquad (6.24')$$

In some cases, it is easier to use one of these formulas to calculate the mean square extrapolation error, instead of formula (6.24).

In this connection, we shall regard the problem of extrapolating a process $\xi(t)$ as solved if we succeed in determining the corresponding spectral characteristic $\Phi_\tau(\lambda)$. Below, we shall illustrate by specific examples how to construct the best extrapolation formula which explicitly expresses the predicted value $\overset{\flat}{\xi}{}^{(1)}(t + \tau)$ of the quantity $\xi^{(1)}(t + \tau)$ in terms of $\xi^{(1)}(t - s), s \geqslant 0$. It

---

[3] This relation also follows from (6.21), (6.18) and (6.13).

turns out that in different cases this formula has a completely different appearance. In fact, in the general case, there does not exist any convenient explicit formula for $\xi^{(1)}(t + \tau)$, so that we have to be satisfied with representing $L_\tau(t)$ as the limit of some sequence (6.9) of finite linear combinations of values of the process, or equivalently, as an integral of the form (6.20). The situation was different in Chap. 4, where in every case of practical interest, $\xi^{(1)}(t + \tau)$ could be represented as a series of the form (4.17), i.e., as a linear combination of all past values of the sequence. In the case of processes, however, the expression for $\xi^{(1)}(t + \tau)$ can contain not only the values of the process at separate times in the past (and in the present), but also terms formed from the past of the process by using various differential and integral operators, e.g.,

$$\frac{d}{dt}\xi^{(1)}(t), \qquad \frac{d^3}{dt^3}\xi^{(1)}(t - a) \qquad (a > 0),$$

$$\int_0^\infty \xi^{(1)}(t - s)K_\tau(s)\,ds, \qquad \int_0^\infty \xi^{(1)}(t - s)\,dK_\tau(s),$$

and terms of a more complicated nature.

At this point it should be noted that from a purely practical point of view, in the theory of extrapolating stationary processes we are more interested in finding the function $\Phi_\tau(\lambda)$ than in finding a simple formula for $\xi^{(1)}(t + \tau)$. The point is that in practice, the random process $\xi(t)$ is either usually given from the outset as a fluctuating current (or voltage), or else can be easily transformed into such a random current by using a special device. Then, the extrapolation of $\xi(t)$ is most easily accomplished by applying this random current to the input of a four-terminal network, i.e., a *(linear) filter* (see p. 42), which is chosen in such a way as to give $\xi^{(1)}(t + \tau)$ at its output. (In this respect, extrapolation of processes differs fundamentally from extrapolation of sequences, which in practice is usually carried out arithmetically, with the aid of some kind of calculating machine.) As is well known, to construct such a filter, we have to know its transfer function (i.e., its "frequency response"), and it is not hard to see that this transfer function will be just the function $\Phi_\tau(\lambda)$. In keeping with this, the case where $\Phi_\tau(\lambda)$ is a rational function in $\lambda$ is of particular practical interest, since filters with rational transfer functions can be "realized" in the form of electrical circuits consisting of a finite number of fixed capacitances, resistances and inductances (so-called "lumped parameters"). Moreover, in cases where $\Phi_\tau(\lambda)$ is not a rational function, it is customary to first approximate $\Phi_\tau(\lambda)$ by a rational function and then use a filter with this approximating rational function as its transfer function. Thus, in this case also, one only needs to know the function $\Phi_\tau(\lambda)$ in order to carry out the extrapolation in practice.

In this way, we have reduced the problem of extrapolating a stationary random process $\xi(t)$ to the problem of finding a function $\Phi_\tau(\lambda)$ which satisfies the conditions (6.23) and (6.21), and which is the limit in the mean with respect to $f(\lambda)\,d\lambda$ of a sequence of finite linear combinations of the functions $e^{-is\lambda}$, $s \geqslant 0$. These conditions uniquely specify the function $\Phi_\tau(\lambda)$, with the proviso mentioned on p. 107, since the perpendicular dropped from the point $\xi(t + \tau)$ onto $H(t)$ is unique.

## 30. Transition to Functions of a Complex Variable

Before considering specific examples of extrapolation, we show how to satisfy the requirements (6.18) and (6.21) by imposing certain restrictions on the behavior of $\Phi_\tau(\lambda)$, regarded as a function of a complex variable $\lambda$. As is easily seen, a sufficient condition for (6.21) to hold is that

$$\Psi_\tau(\lambda) = [e^{i\tau\lambda} - \Phi_\tau(\lambda)]f(\lambda)$$

be an analytic function of $\lambda$ in the upper half-plane and fall off faster than $|\lambda|^{-1-\varepsilon}$, $\varepsilon > 0$, as $|\lambda| \to \infty$ in the upper half-plane. [In particular, this condition guarantees the existence of the integral (6.21).] To see this, consider a closed contour $C_R$ in the complex plane, consisting of the interval $[-R, R]$ of the real axis, completed by a semicircle $L_R$ of radius $R$ lying in the upper half-plane (with $[-R, R]$ as its base). Since the functions $\Psi_\tau(\lambda)$ and $e^{is\lambda}$ are both analytic in the upper-half plane, then, according to the residue theorem,

$$\int_{C_R} e^{is\lambda}\Psi_\tau(\lambda)\,d\lambda = 0,$$

i.e.,

$$\int_{-R}^{R} e^{is\lambda}\Psi_\tau(\lambda)\,d\lambda = \int_{L_R} e^{is\lambda}\Psi_\tau(\lambda)\,d\lambda. \tag{6.25}$$

But $|e^{is\lambda}| \leqslant 1$ if $s \geqslant 0$ and $\lambda$ is in the upper half-plane, and by hypothesis, $|\Psi_\tau(\lambda)| < CR^{-1-\varepsilon}$ for $\lambda = Re^{i\varphi}$, $0 \leqslant \varphi \leqslant \pi$, i.e., on the semicircle $L_R$, so that

$$\lim_{R\to\infty} \left| \int_{L_R} e^{is\lambda}\Psi_\tau(\lambda)\,d\lambda \right| \leqslant \lim_{R\to\infty} \frac{C}{R^{1+\varepsilon}} \int_0^\pi R\,d\varphi = 0, \quad \text{for } s \geqslant 0.$$

According to (6.25), this means that

$$\lim_{R\to\infty} \int_{-R}^{R} e^{is\lambda}\Psi_\tau(\lambda)\,d\lambda = \int_{-\infty}^{\infty} e^{is\lambda}\Psi_\tau(\lambda)\,d\lambda = 0 \quad \text{for } s \geqslant 0 \tag{6.26}$$

as well, which is the same as (6.21).

It is somewhat more complicated to prove that if the behavior of $\Phi_\tau(\lambda)$ for complex $\lambda$ is suitable, then the condition (6.18) holds, i.e., there exists a sequence of functions of the form (6.13) converging to $\Phi_\tau(\lambda)$ in the mean with respect to $f(\lambda)\,d\lambda$. We shall assume that $f(\lambda)$ is bounded (this is the only kind of spectral density considered in this book), and then, (6.18) will certainly hold, provided that we can find a sequence of linear combinations of the form (6.13) converging in the mean to $\Phi_\tau(\lambda)$ in the ordinary sense (i.e., with respect to $d\lambda$). For example, this condition will be met for all functions $\Phi_\tau(\lambda)$ which can be expanded, in the sense of mean convergence, in a one-sided Fourier integral of the form

$$\Phi_\tau(\lambda) = \int_0^\infty e^{-is\lambda} a(s) \, ds. \tag{6.27}$$

In this case, the required linear combinations are just the sums approximating the integral

$$\int_0^T e^{-is\lambda} a(s) \, ds$$

as $T \to \infty$. In particular, the representation (6.27) holds if the function $\Phi_\tau(\lambda)$ is analytic in the lower half-plane and falls off faster than $|\lambda|^{-1-\varepsilon}$, $\varepsilon > 0$. Then we have

$$\int_{-\infty}^\infty |\Phi_\tau(\lambda)|^2 \, d\lambda < \infty, \tag{6.28}$$

which, as is well known (see e.g., T3), implies that $\Phi_\tau(\lambda)$ can be represented as a Fourier integral (in the sense of mean convergence). Moreover, the behavior of $\Phi_\tau(\lambda)$ in the lower half-plane guarantees that the Fourier transform of $\Phi_\tau(\lambda)$, i.e., the function

$$a(s) = \frac{1}{2\pi} \int_{-\infty}^\infty e^{i\lambda s} \Phi_\tau(\lambda) \, d\lambda,$$

vanish for negative values of its argument [cf. the derivation of (6.26)].

Unfortunately, in the majority of cases encountered in practice, the function $\Phi_\tau(\lambda)$ does not satisfy the condition (6.28), and hence cannot be represented as an integral of the form (6.27).[4] For example, as we shall see below, in many cases $\Phi_\tau(\lambda)$ is a polynomial in $\lambda$. However, it turns out that for functions satisfying (6.23), we can often conclude that the condition (6.18) holds just from the fact that $\Phi_\tau(\lambda)$ is analytic in the lower half-plane. In fact, to prove (6.18) it is obviously sufficient to find a sequence of functions $\tilde{\Phi}_{\tau,n}(\lambda)$ converging to $\Phi_\tau(\lambda)$ in the mean with respect to $f(\lambda) \, d\lambda$, each of which is in turn a limit (in the same sense) of a sequence of functions of the form (6.13). This will certainly imply that the function $\Phi_\tau(\lambda)$ itself belongs to the closure of the set of functions $e^{-is\lambda}$, $s \geqslant 0$, in the mean with respect to $f(\lambda) \, d\lambda$.

Thus, let $\Phi_\tau(\lambda)$ be a function of $\lambda$ which is analytic in the lower half-plane, and whose absolute value grows no faster than some power of $|\lambda|$, as $|\lambda| \to \infty$

---

[4] In this regard, it should be noted that the situation is more favorable in Chap. 4, where, as a rule, there is no difficulty in representing the spectral characteristic $\Phi_m(\lambda)$ as a one-sided Fourier series (4.28), analogous to the integral (6.27).

in the lower half-plane (no faster than $|\lambda|^q$, say), and let $\Phi_\tau(\lambda)$ satisfy (6.23). If we write

$$\tilde{\Phi}_{\tau,n}(\lambda) = \frac{\Phi_\tau(\lambda)}{\left(1 + \dfrac{i\lambda}{n}\right)^r}, \qquad (6.29)$$

where $r$ is an integer greater than $q + 1 + \varepsilon$, then all the functions $\tilde{\Phi}_{\tau,n}(\lambda)$ are analytic in the lower half-plane[5] and fall off faster than $|\lambda|^{-1-\varepsilon}$, as $|\lambda| \to \infty$ in the lower half-plane. Therefore, all the functions $\tilde{\Phi}_{\tau,n}(\lambda)$ can be represented as one-sided Fourier integrals of the form (6.27), i.e., they are all limits in the mean [both in the usual sense and with respect to $f(\lambda)\,d\lambda$] of sequences of functions of the form (6.13). However, it is not hard to see that

$$\lim_{n\to\infty} \int_{-\infty}^{\infty} |\Phi_\tau(\lambda) - \tilde{\Phi}_{\tau,n}(\lambda)|^2 f(\lambda)\,d\lambda = 0. \qquad (6.30)$$

In fact, we have

$$
\begin{aligned}
\int_{-\infty}^{\infty} &|\Phi_\tau(\lambda) - \tilde{\Phi}_{\tau,n}(\lambda)|^2 f(\lambda)\,d\lambda \\
&= \int_{|\lambda| \leqslant L} |\Phi_\tau(\lambda) - \tilde{\Phi}_{\tau,n}(\lambda)|^2 f(\lambda)\,d\lambda \\
&\quad + \int_{|\lambda| > L} |\Phi_\tau(\lambda) - \tilde{\Phi}_{\tau,n}(\lambda)|^2 f(\lambda)\,d\lambda \qquad (6.31)\\
&= \int_{|\lambda| \leqslant L} \left|1 - \left(1 + \frac{i\lambda}{n}\right)^{-r}\right|^2 |\Phi_\tau(\lambda)|^2 f(\lambda)\,d\lambda \\
&\quad + \int_{|\lambda| > L} |\Phi_\tau(\lambda) - \tilde{\Phi}_{\tau,n}(\lambda)|^2 f(\lambda)\,d\lambda.
\end{aligned}
$$

The first integral in the right-hand side of (6.31) goes to zero as $n \to \infty$ for any $L > 0$, since for $|\lambda| \leqslant L$, the difference

$$1 - \left(1 + \frac{i\lambda}{n}\right)^{-r}$$

is arbitrarily small in absolute value, for sufficiently large $n$. On the other hand, for the second integral in the right-hand side of (6.31), we have the estimate

$$
\begin{aligned}
\int_{|\lambda| > L} |\Phi_\tau(\lambda) - \tilde{\Phi}_{\tau,n}(\lambda)|^2 f(\lambda)\,d\lambda &\leqslant \int_{|\lambda| > L} [|\Phi_\tau(\lambda)| + |\tilde{\Phi}_{\tau,n}(\lambda)|]^2 f(\lambda)\,d\lambda \\
&\leqslant 4 \int_{|\lambda| > L} |\Phi_\tau(\lambda)|^2 f(\lambda)\,d\lambda,
\end{aligned}
$$

where we have used the obvious inequality $|\tilde{\Phi}_{\tau,n}(\lambda)| \leqslant |\Phi_\tau(\lambda)|$. It follows at once from (6.23) that this integral can be made arbitrarily small by choosing $L$ sufficiently large. Thus, the relation (6.30) is proved, and according to

---

[5] The denominator of (6.29) vanishes for $\lambda = in$, i.e., only in the upper half-plane.

what was said previously, this also proves that the function $\Phi_\tau(\lambda)$ satisfies the condition (6.18).

Finally, therefore, we see that in the case where the spectral density $f(\lambda)$ is bounded, the following conditions are sufficient for a function $\Phi_\tau(\lambda)$ to be the spectral characteristic for extrapolation:

(a) The function $\Phi_\tau(\lambda)$ is analytic in the lower half-plane, and as $|\lambda| \to \infty$ in the lower half-plane, $\Phi_\tau(\lambda)$ grows no faster than some power of $|\lambda|$;

(b) The function $\Psi_\tau(\lambda) = [e^{i\tau\lambda} - \Phi_\tau(\lambda)]f(\lambda)$ is analytic in the upper half-plane, and as $|\lambda| \to \infty$ in the upper half-plane, $\Psi_\tau(\lambda)$ falls off faster than $|\lambda|^{-1-\varepsilon}$, $\varepsilon > 0$;

(c) The integral along the real axis of the function $|\Phi_\tau(\lambda)|^2 f(\lambda)$ is bounded, i.e.,

$$\int_{-\infty}^{\infty} |\Phi_\tau(\lambda)|^2 f(\lambda)\, d\lambda < \infty.$$

In cases where $f(\lambda)$ is a rational function of $\lambda$, a function $\Phi_\tau(\lambda)$ satisfying these three conditions can easily be constructed, just as the corresponding function $\Phi_m(\lambda)$ was constructed in the theory of extrapolation of sequences (see Chap. 4). In the next section, we shall give some specific examples showing how this is done.

## 31. Examples of Linear Extrapolation of Stationary Processes

*Example 1.* Let

$$B(\tau) = Ce^{-\alpha|\tau|}, \qquad \alpha > 0, \tag{6.32}$$

so that

$$f(\lambda) = \frac{C}{\pi}\frac{\alpha}{\alpha^2 + \lambda^2} = \frac{C\alpha}{\pi}\frac{1}{(\lambda - i\alpha)(\lambda + i\alpha)} \tag{6.33}$$

(cf. Example 1, p. 61). In this case,

$$\Psi_\tau(\lambda) = \frac{C\alpha}{\pi}\frac{e^{i\tau\lambda} - \Phi_\tau(\lambda)}{(\lambda - i\alpha)(\lambda + i\alpha)}. \tag{6.34}$$

In order for the function (6.34) to be analytic in the upper half-plane, the zero of the denominator at the point $\lambda = i\alpha$ must be cancelled by a zero of the numerator at the same point, and hence we must have

$$\Phi_\tau(i\alpha) = e^{i\tau(i\alpha)} = e^{-\alpha\tau}.$$

Moreover, since $\Phi_\tau(\lambda)$ has to be analytic in the lower half-plane, while $\Psi_\tau(\lambda)$ has to be analytic in the upper half-plane, it follows that $\Phi_\tau(\lambda)$ can have no singularities other than the point $\lambda = \infty$. In order to satisfy condition (c) as well, it is sufficient to set $\Phi_\tau(\lambda)$ equal to a constant. It is clear that all the conditions concerning the behavior of $\Phi_\tau(\lambda)$ and $\Psi_\tau(\lambda)$ as $|\lambda| \to \infty$ will then be met. Thus, in the present case, the spectral characteristic for extrapolation is just

$$\Phi_\tau(\lambda) = e^{-\alpha\tau} \tag{6.35}$$

The mean square extrapolation error $\sigma_\tau^2$ can now be easily calculated from formula (6.24) by using the residue theorem, or, more simply, by using the second of the formulas (6.24'); the result is

$$\sigma_\tau^2 = C(1 - e^{-2\alpha\tau}). \tag{6.36}$$

Substituting (6.35) into (6.20), and then using (2.40), we find that

$$L_\tau(t) = e^{-\alpha\tau}\xi(t), \tag{6.37}$$

so that the best extrapolation formula has the form

$$\xi^{(1)}(t + \tau) = e^{-\alpha\tau}\xi^{(1)}(t). \tag{6.37'}$$

Thus, in the present example, the best linear prediction of the value of the quantity $\xi(t + \tau)$, which can be made when the values of $\xi(t - s)$ are known for $s \geqslant 0$, uses only the last of these values, and knowledge of the previous history of the process is of no use whatsoever in improving the prediction.

This last fact suggests that $\xi(t)$ is a Markov process, i.e., that the conditional probability distribution for the quantity $\xi(t + \tau)$, $\tau \geqslant 0$, given the values of the quantities $\xi(t - s_1), \xi(t - s_2), \ldots, \xi(t - s_n)$, $0 < s_1 < s_2 < \cdots < s_n$, depends only on the value of $\xi(t - s_1)$, and not on the values of the other random variables $\xi(t - s_2), \ldots, \xi(t - s_n)$. It can be shown that this will indeed be the case, provided that the process $\xi(t)$ is also normal, i.e., that all its distributions (1.3) are normal. Moreover, it is not hard to show that every real normal stationary Markov process has a correlation function of the form (6.32) [see e.g., D4, W2]. Because of this, stationary processes with correlation functions of the form (6.32) will be called *wide-sense Markov processes* or *processes of the Markov type* (cf. Example 1, p. 113).

*Formula (6.36) can easily be explained if we recall that a stationary process $\xi(t)$ with the correlation function (6.32) can be represented as an integral of the form (2.135) [see p. 69]. Then we have

$$\xi(t + \tau) = A \int_{-\infty}^{t+\tau} e^{-\alpha(t+\tau-s)} \, d\zeta(s)$$
$$= Ae^{-\alpha\tau} \int_{-\infty}^{t} e^{-\alpha(t-s)} \, d\zeta(s) + A \int_{t}^{t+\tau} e^{-\alpha(t+\tau-s)} \, d\zeta(s). \tag{6.38}$$

The second term on the right is a random variable with mean value zero, which is uncorrelated with all past values of $\xi(t)$. From this it is clear that the best linear prediction of the value of $\xi(t + \tau)$ in terms of the known past values of

$\xi(t - s)$, $s \geqslant 0$, must be just the value of the first term on the right in (6.38), i.e., the value of the quantity

$$Ae^{-\alpha\tau} \int_{-\infty}^{t} e^{-\alpha(t-s)} \, d\zeta(s) = e^{-\alpha\tau}\xi(t),$$

in complete agreement with formula (6.37).

Another explanation of formula (6.37) follows from the representation of the random process $\xi(t)$ as a solution of a differential equation like (2.137), where the right-hand side is a "purely random" process (see p. 64). In fact, the expression (6.37) is the same as the value at the time $t + \tau$ of the solution of the corresponding homogeneous equation

$$\frac{d\xi}{dt} + \alpha\xi = 0, \tag{6.39}$$

subject to the initial condition

$$\xi|_t = \xi(t).$$

Thus, for example, in order to obtain the best approximation to the velocity of a Brownian particle after the time $\tau$ has elapsed, we should take into account only the friction force, while assuming that in the future there will be no random collisions. In other words, in predicting future values of the particle's velocity we replace the actual perturbing force which the medium exerts on the particle by the mean value of this force. This is only natural, since we can say nothing in advance concerning the future values of the fluctuating "purely random" component of the force.*

*Example 2.* Now let

$$f(\lambda) = \frac{C}{\lambda^4 + \alpha^4}$$

$$= \frac{C}{\left(\lambda + \dfrac{1+i}{\sqrt{2}}\alpha\right)\left(\lambda - \dfrac{1+i}{\sqrt{2}}\alpha\right)\left(\lambda + \dfrac{1-i}{\sqrt{2}}\alpha\right)\left(\lambda - \dfrac{1-i}{\sqrt{2}}\alpha\right)}, \tag{6.40}$$

where $C > 0$ and $\alpha > 0$. According to (2.150), the corresponding correlation function is

$$B(\tau) = \frac{\pi C}{\sqrt{2}\alpha^3} e^{-\alpha|\tau|/\sqrt{2}} \left(\cos\frac{\alpha\tau}{\sqrt{2}} + \sin\frac{\alpha|\tau|}{\sqrt{2}}\right). \tag{6.41}$$

In this case,

$$\Psi_\tau(\lambda) = C \frac{e^{i\tau\lambda} - \Phi_\tau(\lambda)}{\left(\lambda + \dfrac{1+i}{\sqrt{2}}\alpha\right)\left(\lambda - \dfrac{1+i}{\sqrt{2}}\alpha\right)\left(\lambda + \dfrac{1-i}{\sqrt{2}}\alpha\right)\left(\lambda - \dfrac{1-i}{\sqrt{2}}\alpha\right)}, \tag{6.42}$$

from which it is clear that the difference $e^{i\tau\lambda} - \Phi_\tau(\lambda)$ must vanish at the points

$$\lambda = \frac{1+i}{\sqrt{2}}\,\alpha \quad \text{and} \quad \lambda = -\frac{1-i}{\sqrt{2}}\,\alpha,$$

so that

$$\Phi_\tau\left(\frac{1+i}{\sqrt{2}}\,\alpha\right) = \exp\left(i\tau\,\frac{1+i}{\sqrt{2}}\,\alpha\right) = \exp\left(-\frac{\alpha\tau}{\sqrt{2}}(1-i)\right),$$
$$\Phi_\tau\left(-\frac{1-i}{\sqrt{2}}\,\alpha\right) = \exp\left(-i\tau\,\frac{1-i}{\sqrt{2}}\,\alpha\right) = \exp\left(-\frac{\alpha\tau}{\sqrt{2}}(1+i)\right). \tag{6.43}$$

As in Example 1, we can convince ourselves that the function $\Phi_\tau(\lambda)$ must be entire (i.e., can have no singularities at finite points). Moreover, since the integral of $|\Phi_\tau(\lambda)|^2 f(\lambda)$ along the real axis must converge, it is natural to choose $\Phi_\tau(\lambda)$ to be a linear function

$$\Phi_\tau(\lambda) = A\lambda + B. \tag{6.44}$$

It is clear that if (6.44) holds, then all the conditions concerning the behavior of $\Phi_\tau(\lambda)$ and $\Psi_\tau(\lambda)$ as $|\lambda| \to \infty$ are also satisfied. The coefficients $A$ and $B$ in (6.44) are uniquely determined by the conditions (6.43), and finding these coefficients reduces to solving two linear equations in two unknowns. Solving this system, we obtain

$$\Phi_\tau(\lambda) = \frac{\sqrt{2}\,i\lambda}{\alpha}\,e^{-\alpha\tau/\sqrt{2}}\sin\frac{\alpha\tau}{\sqrt{2}} + e^{-\alpha\tau/\sqrt{2}}\left(\cos\frac{\alpha\tau}{\sqrt{2}} + \sin\frac{\alpha\tau}{\sqrt{2}}\right). \tag{6.45}$$

The mean square extrapolation error $\sigma_\tau^2$ can be calculated from formula (6.24) or (6.24′), and turns out to be

$$\sigma_\tau^2 = \frac{\pi C}{\sqrt{2}\,\alpha^3}\left[1 - 2e^{-\sqrt{2}\alpha\tau} - \sqrt{2}\,e^{-\sqrt{2}\alpha\tau}\sin\left(\sqrt{2}\,\alpha\tau - \frac{\pi}{4}\right)\right]. \tag{6.46}$$

If we now substitute (6.45) into (6.20), taking account of (2.40) and (2.80), we find that

$$L_\tau(t) = \frac{\sqrt{2}}{\alpha}\,e^{-\alpha\tau/\sqrt{2}}\sin\frac{\alpha\tau}{\sqrt{2}}\,\xi'(t)$$
$$+ e^{-\alpha\tau/\sqrt{2}}\left(\cos\frac{\alpha\tau}{\sqrt{2}} + \sin\frac{\alpha\tau}{\sqrt{2}}\right)\xi(t). \tag{6.47}$$

It follows from (6.47) that in the present example, the best extrapolation formula is

$$\overset{\approx}{\xi}{}^{(1)}(t+\tau) = \frac{\sqrt{2}}{\alpha}\,e^{-\alpha\tau/\sqrt{2}}\sin\frac{\alpha\tau}{\sqrt{2}}\,\frac{d\xi^{(1)}(t)}{dt}$$
$$+ e^{-\alpha\tau/\sqrt{2}}\left(\cos\frac{\alpha\tau}{\sqrt{2}} + \sin\frac{\alpha\tau}{\sqrt{2}}\right)\xi^{(1)}(t). \tag{6.48}$$

Thus, we see that the best linear prediction of a future value of $\xi(t)$ uses only the values of the quantities $\xi(t)$ and $\xi'(t)$. Therefore, it is apparent that in making the best linear prediction of $\xi(t)$, we only need to know the values of $\xi(t)$ at the time $t$ and during an arbitrarily small time interval preceding $\xi(t)$; all earlier values are of no use whatsoever here.

　*As in Example 1, this result has an intuitive mechanical interpretation. At the end of Sec. 14, we saw that a process $\xi(t)$ with the spectral density (6.40) can be regarded as the solution of the differential equation

$$\frac{d^2\xi}{dt^2} + \sqrt{2}\alpha \frac{d\xi}{dt} + \alpha^2\xi = \zeta'(t), \qquad (6.49)$$

where the right-hand side is a "purely random" process. It is easily verified that (6.47) is just the value at the time $t + \tau$ of the solution of the corresponding homogeneous equation

$$\frac{d^2\xi}{dt^2} + \sqrt{2}\alpha \frac{d\xi}{dt} + \alpha^2\xi = 0, \qquad (6.49')$$

corresponding to the initial conditions

$$\xi|_t = \xi(t), \qquad \xi'|_t = \xi'(t). \qquad (6.49'')$$

Thus, to predict the position of a harmonic oscillator undergoing Brownian motion as described by equation (6.49), we should assume that in the future the random forces will not act at all. Instead, we calculate the motion of the oscillator under the action of just the "regular forces" (i.e., the restoring force and the friction force), by finding the solution of (6.49') subject to the conditions (6.49''), which give the last known values of the oscillator's position and velocity. The value of this solution at time $t + \tau$ is the best linear approximation to the true value of $\xi(t)$ at time $t + \tau$, and the mean square error of the approximation is given by (6.46).

　We can also explain formula (6.47), starting from the representation of the process $\xi(t)$ in the form

$$\xi(t) = \frac{\sqrt{2}}{\alpha} \int_{-\infty}^{t} e^{-\alpha(t-s)/\sqrt{2}} \sin \frac{\alpha}{\sqrt{2}} (t - s) \, d\zeta(s) \qquad (6.50)$$

[cf. formula (2.156), p. 76]. Differentiating the integral (6.50) by the usual rule, we find that

$$\xi'(t) = \int_{-\infty}^{t} e^{-\alpha(t-s)/\sqrt{2}} \left\{ \cos \frac{\alpha}{\sqrt{2}} (t - s) - \sin \frac{\alpha}{\sqrt{2}} (t - s) \right\} d\zeta(s). \qquad (6.51)$$

But, according to (6.50), we have

$$\xi(t + \tau) = \frac{\sqrt{2}}{\alpha} \int_{-\infty}^{t+\tau} e^{-\alpha(t+\tau-s)/\sqrt{2}} \sin \frac{\alpha}{\sqrt{2}} (t + \tau - s) \, d\zeta(s).$$

Bearing in mind that if $s > t$, the values of $\zeta(s)$ are uncorrelated with the past values of (6.50), it is natural to conclude that the "predictable part" $L_\tau(t)$ of the random variable $\xi(t + \tau)$ is given by the integral

$$L_\tau(t) = \frac{\sqrt{2}}{\alpha} \int_{-\infty}^{t} e^{-\alpha(t+\tau-s)/\sqrt{2}} \sin\frac{\alpha}{\sqrt{2}} (t + \tau - s) \, d\zeta(s)$$

$$= \frac{\sqrt{2}}{\alpha} e^{-\alpha\tau/\sqrt{2}} \left\{ \sin\frac{\alpha\tau}{\sqrt{2}} \int_{-\infty}^{t} e^{-\alpha(t-s)/\sqrt{2}} \cos\frac{\alpha}{\sqrt{2}} (t - s) \, d\zeta(s) \right. \tag{6.52}$$

$$\left. + \cos\frac{\alpha\tau}{\sqrt{2}} \int_{-\infty}^{t} e^{-\alpha(t-s)/\sqrt{2}} \sin\frac{\alpha}{\sqrt{2}} (t - s) \, d\zeta(s) \right\}.$$

Then, formula (6.47) is an immediate consequence of (6.52), (6.50) and (6.51).

*Example 2a.* Just as in Example 2, we can solve the extrapolation problem in the case of the more general spectral density (2.148). In this case, $\Phi_\tau(\lambda)$ will also be a linear function of $\lambda$, and hence $L_\tau(t)$ will be a linear combination of $\xi(t)$ and $\xi'(t)$. The appearance of the resulting formulas will be somewhat different depending on whether $\omega^2 - \alpha^2 > 0$, $\omega^2 - \alpha^2 = 0$ or $\omega^2 - \alpha^2 < 0$. Example 2 (just analyzed) is typical of the case $\omega^2 - \alpha^2 > 0$. When $\omega^2 = \alpha^2$, i.e., in the case of the spectral density

$$f(\lambda) = \frac{C}{(\lambda^2 + \alpha^2)^2}$$

[cf. formula (2.152)], we have

$$\Psi_\tau(\lambda) = C \frac{e^{i\tau\lambda} - \Phi_\tau(\lambda)}{(\lambda + i\alpha)^2(\lambda - i\alpha)^2}.$$

It follows that here the coefficients in the formula

$$\Phi_\tau(\lambda) = A\lambda + B$$

are obtained from the conditions

$$[e^{i\tau\lambda} - \Phi_\tau(\lambda)]_{\lambda=i\alpha} = 0, \qquad \frac{d}{d\lambda} [e^{i\tau\lambda} - \Phi_\tau(\lambda)]_{\lambda=i\alpha} = 0,$$

i.e.,

$$\Phi_\tau(i\alpha) = e^{-\alpha\tau}, \qquad \Phi_\tau'(i\alpha) = i\tau e^{-\alpha\tau}.$$

After some elementary calculations, we obtain

$$\Phi_\tau(\lambda) = \tau e^{-\alpha\tau} i\lambda + e^{-\alpha\tau}(1 + \alpha\tau), \tag{6.53}$$

from which it follows at once that

$$L_\tau(t) = \tau e^{-\alpha\tau} \xi'(t) + e^{-\alpha\tau}(1 + \alpha\tau)\xi(t) \tag{6.54}$$

and

$$\sigma_\tau^2 = \frac{\pi C}{2\alpha^3} [1 - e^{-2\alpha\tau}(1 + 2\alpha\tau + 2\alpha^2\tau^2)]. \tag{6.55}$$

Next, to illustrate the spectral density (2.148) when $\omega^2 - \alpha^2 < 0$, we consider

$$f(\lambda) = \frac{C}{(\lambda^2 + 2\omega^2)\left(\lambda^2 + \frac{\omega^2}{2}\right)}, \tag{6.56}$$

where we find at once that $\alpha^2 = 9\omega^2/8$. It is not hard to see that here the coefficients in formula (6.44) have to be determined from the conditions

$$\Phi_\tau(\sqrt{2}\,i\omega) = e^{-\sqrt{2}\omega\tau}, \qquad \Phi_\tau\left(\frac{i\omega}{\sqrt{2}}\right) = e^{-\omega\tau/\sqrt{2}},$$

from which we easily derive

$$\begin{aligned}
\Phi_\tau(\lambda) = \ &\frac{\sqrt{2}}{\omega} e^{-\omega\tau/\sqrt{2}}(1 - e^{-\omega\tau/\sqrt{2}})i\lambda \\
&+ 2e^{-\omega\tau/\sqrt{2}}\left(1 - \frac{1}{2}e^{-\omega\tau/\sqrt{2}}\right),
\end{aligned} \tag{6.57}$$

$$\begin{aligned}
L_\tau(t) = \ &\frac{\sqrt{2}}{\omega} e^{-\omega\tau/\sqrt{2}}(1 - e^{-\omega\tau/\sqrt{2}})\,\xi'(t) \\
&+ 2e^{-\omega\tau/\sqrt{2}}\left(1 - \frac{1}{2}e^{-\omega\tau/\sqrt{2}}\right)\xi(t),
\end{aligned} \tag{6.58}$$

and

$$\sigma_\tau^2 = \frac{\sqrt{2}\pi C}{3\omega^3}[1 - e^{-\sqrt{2}\omega\tau}(6 - 8e^{-\omega\tau/\sqrt{2}} + 3e^{-\sqrt{2}\omega\tau})]. \tag{6.59}$$

It is clear that the explanation of the meaning of extrapolation given at the end of Example 2 can be carried over almost word for word to the case of the general spectral density (2.148).*

*Example 3.* A generalization of the spectral densities considered in Examples 1 to 2a is the spectral density of the form

$$f(\lambda) = \frac{C}{|\lambda^n + a_1\lambda^{n-1} + \cdots + a_n|^2}, \qquad C > 0, \tag{6.60}$$

where $\lambda^n + a_1\lambda^{n-1} + \cdots + a_n$ is a polynomial all of whose zeros lie in the upper half-plane. It is easy to see that in this case, the function $\Phi_\tau(\lambda)$ satisfying all the conditions (a) to (c) given on p. 153 will be a polynomial of degree $n - 1$, with the property that at every zero of the polynomial $\lambda^n + a_1\lambda^{n-1} + \cdots + a_n$, the function $e^{i\tau\lambda} - \Phi_\tau(\lambda)$ has a zero of the same order. It follows at once that in the case of a spectral density of the form (6.60), the best prediction of the value of $\xi(t + \tau)$ which depends linearly on the values of the process $\xi(t)$ "in the past" is given by a linear combination of the values at time $t$ of the process $\xi(t)$ itself and of its derivatives up to order

$n - 1$ inclusive. [We recall that according to the condition (2.86), a process with a spectral density of the form (6.60) is differentiable $n - 1$ times.]

*It is clear that the interpretation of the meaning of extrapolation given at the end of Examples 1 and 2, in which we regard the corresponding random processes as solutions of the equations (2.137) and (2.155), can be extended without difficulty to the case of the more general spectral density (6.60). The only difference is that now we have to find the solution of the homogeneous differential equation

$$\frac{d^n\xi}{dt^n} + ia_1 \frac{d^{n-1}\xi}{dt^{n-1}} + \ldots + i^n a_n \xi = 0$$

of order $n$, subject to the initial conditions

$$\xi|_t = \xi(t), \qquad \xi'|_t = \xi'(t), \ldots, \xi^{(n-1)}|_t = \xi^{(n-1)}(t).$$

Then, the value of this solution at time $t + \tau$ is the best linear approximation to the true value of $\xi(t)$ at time $t + \tau$, so that the extrapolation formula involves the derivatives of $\xi(t)$ up to order $n - 1$ inclusive.*

***Example 4.*** As a last example, let

$$f(\lambda) = C\,\frac{\lambda^2 + \alpha^2}{\lambda^4 + \alpha^4}, \tag{6.61}$$

so that

$$\Psi_\tau(\lambda) = C\,\frac{[e^{i\tau\lambda} - \Phi_\tau(\lambda)](\lambda + i\alpha)(\lambda - i\alpha)}{\left(\lambda + \dfrac{1+i}{\sqrt{2}}\alpha\right)\left(\lambda - \dfrac{1+i}{\sqrt{2}}\alpha\right)\left(\lambda + \dfrac{1-i}{\sqrt{2}}\alpha\right)\left(\lambda - \dfrac{1-i}{\sqrt{2}}\alpha\right)}. \tag{6.62}$$

In view of the conditions (a) to (c) imposed on the function $\Phi_\tau(\lambda)$, we conclude that the difference $e^{i\tau\lambda} - \Phi_\tau(\lambda)$ must vanish at the points

$$\lambda = \frac{1+i}{\sqrt{2}}\alpha, \qquad \lambda = -\frac{1-i}{\sqrt{2}}\alpha,$$

and $\Phi_\tau(\lambda)$ can have no singularities other than a simple pole at the point $\lambda = i\alpha$. Therefore

$$\Phi_\tau(\lambda) = \frac{\gamma_\tau(\lambda)}{\lambda - i\alpha}, \tag{6.63}$$

where $\gamma_\tau(\lambda)$ is an entire function, which according to condition (c) cannot grow faster than the first power of $|\lambda|$, as $|\lambda| \to \infty$. It follows that in order to satisfy conditions (a) to (c), we have to set

$$\Phi_\tau(\lambda) = \frac{A\lambda + B}{\lambda - i\alpha}, \tag{6.64}$$

where the coefficients $A$ and $B$ are determined from the conditions

$$\Phi_\tau\left(\frac{1+i}{\sqrt{2}}\alpha\right) = e^{-\alpha\tau(1-i)/\sqrt{2}}, \qquad \Phi_\tau\left(-\frac{1-i}{\sqrt{2}}\alpha\right) = e^{-\alpha\tau(1+i)/\sqrt{2}}.$$

After making some simple calculations, we find that

$$A = e^{-\alpha\tau/\sqrt{2}}\left[\cos\frac{\alpha\tau}{\sqrt{2}} + (\sqrt{2}-1)\sin\frac{\alpha\tau}{\sqrt{2}}\right],$$

$$B = i\alpha e^{-\alpha\tau/\sqrt{2}}\left[-\cos\frac{\alpha\tau}{\sqrt{2}} + (\sqrt{2}-1)\sin\frac{\alpha\tau}{\sqrt{2}}\right] = -iB_1. \tag{6.65}$$

The corresponding mean square error turns out to be

$$\sigma_\tau^2 = \frac{\sqrt{2}\pi C}{\alpha}[1 - (\sqrt{2}-1)e^{-\sqrt{2}\alpha\tau}(\sqrt{2} + \cos\sqrt{2}\alpha\tau)]. \tag{6.66}$$

It is convenient to rewrite (6.64) in the form

$$\Phi_\tau(\lambda) = A + \frac{i\alpha A + B}{\lambda - i\alpha} = A - \frac{\alpha A - iB}{\alpha + i\lambda} = A - \frac{\alpha A - B_1}{\alpha + i\lambda}. \tag{6.64'}$$

Then

$$L_\tau(t) = A\int_{-\infty}^{\infty} e^{i\lambda t}\,dZ(\lambda) - (\alpha A - B_1)\int_{-\infty}^{\infty}\frac{e^{i\lambda t}}{\alpha + i\lambda}\,dZ(\lambda),$$

so that, according to (2.95),

$$L_\tau(t) = A\xi(t) - (\alpha A - B_1)\int_0^{\infty} e^{-\alpha\tau}\xi(t-\tau)\,d\tau, \tag{6.67}$$

and

$$\xi^{(1)}(t+\tau) = A\xi^{(1)}(t) - (\alpha A - B_1)\int_0^{\infty} e^{-\alpha\tau}\xi^{(1)}(t-\tau)\,d\tau. \tag{6.68}$$

Thus, we see that to predict the future value of the process $\xi(t)$, we have to take its value at the last moment available to us (with the finite weight $A$), and then add an extra term, equal to a certain integral extending over the entire past of the process.

## *32. The Case of a General Rational Spectral Density in $\lambda$

It is not hard to show that the most general spectral density which is a rational function of $\lambda$ can be represented in the form

$$f(\lambda) = \frac{|B_0\lambda^M + B_1\lambda^{M-1} + \cdots + B_M|^2}{|A_0\lambda^N + A_1\lambda^{N-1} + \cdots + A_N|^2} = \frac{|B(\lambda)|^2}{|A(\lambda)|^2}, \tag{6.69}$$

where the coefficients $A_0$, $B_0$, $A_N$ and $B_M$ are nonzero, all the zeros of $A(\lambda)$

have positive imaginary parts, all the zeros of $B(\lambda)$ have nonnegative imaginary parts, and $M < N$ (see D6, p. 542). (See also the remarks made at the beginning of Sec. 24.[5])   As in Sec. 24, we make the additional assumption that $f(\lambda)$ does not vanish for any value of $\lambda$, in which case all the zeros of $B(\lambda)$ have strictly positive imaginary parts.   The class of spectral densities (6.69), even with this restriction, is sufficiently broad to encompass almost all cases of practical interest.   In this section, we shall consider the problem of extrapolating stationary processes with spectral densities of the form (6.69).

In cases where the correlation function $B(\tau)$ of the process $\xi(t)$ to be extrapolated is determined empirically, it is often not hard to approximate $B(\tau)$ by a function whose Fourier transform is rational; some practical methods of making such approximations are given in S12.   Moreover, the rational character of $f(\lambda)$ follows from theoretical considerations in the case of electrical fluctuations at the output of any "lumped-parameter" electrical circuit[6] (in general, one with many meshes), some part of which contains a source of "white noise" (see p. 64). Another situation in which it can be asserted that the spectral density (if it exists) is rational is the case of a normal stationary random process $\xi(t)$, for which we can find a finite number of subsidiary random "coordinates" $\xi_1(t), \xi_2(t), \ldots, \xi_{N-1}(t)$, which are stationary and stationarily correlated with $\xi(t)$ and are such that the multidimensional random process

$$\boldsymbol{\xi}(t) = (\xi(t), \xi_1(t), \ldots, \xi_{N-1}(t))$$

is a Markov process.   The converse assertion is also true: For every normal stationary random process $\xi(t)$ which has a rational spectral density, we can find a finite number of subsidiary random processes with the properties just indicated (see Theorem 4.9 of D4).

We now factor the polynomials $B(\lambda)$ and $A(\lambda)$, obtaining

$$\begin{aligned}
B(\lambda) &= B_0(\lambda - \beta_1)(\lambda - \beta_2)\ldots(\lambda - \beta_M), \quad \text{where} \quad \operatorname{Im}\beta_k > 0 \\
& \hspace{5.5cm} (k = 1, 2, \ldots, M), \\
A(\lambda) &= A_0(\lambda - \alpha_1)(\lambda - \alpha_2)\ldots(\lambda - \alpha_N), \quad \text{where} \quad \operatorname{Im}\alpha_l > 0 \\
& \hspace{5.5cm} (l = 1, 2, \ldots, N).
\end{aligned} \quad (6.70)$$

For the time being, we shall assume that all the zeros $\alpha_1, \alpha_2, \ldots, \alpha_N$ are different.   In this case, a function $\Phi_\tau(\lambda)$ satisfying conditions (a) to (c) on p. 153 can be found in just the same way as in Example 4 of Sec. 31.   Since $\Phi_\tau(\lambda)$ can have no singularities in the lower half-plane, and since $\Psi_\tau(\lambda) = [e^{i\lambda\tau} - \Phi_\tau(\lambda)]f(\lambda)$ can have no singularities in the upper half-plane, it follows at once that $\Phi_\tau(\lambda)$ can have singularities only at the points $\beta_1, \beta_2, \ldots, \beta_M$,

---

[5] We leave it to the reader to prove that in the case where the process $\xi(t)$ is real, so that $f(\lambda)$ is even, $A(\lambda)$ and $B(\lambda)$ have real coefficients when written as polynomials in $i\lambda$.

[6] See p. 149.

where the order of each pole is at most equal to the order of the corresponding zero of $B(\lambda)$. Thus,

$$\Phi_\tau(\lambda) = \frac{\gamma_\tau(\lambda)}{(\lambda - \beta_1)(\lambda - \beta_2)\ldots(\lambda - \beta_M)}, \qquad (6.71)$$

where $\gamma_\tau(\lambda)$ is an entire function. Then, according to condition (c), we can choose $\gamma_\tau(\lambda)$ to be a polynomial of degree $N - 1$, but not of greater degree.

Using (6.71), we can write the expression for $\Psi_\tau(\lambda)$ in the form

$$\Psi_\tau(\lambda) = \frac{|B_0|^2}{|A_0|^2}$$

$$\times \frac{[e^{i\tau\lambda}(\lambda - \beta_1)(\lambda - \beta_2)\ldots(\lambda - \beta_M) - \gamma_\tau(\lambda)](\lambda - \bar\beta_1)(\lambda - \bar\beta_2)\ldots(\lambda - \bar\beta_M)}{(\lambda - \alpha_1)(\lambda - \alpha_2)\ldots(\lambda - \alpha_N)(\lambda - \bar\alpha_1)(\lambda - \bar\alpha_2)\ldots(\lambda - \bar\alpha_N)}.$$
$$(6.72)$$

In order for this function to have no singularities in the upper half-plane, the following $N$ equations must be satisfied:

$$e^{i\tau\lambda}(\lambda - \beta_1)(\lambda - \beta_2)\ldots(\lambda - \beta_M) - \gamma_\tau(\lambda) = 0$$
$$\text{for} \quad \lambda = \alpha_1, \lambda = \alpha_2, \ldots, \lambda = \alpha_N. \qquad (6.73)$$

But a general polynomial of degree $N - 1$ has just $N$ arbitrary coefficients. Thus, if $\gamma_\tau(\lambda)$ is such a polynomial, (6.73) leads to a system of $N$ linear equations for the coefficients of $\gamma_\tau(\lambda)$, and this system uniquely determines $\gamma_\tau(\lambda)$. Substituting $\gamma_\tau(\lambda)$ into (6.71), we find a function $\Phi_\tau(\lambda)$ satisfying all the conditions (a) to (c).

Thus, we see that to find the spectral characteristic for extrapolation, we have only to solve a system of $N$ linear equations. We now show how to obtain the general formula for $\Phi_\tau(\lambda)$. First, we note that equation (6.73) can be replaced by the single condition that the function

$$\frac{e^{i\tau\lambda}(\lambda - \beta_1)(\lambda - \beta_2)\ldots(\lambda - \beta_M) - \gamma_\tau(\lambda)}{(\lambda - \alpha_1)(\lambda - \alpha_2)\ldots(\lambda - \alpha_N)} \qquad (6.74)$$

should have no poles at the points $\alpha_1, \alpha_2, \ldots, \alpha_N$. Then, we expand the following two rational functions $g(\lambda)$ and $h_\tau(\lambda)$ in partial fractions:

$$g(\lambda) = \frac{(\lambda - \beta_1)(\lambda - \beta_2)\ldots(\lambda - \beta_M)}{(\lambda - \alpha_1)(\lambda - \alpha_2)\ldots(\lambda - \alpha_N)} = \sum_{k=1}^{N} \frac{c_k}{\lambda - \alpha_k}, \qquad (6.75)$$

$$h_\tau(\lambda) = \Phi_\tau(\lambda)g(\lambda) = \frac{\gamma_\tau(\lambda)}{(\lambda - \alpha_1)(\lambda - \alpha_2)\ldots(\lambda - \alpha_N)} = \sum_{k=1}^{N} \frac{d_k}{\lambda - \alpha_k}. \qquad (6.76)$$

[We use the fact that in both $g(\lambda)$ and $h_\tau(\lambda)$, the degree of the denominator exceeds that of the numerator, and that all the roots $\alpha_1, \alpha_2, \ldots, \alpha_N$ are different.] Using (6.75) and (6.76), we can write the function (6.74) in the form

$$\sum_{k=1}^{N} \frac{c_k e^{i\tau\lambda} - d_k}{\lambda - \alpha_k},$$

from which it is clear that (6.74) will have no poles if and only if

$$d_k = c_k e^{i\alpha_k\tau}, \qquad h_\tau(\lambda) = \sum_{k=1}^{N} \frac{c_k e^{i\alpha_k\tau}}{\lambda - \alpha_k}. \tag{6.77}$$

Therefore, we finally have the required general formula

$$\Phi_\tau(\lambda) = \frac{1}{g(\lambda)} \sum_{k=1}^{N} \frac{c_k e^{i\alpha_k\tau}}{\lambda - \alpha_k} = \frac{\displaystyle\sum_{k=1}^{N} \frac{c_k e^{i\alpha_k\tau}}{\lambda - \alpha_k}}{\displaystyle\sum_{k=1}^{N} \frac{c_k}{\lambda - \alpha_k}}, \tag{6.78}$$

where the coefficients $c_k$ are uniquely determined by (6.75).

In the case where the zeros $\alpha_1, \alpha_2, \ldots, \alpha_N$ are not all different, an analogous construction of the spectral characteristic for extrapolation can be carried out. For example, suppose that among the zeros $\alpha_1, \alpha_2, \ldots, \alpha_N$, the zero $\alpha_1$ appears $j_1$ times, the zero $\alpha_2$ appears $j_2$ times, ..., and the zero $\alpha_m$ appears $j_m$ times, where $j_1 + j_2 + \cdots + j_m = N$. We again look for a function $\Phi_\tau(\lambda)$ of the form (6.71), where $\gamma_\tau(\lambda)$ is a polynomial of degree no greater than $N - 1$. However, now the coefficients of $\gamma_\tau(\lambda)$ are determined by the system

$$\frac{d^j}{d\lambda^j} [e^{i\tau\lambda}(\lambda - \beta_1)(\lambda - \beta_2)\ldots(\lambda - \beta_M) - \gamma_\tau(\lambda)]_{\lambda=\alpha_k} = 0$$
$$(j = 0, 1, \ldots, j_{k-1}; \quad k = 1, 2, \ldots, m), \tag{6.79}$$

rather than by the system (6.73). Thus, in this case as well, the solution of the extrapolation problem reduces to the solution of a system of $N$ linear equations.

Formula (6.78) can also be generalized to the case where multiple zeros are present. Instead of the expansions (6.75) and (6.76), we now have the expansions

$$g(\lambda) = \frac{(\lambda - \beta_1)(\lambda - \beta_2)\ldots(\lambda - \beta_M)}{(\lambda - \alpha_1)^{j_1}(\lambda - \alpha_2)^{j_2}\ldots(\lambda - \alpha_m)^{j_m}} = \sum_{k=1}^{m} \sum_{l=1}^{j_k} \frac{c_{kl}}{(\lambda - \alpha_k)^l}. \tag{6.80}$$

and

$$h_\tau(\lambda) = \Phi_\tau(\lambda) g(\lambda) = \frac{\gamma_\tau(\lambda)}{(\lambda - \alpha_1)^{j_1}(\lambda - \alpha_2)^{j_2}\ldots(\lambda - \alpha_m)^{j_m}}$$
$$= \sum_{k=1}^{m} \sum_{l=1}^{j_k} \frac{d_{kl}}{(\lambda - \alpha_k)^l}. \tag{6.81}$$

Then, the coefficients of $\gamma_\tau(\lambda)$ must be chosen in such a way that the difference

$$g(\lambda) - h_\tau(\lambda) = \sum_{k=1}^{m} \sum_{l=1}^{j_k} \frac{c_{kl}e^{i\tau\lambda} - d_{kl}}{(\lambda - \alpha_k)^l}$$

has no poles at the points $\alpha_1, \alpha_2, \ldots, \alpha_m$. As is easily seen, this leads to the requirement that

$$d_{kj_k} = [c_{kj_k}e^{i\tau\lambda}]_{\lambda=\alpha_k},$$

$$d_{kj_{k-1}} = \left[c_{kj_{k-1}}e^{i\tau\lambda} + c_{kj_k}\frac{d}{d\lambda}e^{i\tau\lambda}\right]_{\lambda=\alpha_k},$$

$$\cdots\cdots\cdots\cdots\cdots\cdots\cdots \tag{6.82}$$

$$d_{k1} = \left[c_{k1}e^{i\tau\lambda} + c_{k2}\frac{d}{d\lambda}e^{i\tau\lambda} + c_{k3}\frac{1}{2!}\frac{d^2}{d\lambda^2}e^{i\tau\lambda}\right.$$

$$\left. + \cdots + c_{kj_k}\frac{1}{(j_k-1)!}\frac{d^{j_k-1}}{d\lambda^{j_k-1}}e^{i\tau\lambda}\right]_{\lambda=\alpha_k} \quad (k = 1, 2, \ldots, m),$$

from which we obtain

$$d_{kj_k} = c_{kj_k}e^{i\alpha_k\tau},$$

$$d_{kj_{k-1}} = [c_{kj_{k-1}} + (i\tau)c_{kj_k}]e^{i\alpha_k\tau},$$

$$\cdots\cdots\cdots\cdots\cdots\cdots\cdots \tag{6.83}$$

$$d_{k1} = \left[c_{k1} + (i\tau)c_{k2} + \frac{(i\tau)^2}{2!}c_{k3} + \cdots + \frac{(i\tau)^{j_k-1}}{(j_k-1)!}c_{kj_k}\right]e^{i\alpha_k\tau}$$

$$(k = 1, 2, \ldots, m).$$

Substituting (6.83) into (6.81), we finally obtain

$$\Phi_\tau(\lambda) = \frac{1}{g(\lambda)}\sum_{k=1}^{m}\sum_{l=1}^{j_k}c_{kl}e^{i\alpha_k\tau}\sum_{r=1}^{l}\frac{(i\tau)^{l-r}}{(l-r)!(\lambda-\alpha_k)^r}$$

$$= \frac{\sum_{k=1}^{m}\sum_{l=1}^{j_k}c_{kl}e^{i\alpha_k\tau}\sum_{r=1}^{l}\frac{(i\tau)^{l-r}}{(l-r)!(\lambda-a_k)^r}}{\sum_{k=1}^{m}\sum_{l=1}^{j_k}\frac{c_{kl}}{(\lambda-\alpha_k)^l}}. \tag{6.84}$$

This is the general formula which takes account of the possibility that $A(\lambda)$ may have multiple zeros.

The spectral characteristic (6.84) is a rational function whose numerator contains a polynomial of degree $N - 1$, and whose denominator contains a polynomial of degree $M$ [see equation (6.71)]. From this, it is not hard to deduce that the corresponding best extrapolation formula will in general be the sum of a linear combination of the values of the process $\xi(t)$ itself and its derivatives up to order $N - M - 1$ inclusive (all evaluated at the "last available" time $t$) and an integral extending over the entire past of the process. Moreover, it follows at once from the fact that all the numbers $\alpha_k$ ($k = 1, 2, \ldots, m$) have positive imaginary parts that

$$\lim_{\tau \to \infty} \int_{-\infty}^{\infty} |\Phi_\tau(\lambda)|^2 f(\lambda) \, d\lambda = 0.$$

According to (6.24'), this means that

$$\lim_{\tau \to \infty} \sigma_\tau^2 = \int_{-\infty}^{\infty} f(\lambda) \, d\lambda = B(0). \tag{6.85}$$

Thus, here, as in the case of the problem analyzed in Sec. 24, the "predictable part" of $\xi(t + \tau)$ approaches zero as the prediction time is increased without limit.

Finally, we call attention once more to the important fact that in all cases the spectral characteristic for extrapolation (6.84) is a rational function of $\lambda$. This leads to great simplifications in practical implementation of the results obtained here (see p. 149).

# 7

---

# LINEAR FILTERING
# OF STATIONARY
# RANDOM PROCESSES

---

## 33. Statement of the Problem

Let $\xi(t)$ be a "signal," e.g., a telephone message, a radio broadcast, etc. If $\xi(t)$ is a known function of $t$, then obviously transmission of the signal is superfluous. In fact, the only situation in which it makes sense to use the available means of communication is when the future values of $\xi(t)$ are not known in advance. Here, we consider only the simplest case, where $\xi(t)$ can be regarded as a stationary random process. Because of the unavoidable presence of noise, the signal $\xi(t)$ is "corrupted" to a certain extent during transmission, so that what is received is no longer $\xi(t)$, but rather the sum

$$\zeta(t) = \xi(t) + \eta(t),$$

where $\eta(t)$ is some extraneous "noise," which is also assumed to be a stationary random process.

In this chapter, we consider the problem of reconstruction of the value of the signal $\xi(t)$ at the time $t + \tau$ (where $\tau$ is positive, negative, or zero) by using the known values of the process $\zeta(t)$ "in the past," i.e., at the times $t - s, s \geqslant 0$. As elsewhere in this book, by "reconstruction" of the value of $\xi(t)$ at the time $t + \tau$, we mean finding the best approximation to $\xi^{(1)}(t + \tau)$, in the sense of the method of least squares, which depends linearly on $\zeta^{(1)}(t - s), s \geqslant 0$. It is clear that the problem of finding such an approximation is completely analogous to the problem studied in Chap. 5, except that now we are dealing with stationary random processes instead of stationary random sequences. Henceforth, this problem will be referred to as the problem of *linear filtering of stationary processes*. As in Chap. 5, we assume that we know the correlation functions $B_{\zeta\zeta}(\tau)$ and $B_{\eta\eta}(\tau)$, and that the signal and the noise are not correlated with each other, i.e.,

$$\mathbf{E}\xi(t)\overline{\eta(s)} = 0 \tag{7.1}$$

for any $t$ and $s$.[1] We also assume that the correlation functions $B_{\zeta\zeta}(\tau)$ and $B_{\eta\eta}(\tau)$ fall off fast enough at infinity to have spectral densities $f_{\zeta\zeta}(\lambda)$ and $f_{\eta\eta}(\lambda)$. Then, because of (7.1), it is obvious that

$$B_{\xi\xi}(\tau) = B_{\zeta\zeta}(\tau) - B_{\eta\eta}(\tau) \tag{7.2}$$

and

$$f_{\xi\xi}(\lambda) = f_{\zeta\zeta}(\lambda) - f_{\eta\eta}(\lambda). \tag{7.2'}$$

It is easy to see that the linear filtering problem is equivalent to the geometric problem of dropping a perpendicular from the point $\xi(t + \tau)$ of the space $H$ onto the (linear) subspace $H^{\zeta}(t)$ spanned by the set of all vectors $\zeta(t - s), s \geqslant 0$. To solve this latter problem, we need only find the vector $L_{\tau}(t)$ [the "foot" of the perpendicular], belonging to the subspace $H^{\zeta}(t)$, such that

$$(\xi(t + \tau) - L_{\tau}(t), \zeta(t - s)) = 0 \quad \text{for} \quad s \geqslant 0; \tag{7.3}$$

the vector $L_{\tau}(t)$ satisfying these conditions is uniquely defined. Then, the random variable $L_{\tau}(t)$ is the best approximation to $\xi(t + \tau)$ which depends linearly on $\zeta(t - s), s \geqslant 0$, and hence the *mean square filtering error* $\sigma_{\tau}^2$ equals

$$\sigma_{\tau}^2 = \mathbf{E}|\xi(t + \tau) - L_{\tau}(t)|^2. \tag{7.4}$$

Moreover, bearing in mind that the difference $\xi(t + \tau) - L_{\tau}(t)$ is orthogonal to the vector $L_{\tau}(t)$, since $L_{\tau}(t)$ belongs to $H^{\zeta}(t)$, we find that

$$\begin{aligned}
\sigma_{\tau}^2 &= \mathbf{E}[\xi(t + \tau) - L_{\tau}(t)]\overline{\xi(t + \tau)} \\
&= \mathbf{E}|\xi(t + \tau)|^2 - \mathbf{E}L_{\tau}(t)\overline{[(\xi(t + \tau) - L_{\tau}(t)) + L_{\tau}(t)]} \\
&= \mathbf{E}|\xi(t + \tau)|^2 - \mathbf{E}|L_{\tau}(t)|^2.
\end{aligned} \tag{7.4'}$$

The fact that $L_{\tau}(t)$ belongs to $H^{\zeta}(t)$ means that there exists a sequence

$$\sum_{k=1}^{n_1} \alpha_k^{(1)}\zeta(t - s_k^{(1)}), \quad \sum_{k=1}^{n_2} \alpha_k^{(2)}\zeta(t - s_k^{(2)}), \ldots, \quad \sum_{k=1}^{n_m} \alpha_k^{(m)}\zeta(t - s_k^{(m)}), \ldots \tag{7.5}$$

of finite linear combinations of the quantities $\zeta(t - s), s \geqslant 0$, which con-

---

[1] As noted in Chap. 5, the theory can easily be extended to the case where (7.1) is replaced by the condition that $\xi(t)$ and $\eta(t)$ be stationarily correlated. Cf. footnote 1, p. 127, and the remark at the end of Sec. 35, p. 181.

verges in the mean (square) to $L_\tau(t)$. As was shown in Sec. 29, this implies that $L_\tau(t)$ can be represented as an integral of the form

$$L_\tau(t) = \int_{-\infty}^{\infty} e^{it\lambda} \Phi_\tau(\lambda)\, dZ_\zeta(\lambda), \tag{7.6}$$

where $Z_\zeta(\lambda)$ is the random function with uncorrelated increments appearing in the spectral representation (2.40) of the process $\zeta(t)$, and where

$$\lim_{m\to\infty} \int_{-\infty}^{\infty} |\Phi_\tau(\lambda) - \Phi_{\tau,m}(\lambda)|^2 f_{\zeta\zeta}(\lambda)\, d\lambda = 0, \tag{7.7}$$

$$\Phi_{\tau,m}(\lambda) = \sum_{k=1}^{n_m} a_k^{(m)} e^{-is_k^{(m)}\lambda}. \tag{7.8}$$

We now note that according to (7.1) and the formula $\zeta(t) = \xi(t) + \eta(t)$, we can rewrite (7.3) as

$$(\xi(t + \tau), \xi(t - s)) - (L_\tau(t), \zeta(t - s)) = 0 \quad \text{for} \quad s \geq 0. \tag{7.3'}$$

Substituting (7.6) and the spectral representations of the processes $\xi(t)$ and $\zeta(t)$ into (7.3'), we obtain

$$\int_{-\infty}^{\infty} e^{is\lambda} [e^{i\tau\lambda} f_{\xi\xi}(\lambda) - \Phi_\tau(\lambda) f_{\zeta\zeta}(\lambda)]\, d\lambda = 0 \quad \text{for} \quad s \geq 0. \tag{7.9}$$

Thus, the solution of the linear filtering problem reduces to finding a function $\Phi_\tau(\lambda)$ which can be represented as the limit in the mean (square) with respect to $f_{\zeta\zeta}(\lambda)\, d\lambda$ of a sequence of functions $\Phi_{\tau,m}(\lambda)$ of the form (7.8), and which satisfies the condition (7.9). The function $\Phi_\tau(\lambda)$ will be called the *spectral characteristic for filtering*. From the point of view of radio engineering, $\Phi_\tau(\lambda)$ is the transfer function of the filter which performs the required filtering (cf. p. 149). It follows from formulas (7.4') and (7.6) that the mean square filtering error can be expressed in terms of $\Phi_\tau(\lambda)$ as follows:

$$\sigma_\tau^2 = \int_{-\infty}^{\infty} f_{\xi\xi}(\lambda)\, d\lambda - \int_{-\infty}^{\infty} |\Phi_\tau(\lambda)|^2 f_{\zeta\zeta}(\lambda)\, d\lambda. \tag{7.10}$$

To find $\Phi_\tau(\lambda)$, all the considerations of Sec. 30 are still applicable, except that now the function $e^{i\tau\lambda} f_{\xi\xi}(\lambda) - \Phi_\tau(\lambda) f_{\zeta\zeta}(\lambda)$ plays the role of the function $[e^{i\tau\lambda} - \Phi_\tau(\lambda)]f(\lambda)$. Therefore, in the case where the spectral density $f_{\zeta\zeta}(\lambda)$ is bounded, the following conditions are sufficient for a function $\Phi_\tau(\lambda)$ to be the spectral characteristic for filtering:

(a) The function $\Phi_\tau(\lambda)$ is analytic in the lower half-plane, and as $|\lambda| \to \infty$ in the lower half-plane, $\Phi_\tau(\lambda)$ grows no faster than some power of $|\lambda|$;

(b) The function $\Psi_\tau(\lambda) = e^{i\tau\lambda} f_{\xi\xi}(\lambda) - \Phi_\tau(\lambda) f_{\zeta\zeta}(\lambda)$ is analytic in the upper half-plane, and as $|\lambda| \to \infty$ in the upper half-plane, $\Psi_\tau(\lambda)$ falls off faster than $|\lambda|^{-1-\varepsilon}$, where $\varepsilon > 0$;

(c) The integral along the real axis of the function $|\Phi_\tau(\lambda)|^2 f_{\zeta\zeta}(\lambda)$ is bounded, i.e.,

$$\int_{-\infty}^{\infty} |\Phi_\tau(\lambda)|^2 f_{\zeta\zeta}(\lambda) \, d\lambda < \infty. \tag{7.11}$$

In the next section, we shall give some specific examples showing how to use these conditions to find $\Phi_\tau(\lambda)$.

## 34. Examples of Linear Filtering of Stationary Processes

In this section, we consider some examples of filtering of stationary processes which are the analogs of the examples considered in Sec. 27 for the case of stationary sequences.

**Example 1.** We begin by considering the case where the values of the sum $\zeta(t) = \xi(t) + \eta(t)$ of the "signal" and the "noise" are known at all moments of time, and it is required to use these values to reconstruct the value of the signal $\xi(t)$ at the time $t$. In this case, the spectral characteristic $\Phi_0(\lambda)$ for filtering must satisfy the condition

$$\int_{-\infty}^{\infty} e^{is\lambda}[f_{\xi\xi}(\lambda) - \Phi_0(\lambda)f_{\zeta\zeta}(\lambda)] \, d\lambda = 0 \quad \text{for} \quad -\infty < s < \infty, \tag{7.12}$$

instead of the condition (7.9). Since the Fourier transform of the function $f_{\xi\xi}(\lambda) - \Phi_0(\lambda)f_{\zeta\zeta}(\lambda)$ vanishes identically, we must have

$$f_{\xi\xi}(\lambda) - \Phi_0(\lambda)f_{\zeta\zeta}(\lambda) = 0,$$

so that

$$\Phi_0(\lambda) = \frac{f_{\xi\xi}(\lambda)}{f_{\eta\eta}(\lambda)}. \tag{7.13}$$

From (7.10) and (7.2'), we find that the mean square filtering error is given by the formula

$$\sigma_0^2 = \int_{-\infty}^{\infty} \frac{f_{\xi\xi}(\lambda)f_{\eta\eta}(\lambda)}{f_{\xi\xi}(\lambda) + f_{\eta\eta}(\lambda)} \, d\lambda. \tag{7.14}$$

This shows that complete suppression of the noise is possible only when the spectra of the signal and the noise do not overlap.

**Example 2.** As an example of filtering by using only past known values of $\zeta(t)$, we consider the case where the signal and the noise are both stationary processes of the Markov type (see Example 1, p. 153). Thus, let

$$f_{\xi\xi}(\lambda) = \frac{C_1}{\lambda^2 + \alpha_1^2}, \qquad f_{\eta\eta}(\lambda) = \frac{C_2}{\lambda^2 + \alpha_2^2}, \tag{7.15}$$

where $C_1$, $C_2$, $\alpha_1$ and $a_2$ are positive constants. Then, by (7.2'), we have

$$f_{\zeta\zeta}(\lambda) = \frac{C_1}{\lambda^2 + \alpha_1^2} + \frac{C_2}{\lambda^2 + \alpha_2^2} = \frac{C_3(\lambda^2 + \beta^2)}{(\lambda^2 + \alpha_1^2)(\lambda^2 + \alpha_2^2)}, \tag{7.16}$$

where

$$C_3 = C_1 + C_2, \qquad \beta^2 = \frac{C_1\alpha_2^2 + C_2\alpha_1^2}{C_1 + C_2}. \tag{7.17}$$

We now show that starting from (7.15) and (7.16), we can easily find a function $\Phi_\tau(\lambda)$ satisfying all the conditions (a) to (c) given at the end of Sec. 33.

1. First, we consider the case $\tau \geqslant 0$ (filtering with prediction). According to condition (b), the function

$$\begin{aligned}\Psi_\tau(\lambda) &= e^{i\lambda\tau}f_{\xi\xi}(\lambda) - \Phi_\tau(\lambda)f_{\zeta\zeta}(\lambda) \\ &= \frac{C_1 e^{i\tau\lambda}(\lambda_2 + \alpha_2^2) - C_3\Phi_\tau(\lambda)(\lambda^2 + \beta^2)}{(\lambda^2 + \alpha_1^2)(\lambda^2 + \alpha_2^2)}\end{aligned} \tag{7.18}$$

must be analytic in the upper half-plane. This implies that the zeros of the denominator of $\Psi_\tau(\lambda)$ at the points $\lambda = i\alpha_1$ and $\lambda = i\alpha_2$ must be cancelled by zeros of the numerator at the same points, so that

$$\Phi_\tau(i\alpha_1) = \frac{C_1}{C_3}\frac{\alpha_2^2 - \alpha_1^2}{\beta^2 - \alpha_1^2}e^{-i\alpha_1\tau}, \qquad \Phi_\tau(i\alpha_2) = 0. \tag{7.19}$$

Moreover, since the function $\Phi_\tau(\lambda)$ must be analytic in the lower half-plane and the function $\Psi_\tau(\lambda)$ must be analytic in the upper half plane, it follows that $\Phi_\tau(\lambda)$ and $\Psi_\tau(\lambda)$ can have no singularities in common. Because of the form of (7.18), this implies that $\Phi_\tau(\lambda)$ can have no singularities other than a simple pole at the point $\lambda = i\beta$. Therefore

$$\Phi_\tau(\lambda) = \frac{\gamma_\tau(\lambda)}{\beta + i\lambda}, \tag{7.20}$$

where $\gamma_\tau(\lambda)$ is an entire function. In order for condition (c) to hold as well, the function $\gamma_\tau(\lambda)$ must be linear, i.e.,

$$\gamma_\tau(\lambda) = A\lambda + B.$$

The coefficients $A$ and $B$ are easily determined from the relations (7.19). Substituting these coefficients into the expression for $\gamma_\tau(\lambda)$, and then substituting the resulting expression for $\gamma_\tau(\lambda)$ into (7.20), we obtain

$$\Phi_\tau(\lambda) = \frac{C_1}{C_3}e^{-\alpha_1\tau}\frac{\alpha_2 + \alpha_1}{\beta + \alpha_1}\frac{\alpha_2 + i\lambda}{\beta + i\lambda}. \tag{7.21}$$

It is easily verified that for $\tau \geqslant 0$, the function (7.21) satisfies all the conditions (a) to (c) of Sec. 33, and hence (7.21) is the required spectral characteristic for filtering.

2. Next, we consider the case $\tau < 0$ (filtering with lag). In this case, the function (7.21) no longer satisfies all the required conditions. In fact, substituting (7.21) into (7.18), we obtain a function which contains a term proportional to $e^{i\tau\lambda}$ and grows exponentially in the upper half-plane (recall that $\tau < 0$). Again, we make use of the fact that if condition (b) is to be satisfied, then the function

$$\chi_\tau(\lambda) = C_1 e^{i\tau\lambda}(\lambda^2 + \alpha_2^2) - C_3 \Phi_\tau(\lambda)(\lambda^2 + \beta^2)$$
$$= C_1 e^{-i|\tau|\lambda}(\lambda^2 + \alpha_2^2) - C_3 \Phi_\tau(\lambda)(\lambda^2 + \beta^2) \tag{7.22}$$

must be analytic in the upper half-plane and vanish at the points $\lambda = i\alpha_1$ and $\lambda = i\alpha_2$. Moreover, since the function

$$\Phi_\tau(\lambda) = \frac{C_1 e^{-i|\tau|\lambda}(\lambda^2 + \alpha_2^2) - \chi_\tau(\lambda)}{C_3(\lambda^2 + \beta^2)} \tag{7.23}$$

must be analytic in the lower half-plane, $\chi_\tau(\lambda)$ cannot have any singularities in the lower half-plane either (and hence must be an entire function), and the relation

$$\chi_\tau(-i\beta) = C_1 e^{-\beta|\tau|}(\alpha_2^2 - \beta^2) \tag{7.24}$$

must hold. Then, taking into account that $\chi_\tau(i\alpha_1) = \chi_\tau(i\alpha_2) = 0$, we can write $\chi_\tau(\lambda)$ in the form

$$\chi_\tau(\lambda) = A_\tau(\lambda)(\lambda - i\alpha_1)(\lambda - i\alpha_2), \tag{7.25}$$

where $A_\tau(\lambda)$ is another entire function and

$$A_\tau(-i\beta) = -C_1 \frac{\alpha_2 - \beta}{\alpha_1 + b} e^{-\beta|\tau|}, \tag{7.26}$$

because of (7.24). To satisfy condition (c) as well, it is sufficient to set the function $A_\tau(\lambda)$ equal to a constant, and then using (7.26), (7.25) and (7.23), we finally obtain

$$\Phi_\tau(\lambda) = \frac{C_1}{C_3} \frac{(\alpha_1 + \beta)(\lambda^2 + \alpha_2^2)e^{-i|\tau|\lambda} + (\alpha_2 - \beta)(\lambda - i\alpha_1)(\lambda - i\alpha_2)e^{-\beta|\tau|}}{(\alpha_1 + \beta)(\lambda^2 + \beta^2)}$$
$$= \frac{C_1}{C_3} \frac{(\alpha_1 + \beta)(\lambda^2 + \alpha_2^2)e^{i\tau\lambda} + (\alpha_2 - \beta)(\lambda - i\alpha_1)(\lambda - i\alpha_2)e^{\beta\tau}}{(\alpha_1 + \beta)(\lambda^2 + \beta^2)}. \tag{7.27}$$

It is not hard to see that when $\tau < 0$, the function (7.27) satisfies all the conditions (a) to (c), and hence it is the required spectral characteristic for filtering. However, for $\tau > 0$, the function (7.27) will no longer satisfy all the conditions (a) to (c), and in fact, it will grow exponentially in the lower half-plane. This is understandable, since the spectral characteristic for filtering is unique, and as we have seen, the spectral characteristic has a quite different form for $\tau > 0$. Of course, in the case $\tau = 0$, formulas (7.21) and (7.27) lead to the same function, as is easily verified.

It should be noted that for $\tau > 0$, the spectral characteristic (7.21) turns out to be a rational function of $\lambda$, so that the filter with this function as its transfer function can easily be constructed. However, in the case $\tau < 0$, the corresponding spectral characteristic (7.27) is no longer rational, and contains the exponential $e^{i\tau\lambda} = e^{-i|\tau|\lambda}$. Therefore, to construct a "lumped-parameter" filter (see p. 149) which performs filtering with lag, we must first approximate the function (7.27) by a rational function. For example, we can replace $e^{-i|\tau|\lambda}$ in (7.27) by the approximation

$$\left(\frac{1 - (i|\tau|\lambda/2n)}{1 + (i|\tau|\lambda/2n)}\right)^n \approx e^{-i|\tau|\lambda}. \tag{7.28}$$

The increase in the mean square filtering error produced by this change can easily be calculated by using formula (7.4) and (7.6). It is clear that as we make the number $n$ larger, the filtering error becomes smaller, but, on the other hand, the filter becomes more complicated and hence more expensive. For further remarks on the approximation of the function (7.27) by a rational function, we refer the reader to Wiener's book (W4).

The rather complicated formulas (7.21) and (7.27) obtained above become much simpler in the case (which is encountered very often in practice) where the correlation function $B_{\eta\eta}(\tau)$ of the "noise" falls off much more rapidly than the correlation function $B_{\xi\xi}(\tau)$ of the "signal," so that $\alpha_2$ is much larger than $\alpha_1$. We make the additional assumption that the spectral density $f_{\eta\eta}(\lambda)$ of the noise is not negligible with respect to the spectral density $f_{\xi\xi}(\lambda)$ of the signal, for otherwise we would not have to take the noise into account at all. Because of (7.15), this implies that $C_2/\alpha_2^2$ is of the same order of magnitude as $C_1/\alpha_1^2$, and hence that $C_2$ is much larger than $C_1$. In this case, the main terms in (7.21) and (7.27) are the limiting expressions obtained as

$$\alpha_2 \to \infty, \qquad C_2 \to \infty, \qquad \frac{C_2}{\alpha_2^2} \to \varepsilon^2 = \text{const.} \tag{7.29}$$

(This is the transition to "purely random" noise; cf. Example 1 of Sec. 14.) It follows from formula (7.17) that if we pass to the limit in this way, then

$$\beta^2 \to \beta_1^2 = \frac{C_1}{\varepsilon^2} + \alpha_1^2 = \alpha_1^2(1 + \rho^2), \tag{7.30}$$

where

$$\rho = \sqrt{\frac{f_{\xi\xi}(0)}{f_{\eta\eta}(0)}} = \sqrt{\frac{C_1}{\alpha_1^2\varepsilon^2}} \tag{7.31}$$

is a dimensionless constant characterizing the "signal-to-noise ratio" (cf. p. 138).[2]

---

[2] The quantity $\rho$ in (7.31) plays the same role as the quantity $\rho$ in (5.46), but is defined somewhat differently, since $\mathbf{E}|\eta(t)|^2 = \infty$ for white noise.

Substituting (7.17) into (7.21) and taking the limit (7.27) and (7.30), we easily find that

$$\Phi_\tau(\lambda) \to \Phi_\tau^{(1)}(\lambda) = \frac{C_1}{\varepsilon^2} \frac{e^{-\alpha_1 \tau}}{\beta_1 + \alpha_1} \frac{1}{\beta_1 + i\lambda}$$

$$= \frac{\beta_1 - \alpha_1}{\beta_1 + i\lambda} e^{-\alpha_1 \tau}, \quad \text{where} \quad \tau \geqslant 0. \tag{7.32}$$

Carrying out the same passage to the limit in formula (7.27), we obtain

$$\Phi_\tau(\lambda) \to \Phi_\tau^{(1)}(\lambda) = \frac{C_1}{\varepsilon^2} \frac{e^{-i|\tau|\lambda}}{\lambda^2 + \beta_1^2} - \frac{C_1}{\varepsilon^2} \frac{\alpha_1 + i\lambda}{(\alpha_1 + \beta_1)(\lambda^2 + \beta_1^2)} e^{-\beta_1|\tau|}$$

$$= \frac{\beta_1^2 - \alpha_1^2}{\beta_1^2 + \lambda^2} e^{-i|\tau|\lambda} - (\beta_1 - \alpha_1) \frac{\alpha_1 + i\lambda}{\beta_1^2 + \lambda^2} e^{-\beta_1|\tau|}, \tag{7.33}$$

$$\text{where} \quad \tau \leqslant 0.$$

The functions (7.32) and (7.33) are the spectral characteristics for filtering (for $\tau \geqslant 0$ and $\tau \leqslant 0$, respectively) when the "signal" is of Markov type and the noise is "purely random," i.e., when

$$f_{\xi\xi}(\lambda) = \frac{C_1}{\lambda^2 + \alpha_1^2}, \qquad f_{\eta\eta}(\lambda) = \varepsilon^2 = \text{const}; \tag{7.34}$$

this limiting case is the analog of the case analyzed in Example 2 of Sec. 27. In the case of the spectral densities (7.34), by using formulas (7.10), (7.32) and (7.33), we easily obtain the expression

$$\sigma_\tau^2 = \begin{cases} \dfrac{C_1\pi}{\alpha_1} \left[ 1 - \dfrac{\beta_1 - \alpha_1}{\beta_1 + \alpha_1} e^{-2\alpha_1\tau} \right] & \text{for} \quad \tau \geqslant 0, \\[3mm] \dfrac{C_1\pi}{\beta_1} \left[ 1 + \dfrac{\beta_1 - \alpha_1}{\beta_1 + \alpha_1} e^{-2\beta_1|\tau|} \right] & \text{for} \quad \tau \leqslant 0 \end{cases} \tag{7.35}$$

for the mean square error. In particular,

$$\sigma_\infty^2 = \lim_{\tau \to \infty} \sigma_\tau^2 = \frac{C_1\pi}{\alpha_1} = B_{\xi\xi}(0),$$

$$\sigma_0^2 = \frac{2C_1\pi}{\alpha_1 + \beta_1} = \frac{2\alpha_1}{\alpha_1 + \beta_1} B_{\xi\xi}(0),$$

and

$$\sigma_{-\infty}^2 = \lim_{\tau \to -\infty} \sigma_\tau^2 = \frac{C_1\pi}{\beta_1} = \frac{\alpha_1}{\beta_1} B_{\xi\xi}(0).$$

This formula for $\sigma_{-\infty}^2$ can also be obtained from (7.14).

A simpler derivation of formulas (7.32) and (7.33) is possible if we use the spectral densities (7.34) from the very beginning. However, the more complicated formulas (7.21) and (7.27) are needed, if we wish to calculate the error due to the assumption that the noise spectral density is constant over the whole spectrum.

It should be noted that formulas (7.32) and (7.6) imply that in the case of the spectral densities (7.34), we have

$$L_\tau(t) = (\beta_1 - \alpha_1)e^{-\alpha_1\tau} \int_{-\infty}^{\infty} \frac{e^{it\lambda}}{\beta_1 + i\lambda} \, dZ_\zeta(\lambda) \qquad (7.36)$$

if $\tau \geqslant 0$.   Using (2.95), we can rewrite (7.36) in the form

$$L_\tau(t) = (\beta_1 - \alpha_1)e^{-\alpha_1\tau} \int_{0}^{\infty} e^{-\beta_1 s} \zeta(t - s) \, ds. \qquad (7.37)$$

Formula (7.37) shows that in this case we should take

$$\xi^{(1)}(t + \tau) = (\beta_1 - \alpha_1)e^{-\alpha_1\tau} \int_{0}^{\infty} e^{-\beta_1 s} \zeta^{(1)}(t - s) \, ds \qquad (7.38)$$

as the approximation to $\xi^{(1)}(t + \tau)$.

It is interesting to compare formula (7.38) with formula (6.37'), which gives the best prediction for a process of Markov type in the absence of any "corrupting noise" $\eta(t)$.   We see that when noise is present, to approximate $\xi^{(1)}(t + \tau)$ we have to use the smoothed quantity

$$\hat{\xi}^{(1)}(t) = \beta_1 \int_{0}^{\infty} e^{-\beta_1 s} \zeta^{(1)}(t - s) \, ds \qquad (7.39)$$

instead of $\xi^{(1)}(t)$.   The factor $\beta_1$ is put before the integral in (7.39) in order to make the total "smoothing weight" equal to unity:

$$\beta_1 \int_{0}^{\infty} e^{-\beta_1 s} \, ds = 1.$$

According to (7.30), the "smoothing period" $T = 1/\beta_1$ in (7.39) depends on the signal-to-noise ratio $\rho$.   As $\rho \to \infty$, $T$ approaches zero, and as $\rho \to 0$, $T$ approaches $1/\alpha_1$, i.e., the "decay time" of the signal correlation function.[3]

Using (7.39), we can write formula (7.38) as

$$\xi^{(1)}(t + \tau) = \left(1 - \frac{\alpha_1}{\beta_1}\right)e^{-\alpha_1\tau} \hat{\zeta}^{(1)}(t). \qquad (7.40)$$

Thus, in the presence of noise, the factor $e^{-\alpha_1\tau}$ has to be multiplied by the extra factor

$$k = 1 - \frac{\alpha_1}{\beta_1},$$

which is always less than 1.   If $\rho \gg 1$, we have $k \approx 1 - \rho^{-1}$, so that $k$ is close to 1, while if $\rho \ll 1$, we have $k \approx \rho^2/2$, so that $k$ is close to 0.

---

[3] Synonymously, the "correlation time" of the signal.   Cf. footnote 12, p. 62.

Obviously, if $\rho \ll 1$, the mean square filtering error $\sigma_\tau^2$ is close to $B_{\xi\xi}(0)$ for any $\tau$, while if $\rho \gg 1$ and $\tau \geqslant 0$, $\sigma_\tau^2$ is close to the quantity (6.36). In particular, for $\tau = 0$ we easily find that

$$\sigma_0^2 \approx \frac{2}{\rho} B_{\xi\xi}(0) \quad \text{for} \quad \rho \gg 1,$$

and

$$\sigma_0^2 \approx \left(1 - \frac{1}{4}\rho^2\right)B_{\xi\xi}(0) \quad \text{for} \quad \rho \ll 1.$$

Introduction of lag decreases the error further, and the expression for $\sigma_{-\infty}^2$ characterizes the limiting improvement of the filtering formula which can be obtained in this way. In the two extreme cases of a very high noise level and a very low noise level, this error, which is unavoidable no matter how much lag is introduced, turns out to be

$$\sigma_{-\infty}^2 \approx \frac{1}{\rho} B_{\xi\xi}(0) \quad \text{for} \quad \rho \gg 1,$$

and

$$\sigma_{-\infty}^2 \approx \left(1 - \frac{1}{2}\rho^2\right)B_{\xi\xi}(0) \quad \text{for} \quad \rho \ll 1.$$

## *35. The Case of a General Rational Spectral Density in $\lambda$

The method used to solve the filtering problem in Example 2 of the preceding section can also be applied to the case of arbitrary rational spectral densities $f_{\zeta\zeta}(\lambda)$ and $f_{\xi\xi}(\lambda)$, provided that the spectral density $f_{\zeta\zeta}(\lambda)$ vanishes nowhere. In fact, let

$$f_{\zeta\zeta}(\lambda) = \frac{|B(\lambda)|^2}{|A(\lambda)|^2} = \frac{C_1|(\lambda - \beta_1)(\lambda - \beta_2)\ldots(\lambda - \beta_M)|^2}{|(\lambda - \alpha_1)(\lambda - \alpha_2)\ldots(\lambda - \alpha_N)|^2} \tag{7.41}$$

and

$$f_{\xi\xi}(\lambda) = \frac{|D(\lambda)|^2}{|C(\lambda)|^2} = \frac{C_2|(\lambda - \delta_1)(\lambda - \delta_2)\ldots(\lambda - \delta_K)|^2}{|(\lambda - \gamma_1)(\lambda - \gamma_2)\ldots(\lambda - \gamma_L)|^2}, \tag{7.42}$$

where $C_1 > 0$, $C_2 > 0$, $M < N$, $K < L$, and all the zeros $\alpha_j$, $\beta_k$ and $\gamma_l$ of the polynomials $A(\lambda)$, $B(\lambda)$ and $C(\lambda)$ have positive imaginary parts (cf. the beginning of Sec. 28). We assume that the first $n$ zeros of the polynomials $A(\lambda)$ and $B(\lambda)$ agree in pairs, i.e.,

$$\alpha_j = \gamma_j \quad (j = 1, 2, \ldots, n; n \leqslant N, n \leqslant L), \tag{7.43}$$

and that the other zeros of $A(\lambda)$ are not zeros of $C(\lambda)$. In this case,

$$\Psi_\tau(\lambda) = e^{i\tau\lambda}f_{\xi\xi}(\lambda) - \Phi_\tau(\lambda)f_{\xi\xi}(\lambda)$$

$$= \frac{1}{(\lambda-\alpha_1)\ldots(\lambda-\alpha_N)(\lambda-\gamma_{n+1})\ldots(\lambda-\gamma_L)(\lambda-\bar{\alpha}_1)\ldots(\lambda-\bar{\alpha}_N)(\lambda-\bar{\gamma}_{n+1})\ldots(\lambda-\bar{\gamma}_L)} \tag{7.44}$$

$$\times [C_2(\lambda-\delta_1)\ldots(\lambda-\delta_K)(\lambda-\alpha_{n+1})\ldots(\lambda-\alpha_N)(\lambda-\bar{\delta}_1)\ldots(\lambda-\bar{\delta}_K)(\lambda-\bar{\alpha}_{n+1})\ldots(\lambda-\bar{\alpha}_N)e^{i\tau\lambda}$$

$$- C_1\Phi_\tau(\lambda)(\lambda-\beta_1)\ldots(\lambda-\beta_M)(\lambda-\gamma_{n+1})\ldots(\lambda-\gamma_L)(\lambda-\bar{\beta}_1)\ldots(\lambda-\bar{\beta}_M)(\lambda-\bar{\gamma}_{n+1})\ldots(\lambda-\bar{\gamma}_L)]$$

We have to find a function $\Phi_\tau(\lambda)$ which is analytic in the lower half-plane and does not grow too rapidly (in the lower half-plane), such that the function $\Psi_\tau(\lambda)$ given by (7.44) is analytic in the upper half-plane and falls off at infinity faster than $|\lambda|^{-1-\varepsilon}$, $\varepsilon > 0$ (in the upper half-plane).

1. First, we consider the case $\tau \geqslant 0$, and for the time being, we assume that all of the zeros $\gamma_1, \gamma_2, \ldots, \gamma_L$ are different. It is clear from formula (7.44) that the only possible singularities of $\Phi_\tau(\lambda)$ are poles at the points $\beta_1, \ldots, \beta_M, \gamma_{n+1}, \ldots, \gamma_L$, since any other singularity would either destroy the analyticity of $\Phi_\tau(\lambda)$ in the lower half-plane or the analyticity of $\Psi_\tau(\lambda)$ in the upper half-plane. Moreover, it follows from (7.44) that $\Psi_\tau(\lambda)$ will certainly have a pole in the upper half-plane unless $\Phi_\tau(\lambda)$ vanishes at the points $\alpha_{n+1}, \alpha_{n+2}, \ldots, \alpha_N$. Consequently,

$$\Phi_\tau(\lambda) = \frac{C_2 \omega_\tau(\lambda)(\lambda - \alpha_{n+1})\ldots(\lambda - \alpha_N)}{C_1(\lambda - \beta_1)\ldots(\lambda - \beta_M)(\lambda - \gamma_{n+1})\ldots(\lambda - \gamma_L)}, \qquad (7.45)$$

where $\omega_\tau(\lambda)$ is an entire function. Substituting (7.45) into (7.44), and using (7.43), we obtain

$$\Psi_\tau(\lambda) =$$

$$\frac{C_2}{(\lambda - \gamma_1)\ldots(\lambda - \gamma_L)(\lambda - \bar{\gamma}_1)\ldots(\lambda - \bar{\gamma}_L)(\lambda - \bar{\alpha}_{n+1})\ldots(\lambda - \bar{\alpha}_N)} \qquad (7.46)$$
$$\times \{(\lambda - \delta_1)\ldots(\lambda - \delta_K)(\lambda - \bar{\delta}_1)\ldots(\lambda - \bar{\delta}_K)(\lambda - \bar{\alpha}_{n+1})\ldots(\lambda - \bar{\alpha}_N)e^{i\tau\lambda}$$
$$- \omega_\tau(\lambda)(\lambda - \bar{\beta}_1)\ldots(\lambda - \bar{\beta}_M)(\lambda - \bar{\gamma}_{n+1})\ldots(\lambda - \bar{\gamma}_L)\}$$

In order for the function (7.46) to be analytic in the upper half-plane, the difference appearing in braces in (7.46) must vanish at the $L$ points $\gamma_1, \gamma_2, \ldots, \gamma_L$. This can be achieved by choosing $\omega_\tau(\lambda)$ to be a polynomial of degree $L - 1$, whose $L$ coefficients are determined from the system of $L$ linear equations

$$\frac{(\lambda - \delta_1)\ldots(\lambda - \delta_K)(\lambda - \bar{\delta}_1)\ldots(\lambda - \bar{\delta}_K)(\lambda - \bar{\alpha}_{n+1})\ldots(\lambda - \bar{\alpha}_N)}{(\lambda - \bar{\beta}_1)\ldots(\lambda - \bar{\beta}_M)(\lambda - \bar{\gamma}_{n+1})\ldots(\lambda - \bar{\gamma}_L)} e^{i\tau\lambda}$$
$$- \omega_\tau(\lambda) = 0 \quad \text{for} \quad \lambda = \gamma_1, \gamma_2, \ldots, \gamma_L, \qquad (7.47)$$

analogous to the system (6.73). Moreover, the condition (7.11) only allows us to choose $\omega_\tau(\lambda)$ to be such a polynomial of degree $L - 1$ (and not of greater degree). For such a choice of $\omega_\tau(\lambda)$, the function $\Phi_\tau(\lambda)$ will be rational, and hence as $|\lambda| \to \infty$, $\Phi_\tau(\lambda)$ will grow no faster than some power of $|\lambda|$, while as $|\lambda| \to \infty$ in the upper half-plane, the function $\Psi_\tau(\lambda)$ will fall off no slower than $|\lambda|^{-2}$ (recall that $\tau \geqslant 0$). This means that the corresponding function (7.45) is the required spectral characteristic for filtering, and finding (7.45) reduces to solving the system (7.47).

Just as in the case of extrapolation, by making certain expansions in partial fractions, we can easily give a general formula for $\Phi_\tau(\lambda)$. In fact, the condition (7.47) can be replaced by the single condition that the function

$$\frac{(\lambda - \delta_1)\ldots(\lambda - \delta_K)(\lambda - \bar{\delta}_1)\ldots(\lambda - \bar{\delta}_K)(\lambda - \bar{\alpha}_{n+1})\ldots(\lambda - \bar{\alpha}_N)e^{i\tau\lambda}}{(\lambda - \gamma_1)\ldots(\lambda - \gamma_L)(\lambda - \bar{\gamma}_{n+1})\ldots(\lambda - \bar{\gamma}_L)(\lambda - \bar{\beta}_1)\ldots(\lambda - \bar{\beta}_M)}$$
$$+ \frac{\omega_\tau(\lambda)}{(\lambda - \gamma_1)\ldots(\lambda - \gamma_L)} \tag{7.48}$$

should have no poles at the points $\gamma_1, \gamma_2, \ldots, \gamma_L$. Now let[4]

$$g(\lambda) = \frac{(\lambda - \delta_1)\ldots(\lambda - \delta_K)(\lambda - \bar{\delta}_1)\ldots(\lambda - \bar{\delta}_K)(\lambda - \bar{\alpha}_{n+1})\ldots(\lambda - \bar{\alpha}_N)}{(\lambda - \gamma_1)\ldots(\lambda - \gamma_L)(\lambda - \bar{\gamma}_{n+1})\ldots(\lambda - \bar{\gamma}_L)(\lambda - \bar{\beta}_1)\ldots(\lambda - \bar{\beta}_M)}$$
$$\tag{7.49}$$

$$= \sum_{k=1}^{L} \frac{c_k}{\lambda - \gamma_k} + \sum_{k=n+1}^{L} \frac{c_k^{(1)}}{\lambda - \bar{\gamma}_k} + \sum_{k,j} \frac{c_{kj}^{(2)}}{(\lambda - \bar{\beta}_k)^k},$$

and

$$h(\lambda) = \frac{\omega_\tau(\lambda)}{(\lambda - \gamma_1)\ldots(\lambda - \gamma_L)} = \sum_{k=1}^{L} \frac{d_k}{\lambda - \gamma_k}. \tag{7.50}$$

It is clear that we must have

$$d_k = c_k e^{i\gamma_k\tau}, \qquad h(\lambda) = \frac{\omega_\tau(\lambda)}{(\lambda - \gamma_1)\ldots(\lambda - \gamma_L)} = \sum_{k=1}^{L} \frac{c_k e^{i\gamma_k\tau}}{\lambda - \gamma_k}, \tag{7.51}$$

in order for the function (7.48) to have no poles at the points $\gamma_1, \gamma_2, \ldots, \gamma_L$. Substituting (7.51) into (7.45), we obtain

$$\Phi_\tau(\lambda) = \frac{C_2}{C_1} \frac{(\lambda - \alpha_1)(\lambda - \alpha_2)\ldots(\lambda - \alpha_N)}{(\lambda - \beta_1)(\lambda - \beta_2)\ldots(\lambda - \beta_M)} \sum_{k=1}^{L} \frac{c_k e^{i\gamma_k\tau}}{\lambda - \gamma_k}, \tag{7.52}$$

which is the desired general formula for $\Phi_\tau(\lambda)$, expressed in terms of the coefficients $c_k$ of the expansion (7.49).

In the case where some of the zeros $\gamma_1, \gamma_2, \ldots, \gamma_L$ are multiple zeros, the spectral characteristic $\Phi_\tau(\lambda)$ still has the form (7.45), where $\omega_\tau(\lambda)$ is a polynomial of degree no greater than $L - 1$, except that now the coefficients of this polynomial are determined not from a system similar to (6.73), but from

---

[4] In the expansion (7.49), we have taken account of the possibility that there may be multiple zeros among the zeros $\beta_1, \beta_2, \ldots, \beta_M$, and also that the degree of the denominator of $g(\lambda)$ is greater than the degree of the numerator of $g(\lambda)$. This latter fact is a consequence of the inequality $f_{\xi\xi}(\lambda) \leqslant f_{\zeta\zeta}(\lambda) = f_{\xi\xi}(\lambda) + f_{\eta\eta}(\lambda)$; since this inequality must also hold for very large $|\lambda|$, it follows that $L - K \geqslant N - M \geqslant 1$, which means that $2L + M > 2K + N$.

a system similar to (6.79).   In this case, the expansions (7.49) and (7.50) take the form

$$g(\lambda) = \sum_{k,j} \frac{c_{kj}}{(\lambda - \gamma_k)^j} + \sum_{k,j} \frac{c_{kj}^{(1)}}{(\lambda - \bar{\gamma}_k)^j} + \sum_{k,j} \frac{c_{kj}^{(2)}}{(\lambda - \bar{\beta}_k)^j}, \qquad (7.49')$$

$$h(\lambda) = \sum_{k,j} \frac{d_{kj}}{(\lambda - \gamma_k)^j}, \qquad (7.50')$$

and the condition that the function (7.48) should have no poles at the points $\gamma_1, \gamma_2, \ldots, \gamma_L$ becomes the condition that the function

$$\sum_{k,j} \frac{c_{kj}e^{i\tau\lambda} - d_{kj}}{(\lambda - \gamma_k)^j}$$

should have no poles at these points.   The rest of the argument is completely analogous to the derivation of formula (6.84), and the final expression for $\Phi_\tau(\lambda)$ is found to be

$$\Phi_\tau(\lambda) = \frac{C_2}{C_1} \frac{(\lambda - \alpha_1)(\lambda - \alpha_2)\ldots(\lambda - \alpha_N)}{(\lambda - \beta_1)(\lambda - \beta_2)\ldots(\lambda - \beta_M)}$$

$$\times \sum_{k,j} c_{kj}e^{i\gamma_k\tau} \sum_{r=1}^{j} \frac{(i\tau)^{j-r}}{(j-r)!(\lambda - \gamma_k)^r}. \qquad (7.52')$$

2. Next, we consider the case of filtering with lag, i.e., we let $\tau < 0$. Then, the function (7.52) or (7.52') will no longer be the spectral characteristic for filtering, since the corresponding $\Psi_\tau(\lambda)$ will grow exponentially in the upper half-plane.   However, it turns out that in this case, all the requirements will be satisfied if we choose $\Psi_\tau(\lambda)$ to be a suitable rational function; of course, then $\Phi_\tau(\lambda)$ will no longer be rational.   According to (7.44), the function

$$\begin{aligned}
\chi_\tau(\lambda) = {}& C_2(\lambda - \delta_1)\ldots(\lambda - \delta_K)(\lambda - \alpha_{n+1})\ldots(\lambda - \alpha_N) \\
& \times (\lambda - \bar{\delta}_1)\ldots(\lambda - \bar{\delta}_K)(\lambda - \bar{\alpha}_{n+1})\ldots(\lambda - \bar{\alpha}_N)e^{i\tau\lambda} \\
& - C_1\Phi_\tau(\lambda)(\lambda - \beta_1)\ldots(\lambda - \beta_M)(\lambda - \gamma_{n+1})\ldots(\lambda - \gamma_L) \\
& \times (\lambda - \bar{\beta}_1)\ldots(\lambda - \bar{\beta}_M)(\lambda - \bar{\gamma}_{n+1})\ldots(\lambda - \bar{\gamma}_L)
\end{aligned} \qquad (7.53)$$

must be analytic in the upper half-plane and must vanish at the points $\alpha_1, \ldots, \alpha_m, \gamma_{n+1}, \ldots, \gamma_L$, and furthermore, the function

$$\Phi_\tau(\lambda)$$

$$= \frac{C_2(\lambda-\delta_1)\ldots(\lambda-\delta_K)(\lambda-\alpha_{n+1})\ldots(\lambda-\alpha_N)(\lambda-\bar{\delta}_1)\ldots(\lambda-\bar{\delta}_K)(\lambda-\bar{\alpha}_{n+1})\ldots(\lambda-\bar{\alpha}_N)e^{i\tau\lambda} - \chi_\tau(\lambda)}{C_1(\lambda-\beta_1)\ldots(\lambda-\beta_M)(\lambda-\gamma_{n+1})\ldots(\lambda-\gamma_L)(\lambda-\bar{\beta}_1)\ldots(\lambda-\bar{\beta}_M)(\lambda-\bar{\gamma}_{n+1})\ldots(\lambda-\bar{\gamma}_L)}$$

$$(7.54)$$

must be analytic in the lower half-plane.   It follows that $\chi_\tau(\lambda)$ must be an entire function, and moreover $\chi_\tau(\lambda)$ must be such that the numerator in (7.54) vanishes at the points $\bar{\beta}_1, \ldots, \bar{\beta}_M, \bar{\gamma}_{n+1}, \ldots, \bar{\gamma}_L$.   All these conditions will be satisfied, if we set

$$\chi_\tau(\lambda) = A_\tau(\lambda)(\lambda - \alpha_1)\ldots(\lambda - \alpha_N)(\lambda - \gamma_{n+1})\ldots(\lambda - \gamma_L), \qquad (7.55)$$

where $A_\tau(\lambda)$ is a polynomial of degree no greater than $M + L - n - 1$, which takes certain definite values when $\lambda$ equals $\bar{\beta}_1, \ldots, \bar{\beta}_M, \bar{\gamma}_{n+1}, \ldots, \bar{\gamma}_L$. Then, $\chi_\tau(\lambda)$ will be a polynomial of degree no greater than $M + N + 2L - 2n - 1$. Under these conditions, it can easily be verified that

(a) $\Psi_\tau(\lambda)$ is a rational function which falls off at infinity no slower than $|\lambda|^{-2}$;

(b) For $\tau < 0, \Phi_\tau(\lambda)$ grows in the upper half-plane no faster than some power of $|\lambda|$;

(c) The integral (7.11) converges.

Therefore, with this choice of $\chi_\tau(\lambda)$, the function (7.54) is the desired spectral characteristic for filtering with lag, and finding (7.54) reduces to solving a system of $M + L - n - 1$ linear equations.

Just as in the case $\tau \geqslant 0$, it is not hard to find a general formula for $\Phi_\tau(\lambda)$, by using partial fraction expansions. It is sufficient to note that the conditions imposed on $A_\tau(\lambda)$ can be replaced by the single condition that the function

$$\frac{C_2(\lambda - \delta_1)\ldots(\lambda - \delta_K)(\lambda - \bar{\delta}_1)\ldots(\lambda - \bar{\delta}_K)(\lambda - \bar{\alpha}_{n+1})\ldots(\lambda - \bar{\alpha}_N)e^{i\tau\lambda}}{(\lambda - \bar{\beta}_1)\ldots(\lambda - \bar{\beta}_M)(\lambda - \bar{\gamma}_{n+1})\ldots(\lambda - \bar{\gamma}_L)(\lambda - \gamma_1)\ldots(\lambda - \gamma_L)}$$
$$- \frac{A_\tau(\lambda)}{(\lambda - \bar{\beta}_1)\ldots(\lambda - \bar{\beta}_M)(\lambda - \bar{\gamma}_{n+1})\ldots(\lambda - \bar{\gamma}_L)}$$

should not have poles at the points $\bar{\beta}_1, \ldots, \bar{\beta}_M, \bar{\gamma}_{n+1}, \ldots, \bar{\gamma}_L$. The rest of the derivation is completely analogous to the derivation of formulas (7.52) and (7.52'), and will not be given here.

If we know $\Phi_\tau(\lambda)$, we can easily obtain the mean square filtering error by using formula (7.10). As $\tau \to \infty$, the function $\Phi_\tau(\lambda)$ approaches zero uniformly, as can be seen from (7.52) and (7.52'). Therefore, we always have

$$\sigma_\infty^2 = \lim_{\tau \to \infty} \sigma_\tau^2 = \int_{-\infty}^{\infty} f_{\xi\xi}(\lambda)\, d\lambda = B_{\xi\xi}(0),$$

as is to be expected. In the other limiting case, as $\tau \to -\infty$, it turns out that the function $\chi_\tau(\lambda)$ approaches zero, and then, according to (7.10), (7.41), (7.42) and (7.54), we have

$$\sigma_{-\infty}^2 = \int_{-\infty}^{\infty} f_{\xi\xi}(\lambda)\, d\lambda$$
$$- \int_{-\infty}^{\infty} \frac{C_2^2|(\lambda - \delta_1)\ldots(\lambda - \delta_K)|^4|(\lambda - \alpha_1)\ldots(\lambda - \alpha_N)^2|}{C_1|(\lambda - \beta_1)\ldots(\lambda - \beta_M)|^2|(\lambda - \gamma_1)\ldots(\lambda - \gamma_L)|^4}\, d\lambda$$
$$= \int_{-\infty}^{\infty} f_{\xi\xi}(\lambda)\, d\lambda - \int_{-\infty}^{\infty} \frac{f_{\xi\xi}^2(\lambda)}{f_{\zeta\zeta}(\lambda)}\, d\lambda = \int_{-\infty}^{\infty} \frac{f_{\xi\xi}(\lambda)f_{\eta\eta}(\lambda)}{f_{\xi\xi}(\lambda) + f_{\eta\eta}(\lambda)}\, d\lambda, \tag{7.56}$$

a result which coincides with (7.14).

In conclusion, we note that the solution of the filtering problem given here can be extended practically without change to the case where the "signal" $\xi(t)$ and the "noise" $\eta(t)$ are processes which are stationarily correlated with each other (see Sec. 15), provided that the cross-spectral distribution function $F_{\xi\zeta}(\lambda)$ is the indefinite integral of a rational spectral density

$$f_{\xi\zeta}(\lambda) = \frac{C_2(\lambda - \delta_1)\ldots(\lambda - \delta_K)}{(\lambda - \gamma_1)\ldots(\lambda - \gamma_{L_1})(\lambda - \gamma_1')\ldots(\lambda - \gamma_{L_2}')}, \qquad (7.42')$$

where $K < L_1 + L_2 - 1$ [this inequality follows from the integrability of $f_{\xi\zeta}(\lambda)$], the zeros $\gamma_1, \ldots, \gamma_{L_1}$ have positive imaginary parts, and the zeros $\gamma_1', \ldots, \gamma_{L_2}'$ have negative imaginary parts. In this case, we need only replace $f_{\xi\xi}(\lambda)$ by $f_{\xi\zeta}(\lambda)$ in all the considerations of Chap. 7.

# 8

---

# FURTHER DEVELOPMENT
# OF THE THEORY OF
# EXTRAPOLATION AND FILTERING[1]

---

In Chaps. 4 to 7, we gave a detailed treatment of extrapolation and filtering, both for stationary random sequences with positive spectral densities which are rational in $e^{i\lambda}$, and for stationary random processes with positive spectral densities which are rational in $\lambda$. In this last chapter, we shall discuss briefly, without proofs, some further results involving the same or related ideas.

## 36. General Theory of Extrapolation and Filtering of Stationary Random Sequences and Processes

So far, we have solved the problems of extrapolation and filtering only for a particular class of random functions, i.e., those with spectral densities of a very special form. However, the historical development of the theory proceeded along different lines: The extrapolation problem was first solved by Kolmogorov (K12, K13) for arbitrary stationary sequences. Then, Kolmogorov's results were extended to the case of arbitrary stationary processes [Krein (K24)], and only later was it shown [Doob (D4), Wiener (W4)], that Kolmogorov's general theory could be simplified and made somewhat more con-

---

[1] For a discussion of important recent developments in the theory of extrapolation and filtering, especially as concerns the multidimensional case, see Appendix II, p. 214, by D. B. Lowdenslager.

crete in the special case of sequences and processes with rational spectral densities (in $e^{i\lambda}$ or in $\lambda$). We now present the general results pertaining to arbitrary stationary random functions, together with some examples clarifying the meaning of these results.

**36.1. Extrapolation of arbitrary stationary sequences.** In studying the extrapolation problem in Chap. 4, emphasis was placed on finding the best extrapolation formula (4.17), or equivalently, the series (4.16). However, it turns out that in the case of an arbitrary stationary sequence $\xi(t)$, the representation of the random variable $L_m(t)$ as a series of the form (4.16) may not exist at all (cf. footnote 6, p. 104), and even if such a representation does exist, in general it will not be unique. Because of this, in the general theory of extrapolation, the problem of finding the mean square error $\sigma_m^2$ receives most attention, and as far as the random variable $L_m(t)$ is concerned, all that is done is to represent it as an integral analogous to (6.20). In the present brief summary, we shall only discuss results pertaining to the quantity $\sigma_m^2$.

In all the examples analyzed in Chap. 4, as the prediction time $m$ is increased without limit, the mean square error $\sigma_m^2$ approaches the mean square of the predicted quantity:

$$\lim_{m \to \infty} \sigma_m^2 = \mathbf{E}|\xi|^2. \tag{8.1}$$

In order to see that this is not always true, it is sufficient to consider the case of a stationary random sequence $\xi(t)$ which can be represented as the sum of a finite number ($N$, say) of uncorrelated harmonic oscillations with random amplitudes and phases. Since

$$\mathbf{E}\left|\xi(t+m) - \sum_{k=1}^{n} \alpha_k \xi(t-k)\right|^2 = \int_{-\pi}^{\pi} \left|e^{im\lambda} - \sum_{k=1}^{n} \alpha_k e^{-ik\lambda}\right|^2 dF(\lambda), \tag{8.2}$$

the problem of extrapolating $\xi(t)$, given a finite number of elements of the sequence $\xi(t)$, can be formulated as follows: Regarding the quantities $\alpha_1, \alpha_2, \ldots, \alpha_n$ as variables, find the minimum of the integral (8.2) and also the values of $\alpha_1, \alpha_2, \ldots, \alpha_n$ for which the minimum is achieved. In the present case, $F(\lambda)$ is a "step function," with its increments entirely "concentrated" at $N$ points. Therefore, if $n \geqslant N$, the coefficients $\alpha_1, \alpha_2, \ldots, \alpha_n$ can always be chosen in such a way that the difference

$$e^{im\lambda} - \sum_{k=1}^{n} \alpha_k e^{-ik\lambda}$$

vanishes at these $N$ points, and hence, for $n \geqslant N$, the mean square extrapolation error will vanish for all $m$. Thus, from a knowledge of the past behavior of the sequence $\xi(t)$ during a certain finite time interval, we can determine its value at any future time with probability 1. This conclusion is true *a fortiori* if we know *all* the past values of the sequence $\xi(t)$. Therefore, in the present case, we have

$$\sigma_m^2 = 0 \quad \text{for all} \quad m \geqslant 0. \tag{8.3}$$

This result is hardly unexpected, since all the $\xi(t)$ are determined once we have specified a finite number of quantities, i.e., the amplitudes and phases of the separate oscillations. Therefore, it is clear that a knowledge of a finite number of elements of the sequence $\xi(t)$ allows us to determine all the other elements of the sequence. It is also not hard to show that the converse assertion is true as well: If the extrapolation error of a random sequence $\xi(t)$ is zero when the values of a finite number of its elements are known, then the increments of its spectral distribution function $F(\lambda)$ are entirely "concentrated" at a finite number of points, and hence $\xi(t)$ can be represented as the sum of a finite number of uncorrelated harmonic oscillations with random amplitudes and phases.

A comparison of (8.1) and (8.3) shows how drastically stationary sequences can differ as far as their extrapolation properties are concerned. Following Kolmogorov, we shall call sequences satisfying the condition (8.1) *regular*, and those satisfying the condition (8.3) *singular*. (Sometimes, singular sequences are called *deterministic*, and regular sequences are called *purely nondeterministic*.) It is clear that the first problem to which extrapolation theory must address itself is that of finding necessary and sufficient conditions for a sequence $\xi(t)$ to be singular (or regular). We now explain the nature of these conditions, which were found in the paper K12.

Since the spectral distribution function $F(\lambda)$ is nondecreasing and bounded, its derivative $f(\lambda) = F'(\lambda)$ exists almost everywhere (i.e., except for a set of measure zero), $f(\lambda)$ is nonnegative, and the integral of $f(\lambda)$ is bounded. So far, we have been concerned only with the case where $F(\lambda)$ coincides with

$$F^{(I)}(\lambda) = \int_{-\pi}^{\lambda} f(\mu) \, d\mu,$$

and in this case, $f(\lambda)$ was called the *spectral density*. However, in the general case, $F(\lambda)$ can differ from $F^{(I)}(\lambda)$ by a "step function" (or "jump function") $F^{(II)}(\lambda)$, equal to the sum of all the jumps of $F(\lambda)$ at its points of discontinuity lying to the left of $\lambda$, and $F(\lambda)$ can also differ from $F^{(I)}(\lambda)$ by a "singular component" $F^{(III)}(\lambda)$, which is a continuous function with a derivative which vanishes almost everywhere (see L8, Sec. 11.1).[2] Since the integral of $f(\lambda)$ is bounded, and since $\log f(\lambda) \leqslant f(\lambda)$, the integral

$$P = \frac{1}{2\pi} \int_{-\pi}^{\pi} \log f(\lambda) \, d\lambda \tag{8.4}$$

---

[2] Note that the word "singular" is used here with two different meanings, so that a "singular random function" need not have a "singular component" in its spectral distribution function.

can be either finite or equal to $-\infty$.[3] It turns out that the value of this integral determines to a considerable extent the extrapolation properties of the random sequence $\xi(t)$. In fact, we have the following possibilities:

(a) If $P = -\infty$, i.e., if the integral (8.4) diverges, then the sequence $\xi(t)$ is singular;

(b) If $P$ is finite and $F(\lambda)$ coincides with $F^{(I)}(\lambda)$, then the sequence $\xi(t)$ is regular;

(c) If $P$ is finite, but $F(\lambda)$ is either discontinuous, or else is continuous but contains a "singular component" $F^{(III)}(\lambda)$, then $\xi(t)$ is neither singular nor regular (K12, K13, K23). (See also A1, App. B and D6, Chap. 12.)

As noted above, if we can "reconstruct" any future value of a stationary random sequence $\xi(t)$ with probability 1 from a knowledge of a finite number of past values of $\xi(t)$, then the spectral distribution function of $\xi(t)$ must be a step function with a finite number of points of discontinuity. This fact might lead us to think that if we can precisely reconstruct any future value of $\xi(t)$ from a knowledge of all the past values of $\xi(t)$, then the spectral distribution function of $\xi(t)$ must be a step function with an arbitrary (finite or infinite) number of points of discontinuity. However, the above results show that this conjecture is quite false, and that the class of singular stationary random sequences is much larger than the class of sequences whose spectral distribution functions are step functions. In fact, the spectral distribution function $F(\lambda)$ of a singular sequence can include not only an arbitrary step function $F^{(II)}(\lambda)$, but also an arbitrary "singular component" $F^{(III)}(\lambda)$,[4] and a function of the form

$$F^{(I)}(\lambda) = \int_{-\pi}^{\lambda} f(\mu)\, d\mu$$

for which the integral (8.4) diverges. Moreover the integral (8.4) will diverge if $f(\lambda)$ vanishes on a set of positive measure, or even if $f(\lambda)$ vanishes at just a single point $\lambda = \lambda_0$ but "sticks very close" to the $\lambda$-axis at $\lambda_0$, e.g., if

$$f(\lambda) = f_1(\lambda) \exp\left\{ -\frac{1}{(\lambda - \lambda_0)^2} \right\},$$

where $f_1(\lambda)$ is a bounded function, which is strictly positive.

If a stationary sequence $\xi(t)$ is regular, then it can be shown (K12) that $\xi(t)$ can be represented as a one-sided moving average

$$\xi(t) = \sum_{k=0}^{\infty} a_k \eta(t - k), \tag{8.5}$$

---

[3] Strictly speaking, the integral (8.4) is meaningless if $f(\lambda)$ vanishes on a set of positive measure, since then the integrand equals $-\infty$ on a set of positive measure. However, in this case it is still appropriate to regard $P$ as being equal to $-\infty$.

[4] Recall that the derivatives of the functions $F^{(II)}(\lambda)$ and $F^{(III)}(\lambda)$ vanish almost everywhere.

where $\eta(t)$ is a sequence of uncorrelated random variables; conversely, every sequence which can be represented in the form (8.5) is regular. This result makes the meaning of the concept of a regular sequence very clear.

If a stationary sequence $\xi(t)$ is neither singular nor regular, then it can be represented as a sum

$$\xi(t) = \xi_s(t) + \xi_r(t), \tag{8.6}$$

of two stationary sequences $\xi_s(t)$ and $\xi_r(t)$ which are not correlated with each other, where $\xi_s(t)$ is singular and $\xi_r(t)$ is regular. In this case, the spectral distribution function of $\xi_r(t)$ is just $F^{(I)}(\lambda)$, and the spectral distribution function of $\xi_s(t)$ is just $F^{(II)}(\lambda) + F^{(III)}(\lambda)$. Moreover, in predicting the values $\xi(t + m)$, $m \geqslant 0$ of the sequence $\xi(t)$, from a knowledge of the past values of the sequence $\xi(t)$, the value of $\xi_s(t + m)$ can be determined exactly, whereas the accuracy of the prediction of $\xi_r(t + m)$ goes to zero as $m \to \infty$. Thus, if the integral (8.4) converges, we find that

$$\sigma_\infty^2 = \lim_{m \to \infty} \sigma_m^2 = \mathbf{E}|\xi_r|^2 = \int_{-\pi}^{\pi} dF^{(I)}(\lambda) = \int_{-\pi}^{\pi} f(\lambda) \, d\lambda. \tag{8.7}$$

It should also be noted that it follows at once from (8.6) and (8.5) that any stationary sequence can be represented in the form

$$\xi(t) = \xi_s(t) + \sum_{k=0}^{\infty} a_k \eta(t - k), \tag{8.8}$$

where $\xi_s(t)$ is a singular stationary sequence, and $\eta(t)$ is a stationary sequence of uncorrelated random variables which is not correlated with $\xi_s(t)$. In general, the representation (8.8) is not unique. However, from all representations of this form, we can select a simplest "canonical representation" (see K12, W6), which plays a basic role in extrapolation theory; for further details we refer the reader to Kolmogorov's paper (K12).

The results given above allow us to immediately write an expression for the quantity $\sigma_\infty^2$ when we know the correlation function of the sequence $\xi(t)$. However, we are also interested in the value of the mean square extrapolation error $\sigma_m^2$ for all finite $m \geqslant 0$. We shall assume that the integral (8.4) converges, for otherwise $\sigma_m^2 = 0$ for any $m$. With this assumption, we can calculate the quantity $\sigma_m^2$ by using the formula

$$\sigma_m^2 = 2\pi e^P (1 + |p_1|^2 + |p_2|^2 + \cdots + |p_m|^2), \tag{8.9}$$

where $p_k$ is defined by the formulas

$$1 + p_1 z + p_2 z^2 + \cdots = \exp\{b_1 z + b_2 z^2 + \cdots\}, \tag{8.10}$$

$$b_k = \frac{1}{2\pi} \int_{-\pi}^{\pi} e^{ik\lambda} \log f(\lambda) \, d\lambda \qquad (k = 1, 2, 3, \ldots) \tag{8.11}$$

(see D6, K12, K13).[5]   Expanding the function $\exp\{b_1 z + b_2 z^2 + \cdots\}$ as a power series (according to the usual rules), we can easily obtain formulas which express $\sigma_m^2$ directly in terms of $f(\lambda) = F'(\lambda)$.   Obviously, for $m = 0$, we have

$$\sigma_0^2 = 2\pi e^P \tag{8.9'}$$

**Example 1.1.**   Let $\xi(t)$ have a positive spectral density of the form (4.76) [see p. 121], which we write as

$$f(\lambda) = \frac{|B_0(1 - \bar{b}_1 e^{i\lambda})\ldots(1 - \bar{b}_M e^{i\lambda})|^2}{|A_0(1 - \bar{a}_1 e^{i\lambda})\ldots(1 - \bar{a}_N e^{i\lambda})|^2}, \tag{8.12}$$

where all the zeros of the polynomials in $e^{i\lambda}$ appearing in the numerator and denominator are greater than 1 in absolute value.   To calculate the integral $P$ in this case, we use the fact that

$$\log|x| = \text{Re}\,\log x$$

to represent $P$ in the form

$$P = \frac{1}{2\pi}\,\text{Re}\int_{-\pi}^{\pi} \log\left[\frac{B_0(1 - \bar{b}_1 e^{i\lambda})\ldots(1 - \bar{b}_M e^{i\lambda})}{A_0(1 - \bar{a}_1 e^{i\lambda})\ldots(1 - \bar{a}_N e^{i\lambda})}\right]^2 d\lambda. \tag{8.13}$$

Introducing the complex variable $z = e^{i\lambda}$, we find that

$$\begin{aligned}
\frac{1}{2\pi}\int_{-\pi}^{\pi} &\log\left[\frac{B_0(1 - \bar{b}_1 e^{i\lambda})\ldots(1 - \bar{b}_M e^{i\lambda})}{A_0(1 - \bar{a}_1 e^{i\lambda})\ldots(1 - \bar{a}_N e^{i\lambda})}\right]^2 d\lambda \\
&= \frac{1}{2\pi i}\oint_{|z|=1}\frac{1}{z}\log\left[\frac{B_0(1 - \bar{b}_1 z)\ldots(1 - \bar{b}_M z)}{A_0(1 - \bar{a}_1 z)\ldots(1 - \bar{a}_N z)}\right]^2 dz.
\end{aligned} \tag{8.14}$$

According to Cauchy's theorem, the integral on the right equals

$$\log\left[\frac{B_0(1 - \bar{b}_1 z)\ldots(1 - \bar{b}_M z)}{A_0(1 - \bar{a}_1 z)\ldots(1 - \bar{a}_N z)}\right]^2\bigg|_{z=0} = \log\left(\frac{B_0}{A_0}\right)^2, \tag{8.15}$$

since its integrand is regular inside the unit circle.   Then, using (8.13) and (8.9'), we find that

$$P = \text{Re}\,\log\left(\frac{B_0}{A_0}\right)^2 = \log\left|\frac{B_0}{A_0}\right|^2, \qquad \sigma_0^2 = 2\pi e^P = 2\pi\left|\frac{B_0}{A_0}\right|^2 \tag{8.16}$$

Naturally, the last result agrees with (4.83).   We leave it to the reader to verify that formula (8.9) can be reduced to the form (4.91) for $m > 0$.

**Example 1.2.**   Let

$$B(\tau) = \begin{cases} \dfrac{m - |\tau|}{m} & \text{for } |\tau| \leqslant m, \\[2mm] 0 & \text{for } |\tau| > m \end{cases} \tag{8.17}$$

---

[5] The results in K13 are for the case where $\xi(t)$ is real.

(cf. Example 2, p. 30). It is not hard to verify that then the spectral density exists and is equal to

$$f(\lambda) = \frac{1}{2\pi m} \frac{|1 - e^{im\lambda}|^2}{|1 - e^{i\lambda}|^2} = \frac{1}{2\pi m} \frac{\sin^2 (m\lambda/2)}{\sin^2 (\lambda/2)}. \qquad (8.18)$$

This spectral density is rational in $e^{i\lambda}$, but is not positive everywhere, since [6]

$$f(\lambda) = 0 \quad \text{for} \quad \lambda = \frac{2\pi k}{m} \quad \left(k = \pm 1, \pm 2, \ldots, \pm \left[\frac{m}{2}\right]\right)$$

In the present case, we have

$$\begin{aligned}
P &= \frac{1}{2\pi} \int_{-\pi}^{\pi} \log f(\lambda) \, d\lambda \\
&= \log \frac{1}{2\pi m} + \frac{1}{2\pi} \int_{-\pi}^{\pi} \log \sin^2 \frac{m\lambda}{2} \, d\lambda - \frac{1}{2\pi} \int_{-\pi}^{\pi} \log \sin^2 \frac{\lambda}{2} \, d\lambda \\
&= \log \frac{1}{2\pi m} + \frac{1}{m} \cdot \frac{1}{2\pi} \int_{-m\pi}^{m\pi} \log \sin^2 \frac{\lambda}{2} \, d\lambda - \frac{1}{2\pi} \int_{-\pi}^{\pi} \log \sin^2 \frac{\lambda}{2} \, d\lambda \\
&= \log \frac{1}{2\pi m},
\end{aligned} \qquad (8.19)$$

since the last two integrals cancel each other. Therefore

$$\sigma_0^2 = 2\pi e^P = \frac{1}{m}, \qquad (8.20)$$

a result verified directly by Kozulyaev (K18, K19), using formula (4.18).

*Example 1.3.* Let

$$B(\tau) = \begin{cases} 1 & \text{for} \quad \tau = 0, \\ \dfrac{2(-1)^k}{2(k+1)\pi} & \text{for} \quad \tau = 2k + 1, \\ 0 & \text{for} \quad \tau = 2k, k \neq 0, \end{cases} \qquad (8.21)$$

so that obviously

$$f(\lambda) = \begin{cases} \dfrac{1}{\pi} & \text{for} \quad 0 \leqslant |\lambda| < \dfrac{\pi}{2}, \\ 0 & \text{for} \quad \dfrac{\pi}{2} \leqslant |\lambda| \leqslant \pi. \end{cases} \qquad (8.22)$$

Therefore, the sequence $\xi(t)$ is singular, and if we know the behavior of $\xi(t)$ in the "past," we can uniquely "reconstruct" any value of $\xi(t)$ in the "future." However, it should be noted that in making this reconstruction, we have to use values of $\xi(t)$ in the arbitrarily remote past. No finite number of past

---

[6] By $[n]$ is meant the largest integer $\leqslant n$.

values of the sequence will suffice for making a precise prediction, since here $F(\lambda)$ is not a step function with a finite number of discontinuities.

**36.2. Extrapolation of arbitrary stationary processes.** The results just given concerning extrapolation of stationary sequences can be extended with slight changes to the case of extrapolation of stationary processes. It turns out that the role of the integral $P$ is taken over by the integral

$$Q = \int_{-\infty}^{\infty} \frac{\log F'(\lambda)}{1 + \lambda^2} \, d\lambda = \int_{-\infty}^{\infty} \frac{\log f(\lambda)}{1 + \lambda^2} \, d\lambda. \tag{8.23}$$

If this integral diverges (i.e., equals $-\infty$), then the process is *singular*, in the sense that the relation (8.3) holds, where now, of course, $m$ is no longer an integer, but any positive real number.[7]   However, if the integral $Q$ converges, extrapolation without error is impossible (see A1, App. B, and D6, K24). The other results, concerning (a) the conditions for a process $\xi(t)$ to be *regular*, (b) the decomposition of an arbitrary process $\xi(t)$ into a "singular component" $\xi_s(t)$ and a "regular component" $\xi_r(t)$, and (c) the spectral distribution functions of the processes $\xi_s(t)$ and $\xi_r(t)$, differ from the corresponding results for stationary sequences only to the extent that the integral $P$ must be replaced everywhere by the integral $Q$.

The representation of a regular sequence as a sum of the form (8.5) is now replaced by the representation of a regular process as an integral of the form

$$\xi(t) = \int_0^{\infty} a(\tau) \, d_\tau \zeta(t - \tau), \tag{8.24}$$

where $\zeta(t)$ is a homogeneous random process with uncorrelated increments (see p. 67). It follows that an arbitrary stationary process $\xi(t)$ has the representation

$$\xi(t) = \xi_s(t) + \int_0^{\infty} a(\tau) \, d_\tau \zeta(t - \tau), \tag{8.25}$$

analogous to (8.8). The proof that the representations (8.24) and (8.25) are possible, and the establishing of conditions necessary to determine these representations uniquely, are much more complicated than the corresponding proofs for stationary sequences, and were found almost simultaneously, by Pinsker (unpublished thesis, Moscow State University, 1949), Karhunen (K3) and Hanner (H2).

If it is now assumed that the integral (8.23) converges, the mean square extrapolation error $\sigma_\tau^2$ will be nonzero. It can be shown that then $\sigma_\tau^2$ is given by the formula

$$\sigma_\tau^2 = \int_0^\tau |p(s)|^2 \, ds, \tag{8.26}$$

---

[7] It is clear that like (8.4), the integral (8.23) will diverge if $f(\lambda)$ vanishes on a set of positive measure, or has a zero of sufficiently high exponential order at some point $\lambda = \lambda_0$, which, in the case of (8.23), may also be infinite (see Example 2.2, p. 190).

where

$$p(s) = \frac{1}{\sqrt{2\pi}} \int_{-\infty}^{\infty} e^{iws} b(w) \, dw,$$

$$b(w) = \exp\left\{ -\frac{1}{2\pi i} \int_{-\infty}^{\infty} \frac{1 + \lambda w}{\lambda - w} \frac{\log f(\lambda)}{1 + \lambda^2} \, d\lambda \right\}.$$

(8.27)

The function $b(w)$ is uniquely determined by the second of the formulas (8.27) for values of $w$ in the lower half-plane, and it can then be determined for real values as well, by letting the imaginary part of $w$ approach zero (D6, K3, K24).

***Example 2.1.*** Let

$$B(\tau) = Ce^{-\alpha|\tau|},$$

so that

$$f(\lambda) = \frac{C\alpha}{\pi(\alpha^2 + \lambda^2)},$$

as we know from Example 1, p. 61. Then, the integral (8.23) obviously converges, and hence the process $\xi(t)$ is regular. We have already found the representation (8.24) for this process on p. 69, and extrapolation of $\xi(t)$ was studied in Example 1, p. 153. By using theorems on the boundary values of integrals of the Cauchy type, it can be shown that formulas (8.26) and (8.27) lead to the same expression (6.36) for the mean square extrapolation error as was derived earlier. The proof is similar to the derivation of (8.16), but is a bit more complicated, and will not be given here.

***Example 2.2.*** Now let

$$B(\tau) = Ce^{-\tau^2/4},$$

(8.28)

so that

$$f(\lambda) = \frac{1}{2\pi} \int_{-\infty}^{\infty} e^{-i\lambda\tau} B(\tau) \, d\tau = \frac{C}{\sqrt{\pi}} e^{-\lambda^2}$$

(8.29)

It is clear that here the integral (8.23) diverges, and hence the process $\xi(t)$ is singular. We emphasize that the correlation function (8.28) falls off much faster than the function $e^{-\alpha|\tau|}$, so that at first one might think that in this case, the value of $\xi(t + \tau)$ for large $\tau$ would be much less dependent on the "past" of the process than in the case of Example 2.1. However, it turns out that the situation is just the opposite: In Example 2.1, knowledge of the behavior of the process in the past allows us to say very little about values of the process in the distant future, whereas in Example 2.2, the value of the process in the arbitrarily distant future can be approximated arbitrarily closely by a linear combination of past values of the process. It should be noted that this seemingly paradoxical state of affairs had already been proved by Slutski, by direct calculation of the extrapolation formulas, before the general theory of extrapolation of stationary processes was developed, and served as a "touchstone" for verifying the general theory.

A simple proof of Slutski's result goes as follows: The correlation function (8.28) is an entire function, and hence

$$B(\tau) = B(0) + \frac{\tau^2}{2!} B''(0) + \frac{\tau^4}{4!} B^{(iv)}(0) + \cdots \qquad (8.30)$$

for all $\tau$. Moreover, according to (2.85), the process $\xi(t)$ with spectral density (8.29) has derivatives of all orders. An elementary calculation shows that (8.30) implies

$$\lim_{n \to \infty} \mathbf{E} \left| \xi(t + \tau) - \xi(t) - \tau\xi'(t) - \cdots - \frac{\tau^n}{n!}\xi^{(n)}(t) \right|^2 = 0$$

for all $\tau$. Thus, knowledge of an arbitrarily small "section" of $\xi(t)$ permits perfect extrapolation indefinitely far ahead, by using the formula[8]

$$\xi(t + \tau) = \xi(t) + \tau\xi'(t) + \cdots + \frac{\tau^n}{n!}\xi^{(n)}(t) + \cdots \qquad (8.31)$$

In this regard, it is interesting to consider the extrapolation of the stationary random process with spectral density

$$f(\lambda) = \frac{C}{\sqrt{\pi}} \left( 1 + \frac{\lambda^2}{n} \right)^{-n}, \qquad (8.29')$$

which approximates (8.29). Such a process will still be regular, and we can calculate the corresponding mean square extrapolation error $\sigma_\tau^2$ by using the formulas of Sec. 32; when $\tau$ is large, it can be shown that to a good approximation, the function $\sigma_\tau^2 = \sigma^2(\tau)$ is given by an "error integral" of the form

$$\frac{2\sigma_\infty^2}{\sqrt{\pi}} \int_0^\tau e^{-2(t - \sqrt{n})^2} \, dt. \qquad (8.32)$$

Thus, $\sigma_\tau^2$ is very small if $\tau \ll \sqrt{n}$, but for $\tau \approx \sqrt{n}$, $\sigma_\tau^2$ begins to grow very rapidly with $\tau$, and approaches $\sigma_\infty^2 = B(0)$ as $\tau \to \infty$. Since the function (8.29') approaches (8.29) very rapidly as $n$ increases, and since $\sqrt{n}$ is not a very rapidly increasing function, it is clear that in making "long-range" extrapolations, a slight deviation of the spectral density from (8.29) requires very considerable changes in the extrapolation formulas. In fact, by using the approximate expression (8.32), and by calculating the deviation of (8.29') from

---

[8] Actually, the realizations of a stationary random process with an entire correlation function are entire with probability 1, provided that the process is *separable* in the sense of D6, Chap. 2 (see Appendix II, p. 214). It follows that perfect extrapolation indefinitely far ahead can be achieved simply by using Taylor's series (8.31), written for the observed realization.

(8.29), it is not hard to show that for $\tau > 1.6$ [in time units corresponding to the frequency units used in (8.29) and (8.29′)], the results of extrapolation can be completely changed by a deviation of the spectral density from (8.29) which nowhere exceeds 10% of the maximum value of (8.29), while for $\tau > 5$, the results of extrapolation can be completely changed by a deviation which nowhere exceeds

$$1\% \approx \left(\frac{1.6}{5}\right)^2 \times 10\%$$

of the maximum value of (8.29), and so on.   [See Sec. 2.2 of Wiener's book W4.]   This makes it clear that in the case of the spectral density (8.29), which is proportional to $e^{-\lambda^2}$, long-range predictions will only be reliable when we are sure that the formula (8.29) is very accurate, and moreover, the required accuracy of (8.29) grows as the square of the prediction time.

**Example 2.3.** Let

$$B(\tau) = \frac{\sin \pi\tau}{\pi\tau},$$

so that

$$f(\lambda) = \begin{cases} \dfrac{1}{2\pi} & \text{for} \quad |\lambda| \leqslant \pi, \\[2ex] 0 & \text{for} \quad |\lambda| > \pi. \end{cases}$$

Clearly, a stationary random process with this spectral density is singular, and hence knowledge of $\xi(t)$ in the past allows us to determine $\xi(t + \tau)$ with probability 1, for any $\tau > 0$. This example is analogous to the example of a stationary random sequence whose correlation function is of the form (8.21). Since in this case $B(\tau)$ is entire, we can again use (8.31) to accomplish the prediction (for any $\tau > 0$). For a discussion of prediction for the general "band-limited" case, where $f(\lambda)$ vanishes outside a finite interval (as here), see W1, p. 70.

**Example 2.4.** Finally, consider a process $\xi(t)$ with correlation function

$$B(\tau) = \frac{C}{1 + \tau^2} \qquad (C > 0)$$

and spectral density

$$f(\lambda) = \frac{C}{2} e^{-|\lambda|}.$$

In this case, the integral (8.23) equals $-\infty$, so that $\xi(t)$ is singular and perfect prediction is possible. However, since the radius of convergence of the series (8.30) is now 1, we can only use formula (8.31) to accomplish the prediction when $\tau < 1$. When $\tau > 1$, we have to apply (8.31) *repeatedly*, as in the case of analytic continuation.

**36.3. Filtering of arbitrary stationary random functions.** The general problem of linear filtering of stationary random functions can be formulated as follows: Let $\xi(t)$ and $\zeta(t)$ be two such functions, which are stationarily correlated with each other (see Sec. 15). It is required to find the best approximation (in the sense of the method of least squares) to the random variable $\xi(t + \tau)$ which depends linearly on the random variables $\zeta(t - s)$, $s \geqslant 0$. The correlation functions $B_{\xi\xi}(\tau)$, $B_{\zeta\zeta}(\tau)$ and $B_{\xi\zeta}(\tau)$, and hence the spectral distribution functions $F_{\xi\xi}(\lambda)$, $F_{\zeta\zeta}(\lambda)$ and $F_{\xi\zeta}(\lambda)$, are assumed to be known.[9]

First, we consider a limiting case of filtering, i.e., we look for the best approximation to $\xi(t)$ which depends linearly on all the quantities $\zeta(t')$, $-\infty < t' < \infty$. It is not hard to see that the solution of the similar problem, given as the first example in both Sec. 27 and Sec. 34, can be extended without difficulty to the more general case considered here; the mean square filtering error turns out to be

$$\sigma^2 = \int_\Lambda dF_{\xi\xi}(\lambda) - \int_\Lambda \left| \frac{dF_{\xi\zeta}(\lambda)}{dF_{\zeta\zeta}(\lambda)} \right|^2 dF_{\zeta\zeta}(\lambda), \tag{8.33}$$

where the region of integration $\Lambda$ is the interval $-\pi \leqslant \lambda \leqslant \pi$ in the case of sequences, and the whole real line $-\infty < \lambda < \infty$ in the case of processes. If $F_{\xi\xi}(\lambda)$, $F_{\xi\zeta}(\lambda)$ and $F_{\zeta\zeta}(\lambda)$ can be represented as integrals of the corresponding spectral densities, then formula (8.33) becomes

$$\sigma^2 = \int_\Lambda \frac{f_{\xi\xi}(\lambda)f_{\zeta\zeta}(\lambda) - |f_{\xi\zeta}(\lambda)|^2}{f_{\zeta\zeta}(\lambda)} \, d\lambda, \tag{8.33'}$$

and if we set $f_{\zeta\zeta} = f_{\xi\xi} + f_{\eta\eta}$, $f_{\xi\zeta} = f_{\xi\xi}$, we obtain our previous formulas (5.31) and (7.14).

In the general case, the random function $\xi(t)$ can be represented in the form

$$\xi(t) = \xi_f(t) + \xi_0(t), \tag{8.34}$$

where $\xi_f(t)$ and $\xi_0(t)$ are two stationary random functions which are not correlated with each other; moreover, $\xi_0(t)$ is not correlated with $\zeta(t)$, and $\xi_f(t)$ lies in the (linear) subspace $H^\zeta$ of the space $H$ spanned by the random variables $\zeta(t')$, $-\infty < t' < \infty$. It is clear that within the context of the correlation theory, knowledge of the values of any of the quantities $\zeta(t')$ gives no information whatsoever about $\xi_0(t)$, and hence the best approximation to $\xi(t + \tau)$ which depends linearly on $\zeta(t')$, $t' \leqslant t$, coincides with the best

---

[9] It is also possible to formulate the linear filtering problem even more generally, as follows: Let $\zeta(t)$ be a stationary random function, and let $\xi$ be a random variable. Then, it is required to find the best approximation to $\xi$ which depends linearly on $\zeta(t - s)$, $s \geqslant 0$, given the correlation function $B_{\zeta\zeta}(\tau)$, the function $B_{\xi\zeta}(t) = \mathrm{E}\xi\overline{\zeta(t)}$, $-\infty < t < \infty$, and the value $B_{\xi\xi} = \mathrm{E}|\xi|^2$. For some results along these lines, see e.g., D6, Chap. 12, Sec. 6, and Y3.

approximation to $\xi_f(t + \tau)$.  As for $\xi_f(t + \tau)$, this quantity can be approximated arbitrarily closely as $\tau \to -\infty$.  Thus, the mathematical expectation of $|\xi_0(t)|^2$, which is just the integral (8.33), gives the limit as $\tau \to -\infty$ of the mean square error $\sigma_\tau^2$ of the best linear approximation to $\xi(t + \tau)$ in terms of the quantities $\zeta(t')$, $t' \leqslant t$, i.e.,

$$\sigma_{-\infty}^2 = \lim_{\tau \to -\infty} \sigma_\tau^2 = \mathbf{E}|\xi_0|^2 = \sigma^2. \tag{8.35}$$

The stationary random function $\xi_f(t)$ can in turn be represented as a sum of the form

$$\xi_f(t) = \xi_{f_s}(t) + \xi_{f_r}(t). \tag{8.36}$$

Here, $\xi_{f_s}(t)$ and $\xi_{f_r}(t)$ are stationary random functions which are not correlated with each other, such that $\xi_{f_s}(t)$ lies in the subspace $H^{\zeta_s}$ spanned by the random variables $\zeta_s(t)$, while $\xi_{f_r}(t)$ lies in the subspace $H^{\zeta_r}$ spanned by the random variables $\zeta_r(t)$ [cf. formula (8.6)].  If we know the values of $\zeta(t')$, $t' \leqslant t$, we can "reconstruct" the quantity $\xi_{f_s}(t + \tau)$ with probability 1 for any $\tau$, whereas the mean square error of the best linear approximation to $\xi_{f_r}(t + \tau)$ approaches $\mathbf{E}|\xi_{f_r}|^2$ as $\tau \to \infty$.  In particular, this implies that

$$\sigma_\infty^2 = \lim_{\tau \to \infty} \sigma_\tau^2 = \mathbf{E}|\xi_{f_r}|^2 + \mathbf{E}|\xi_0|^2. \tag{8.37}$$

The quantity $\sigma_\infty^2$ can be calculated from the formula

$$\sigma_\infty^2 = \int_\Lambda dF_{\xi\xi}(\lambda) - \int_\Lambda \left| \frac{dF_{\xi\zeta}(\lambda)}{dF_{\zeta\zeta}^{(s)}(\lambda)} \right|^2 dF_{\zeta\zeta}^{(s)}(\lambda), \tag{8.38}$$

analogous to formula (8.33), where $F_{\zeta\zeta}^{(s)}(\lambda)$ is the spectral distribution function of $\zeta_s(t)$.  If the stationary random function $\zeta(t)$ is singular, $F_{\zeta\zeta}^{(s)}(\lambda)$ coincides with $F_{\zeta\zeta}(\lambda)$, and otherwise, $F_{\zeta\zeta}^{(s)}(\lambda)$ equals the sum $F_{\zeta\zeta}^{(II)}(\lambda) + F_{\zeta\zeta}^{(III)}(\lambda)$ of the "step function component" and the "singular component" of $F_{\zeta\zeta}(\lambda)$.

The above results allow us to determine $\sigma_{-\infty}^2$ and $\sigma_\infty^2$ immediately from a knowledge of $F_{\xi\xi}(\lambda)$, $F_{\zeta\zeta}(\lambda)$ and $F_{\xi\zeta}(\lambda)$.  It is not hard to derive a formula giving $\sigma_\tau^2$ for finite $\tau$, but this formula is more complicated and will not be discussed here.

### 36.4. The multidimensional case.  Let

$$\boldsymbol{\xi}(t) = (\xi_1(t), \xi_2(t), \ldots, \xi_n(t))$$

be a multidimensional stationary random function:  Then, we can pose the problem of finding the best approximation to the quantity $\xi_1(t + \tau)$ which depends linearly on the quantities $\xi_k(t - s)$, $s \geqslant 0$, $k = 1, 2, \ldots, n$.  Similarly, if

$$\boldsymbol{\zeta}(t) = (\zeta_1(t), \zeta_2(t), \ldots, \zeta_m(t))$$

is another multidimensional stationary random function which is stationarily correlated with $\boldsymbol{\xi}(t)$, we can look for the best approximation to $\xi_1(t + \tau)$ which depends linearly on the quantities $\zeta_l(t - s), s \geqslant 0, l = 1, 2, \ldots, m$. Thus, both the extrapolation problem and the filtering problem generalize naturally to the multidimensional case.

It is not hard to see that the concepts of regularity and singularity of stationary random functions (cf. p. 184), and also the expansion (8.6), carry over almost automatically to the multidimensional case. As early as 1941, Zasukhin (Z2) found an important sufficient condition for a multidimensional stationary sequence to be regular. For a discussion of recent developments in this field, we refer the reader to Appendix II, p. 214.[10]

Some simple examples of extrapolation of multidimensional stationary random functions with spectral density matrices which are rational (in $e^{i\lambda}$ for sequences, and in $\lambda$ for processes) are given in Wiener's book W4. This and related problems are also discussed in the author's paper Y6.

## 37. Problems Related to Extrapolation and Filtering of Stationary Random Functions

**37.1. Approximate differentiation.** Let $\zeta(t)$ be a stationary random process which is the sum of a "signal" $\xi(t)$ and a "noise" $\eta(t)$. It is required to find the derivative $\xi'(t)$ of the signal, in terms of the known values of $\zeta(t - s)$, $s \geqslant 0$. This problem is of great practical importance in the construction of servomechanisms and radar equipment (see J2, S11), and its solution is far from obvious. It is clear that if the "spectral composition" of the noise is much richer in high frequencies than that of the signal (and this is most often the case in practice), then, by forming the ratio

$$\frac{\zeta(t) - \zeta(t - \Delta t)}{\Delta t}$$

for small $\Delta t$, we obtain a quantity which is chiefly due to the noise. On the other hand, if we take $\Delta t$ to be large, the resulting quantity will in general differ greatly from the derivative of the signal at time $t$. However, this problem is just a special case of the filtering problem, in the form in which it was stated in the preceding section. In fact, if the spectral distribution functions $F_{\xi\xi}(\lambda)$, $F_{\eta\eta}(\lambda)$ and $F_{\xi\eta}(\lambda)$ are assumed to be known, then the spectral distribution functions

$$F_{\zeta\zeta}(\lambda) = F_{\xi\xi}(\lambda) + F_{\xi\eta}(\lambda) + \bar{F}_{\xi\eta}(\lambda) + F_{\eta\eta}(\lambda),$$
$$F_{\xi'\xi'}(\lambda) = \lambda^2 F_{\xi\xi}(\lambda),$$
$$F_{\xi'\zeta}(\lambda) = i\lambda[F_{\xi\xi}(\lambda) + F_{\xi\eta}(\lambda)]$$

---

[10] Written by D. B. Lowdenslager.

are also known,[11] and we need only use these data to find the best approximation to $\xi'(t)$ which depends linearly on $\zeta(t - s)$, $s \geqslant 0$.

If the spectral distribution functions $F_{\zeta\zeta}(\lambda)$ and $F_{\xi\zeta}(\lambda)$ correspond to rational spectral densities $f_{\zeta\zeta}(\lambda)$ and $f_{\xi\zeta}(\lambda)$, then the problem of finding the best approximation to $\xi'(t)$ is easily solved by using the method of Chap. 7. For simplicity, we shall assume that $B_{\xi\eta}(\tau) \equiv 0$. In this case, by using the argument given in Sec. 33, we see that to solve our problem it is sufficient to find a function $\Phi(\lambda)$ satisfying the following conditions (cf. pp. 169–170):

(a) The function $\Phi(\lambda)$ is analytic in the lower half-plane, and as $|\lambda| \to \infty$ in the lower half-plane, $\Phi(\lambda)$ grows no faster than some power of $|\lambda|$;

(b) The function $\Psi(\lambda) = i\lambda f_{\xi\xi}(\lambda) - \Phi(\lambda)f_{\zeta\zeta}(\lambda)$ is analytic in the upper half-plane, and as $|\lambda| \to \infty$ in the upper half-plane, $\Psi(\lambda)$ falls off faster than $|\lambda|^{-1-\varepsilon}$, $\varepsilon > 0$;

(c) The integral along the real axis of the function $|\Phi(\lambda)|^2 f_{\zeta\zeta}(\lambda)$ is bounded, i.e.,

$$\int_{-\infty}^{\infty} |\Phi(\lambda)|^2 f_{\zeta\zeta}(\lambda)\, d\lambda < \infty.$$

The function $\Phi(\lambda)$ will be precisely the transfer function of the "differentiator" performing the approximate differentiation, and the mean square error of this differentiator can be determined from the formula

$$\sigma^2 = \int_{-\infty}^{\infty} \lambda^2 f_{\xi\xi}(\lambda)\, d\lambda - \int_{-\infty}^{\infty} |\Phi(\lambda)|^2 f_{\zeta\zeta}(\lambda)\, d\lambda.$$

It should be noted that all these considerations are meaningful only when the spectral density $f_{\xi\xi}(\lambda)$ satisfies the condition

$$\int_{-\infty}^{\infty} \lambda^2 f_{\xi\xi}(\lambda)\, d\lambda$$

[cf. formula (2.85)], for otherwise the process $\xi'(t)$ does not even exist.

***Example 1.1.*** Suppose we have

$$f_{\xi\xi}(\lambda) = \frac{1}{1 + \lambda^4}, \qquad f_{\eta\eta}(\lambda) = \varepsilon^4 = \text{const.} \tag{8.39}$$

As usual, the assumption that $f_{\eta\eta}(\lambda)$ is constant means that we are looking for limiting formulas which are approximately valid in the case where $B_{\eta\eta}(\tau)$, the correlation function of the "noise," falls off much faster that $B_{\xi\xi}(\tau)$, the correlation function of the "signal." To obtain a rigorous solution of this problem, we assume that the spectral density of the noise, rather than being constant, has the form (2.126), and we then pass to the limit

$$\alpha \to \infty, \qquad A \to \infty, \qquad \frac{A}{\alpha^2} \to \varepsilon^4$$

---

[11] The last of these formulas is an immediate consequence of the spectral representation (2.80) for the derivative of a stationary process.

in the resulting formulas (cf. Example 2 of Sec. 34). We shall not go into the details of these calculations, which are completely analogous to calculations carried out many times in Chaps. 6 and 7 (see also W4, Sec. 5.1). Instead, we just write down the final formula for $\Phi(\lambda)$:

$$
\begin{aligned}
\Phi(\lambda) = {} & \frac{1}{(\varepsilon^2 + \sqrt{1 + \varepsilon^4})(\varepsilon + \sqrt[4]{1 + \varepsilon^4})} \\
& \times \frac{(\sqrt[4]{1 + \varepsilon^4} - \varepsilon)i\lambda - \varepsilon\sqrt{2}}{\sqrt{1 + \varepsilon^4} + \varepsilon\sqrt{2}\sqrt[4]{1 + \varepsilon^4}\, i\lambda - \varepsilon^2\lambda^2}.
\end{aligned}
\tag{8.40}
$$

However, we note that in the present case, it is impossible to find a spectral characteristic $\Phi(\lambda)$ which completely satisfies all the conditions (a) to (c), since $\Psi(\lambda) = i\lambda f_{\xi\xi}(\lambda) - \Phi(\lambda)f_{\zeta\zeta}(\lambda)$ falls off only like $|\lambda|^{-1}$ as $|\lambda| \to \infty$, if $\Phi(\lambda)$ has the form (8.40). The spectral characteristics (7.32) and (7.33) exhibit the same peculiarity, which is explained by the fact that $f_{\zeta\zeta}(\lambda)$ is not integrable.

By expanding the rational function (8.40) in partial fractions and using (2.95), it is not hard to show that the best approximation to $\overset{\circ}{\xi}{}'(t)$ has the form

$$
\begin{aligned}
\overset{\circ}{\xi}{}'(t) = {} & \int_0^\infty \left( B\cos\frac{\sqrt[4]{1 + \varepsilon^4}}{\varepsilon\sqrt{2}}\tau + C\sin\frac{\sqrt[4]{1 + \varepsilon^4}}{\varepsilon\sqrt{2}}\tau \right) \\
& \times \exp\left\{ -\frac{\sqrt[4]{1 + \varepsilon^4}}{\varepsilon\sqrt{2}}\tau \right\} \zeta(t - \tau)\,d\tau,
\end{aligned}
\tag{8.41}
$$

where the coefficients $B$ and $C$ also depend on $\varepsilon$, and are given by rather complicated formulas, which we omit here. As $\varepsilon \to 0$, i.e., as the "noise intensity" approaches zero, the function (8.40) approaches $i\lambda$, as must be the case. This implies that as $\varepsilon \to 0$ the integral (8.41) approaches $\zeta'(t)$, a limiting value which does not involve an integral over the past of the process.

It is clear that in the same way, we can solve the problem of finding the second or higher-order derivatives of a process at the time $t$, or more generally, at the time $t + \tau$, where $\tau$ is positive, negative, or zero. All these problems are also special cases of the general filtering problem.

**37.2. Interpolation of stationary random sequences.** Suppose that we are given a series of observations of a quantity whose values constitute a realization of a stationary random sequence $\xi(s)$, and suppose that for some reason or other, one observation, pertaining to the time $t$, is omitted. In order to "reconstruct" this omitted value as accurately as possible, we have to find the best approximation to the quantity $\xi(t)$ which can be formed from the known values of the rest of the elements of the sequence. Confining ourselves to linear approximations, and assuming that the correlation function $B(\tau)$ corresponding to $\xi(t)$ is known, we arrive at the problem of *linear interpolation*, considered in Kolmogorov's papers K12 and K13.

Just as in the case of extrapolation, Kolmogorov's basic aim was to determine the mean square error $\sigma^2$ of the best approximation to $\xi(t)$, where no restrictions whatsoever are imposed on the correlation function $B(\tau)$. The quantity $\sigma^2$ can be defined as follows: Let $\sigma_n^2$ be the minimum value of the mathematical expectation

$$\mathbf{E}|\xi(t) - K_n(t)| = \sigma^2(\alpha_1, \ldots, \alpha_n, \alpha_{-1}, \ldots, \alpha_{-n}),$$

where

$$\begin{aligned} K_n(t) = \alpha_1\xi(t + 1) + \cdots + \alpha_n\xi(t + n) \\ + \alpha_{-1}\xi(t - 1) + \cdots + \alpha_{-n}\xi(t - n). \end{aligned}$$

Then, clearly, the determination of the values of the coefficients of the linear combination $K_n(t)$ for which $\sigma^2(\alpha_1, \ldots, \alpha_n, \alpha_{-1}, \ldots, \alpha_{-n})$ takes its minimum value (and hence the determination of the minimum value of $\sigma_n^2$) reduces to solving a simple system of linear equations (cf. Sec. 20). By definition, the quantity $\sigma_n^2$ cannot increase as $n$ increases. This means that the limit

$$\sigma^2 = \lim_{n \to \infty} \sigma_n^2$$

exists, and it is this limit which is called the *mean square interpolation error* (cf. the end of Sec. 21).

It turns out that in the general case, the quantity $\sigma^2$ can be expressed very simply in terms of the function $f(\lambda) = F'(\lambda)$, where, as always, $F(\lambda)$ denotes the spectral distribution function of the sequence $\xi(t)$. In fact, consider the integral

$$R = \int_{-\pi}^{\pi} \frac{d\lambda}{f(\lambda)}. \tag{8.42}$$

Since $f(\lambda) \geqslant 0$, this integral is either finite or else equals $+\infty$. If $R = +\infty$, i.e., if the integral (8.42) diverges, then $\sigma^2$ equals zero, and if $R$ is finite, then

$$\sigma^2 = \frac{4\pi^2}{R} \tag{8.43}$$

(see K12, K13).

Next, consider the case where the values of several elements of the sequence are unknown (e.g., the elements $\xi(t), \xi(t + k_1), \xi(t + k_2), \ldots, \xi(t + k_m)$, where $k_1, k_2, \ldots, k_m$ are arbitrary positive or negative integers), and suppose that it is required to find the best linear approximation to the value of $\xi(t)$ in terms of the known values of $\xi(t')$ for $t' \neq t, t + k_1, t + k_2, \ldots, t + k_m$. This case was studied in the author's paper Y2, and it was found that instead of (8.42), one must now consider the integral

$$R(\alpha_1, \alpha_2, \ldots, \alpha_m) = \int_{-\pi}^{\pi} \left| 1 + \sum_{j=1}^{m} \alpha_j e^{ik_j\lambda} \right|^2 \frac{d\lambda}{f(\lambda)}, \tag{8.44}$$

where $\alpha_1, \alpha_2, \ldots, \alpha_m$ are arbitrary complex numbers.   If the integral (8.44) diverges for any choice whatsoever of the complex numbers $\alpha_1, \alpha_2, \ldots, \alpha_m$, then the mean square error of the best approximation to the quantity $\xi(t)$ equals zero, i.e., the value of $\xi(t)$ can be "reconstructed" with probability 1. However, if there exist complex numbers $\alpha_1, \alpha_2, \ldots, \alpha_m$ for which the integral (8.44) converges, then the mean square error is not zero, and in fact equals

$$\sigma^2_{(k_1, k_2, \ldots, k_m)} = \frac{4\pi^2}{\min_{\alpha_j} R(\alpha_1, \alpha_2, \ldots, \alpha_m)}. \tag{8.45}$$

In the special case where the values of the quantities $\xi(t)$, $\xi(t + 1)$, $\xi(t + 2)$, $\ldots$, $\xi(t + m)$ are unknown, the integral (8.44) becomes

$$R(\alpha_1, \alpha_2, \ldots, \alpha_m) = \int_{-\pi}^{\pi} \left| 1 + \sum_{k=1}^{m} \alpha_k e^{ik\lambda} \right|^2 \frac{d\lambda}{f(\lambda)}. \tag{8.44'}$$

Then, if the function $f(\lambda)$ vanishes at a single point $\lambda = \lambda_0$, in order for the integral (8.44') to diverge for arbitrary $\alpha_1, \alpha_2, \ldots, \alpha_m$, it is necessary that the order of the zero of $f(\lambda)$ at $\lambda = \lambda_0$ be no less than $2m + 1$, since the coefficients $\alpha_1, \alpha_2, \ldots, \alpha_m$ can be chosen in such a way that the numerator of the integrand in (8.44') has a zero of order $2m$ at the point $\lambda = \lambda_0$.   Similarly, if $f(\lambda)$ vanishes at several isolated points, then, in order for these zeros not to be "cancelled out" by the zeros of the function

$$\left| \sum_{k=1}^{m} \alpha_k e^{ik\lambda} \right|^2,$$

the sum of the orders of these zeros must be sufficiently large.

It is not hard to see that the integral (8.44) diverges for all $m, \alpha_1, \alpha_2, \ldots, \alpha_m$ and $k_1, k_2, \ldots, k_m$ if the integral (8.4) diverges.   This is completely natural, since the possibility of perfect extrapolation for all $\tau > 0$ implies the possibility of perfect interpolation.   However, it is interesting to note that we can easily construct examples of stationary sequences for which interpolation is always possible, with arbitrarily high accuracy, when the values of any finite number of elements are unknown, but for which extrapolation, i.e., approximation to $\xi(t)$ by using a linear combination of the $\xi(t')$ with $t' < t$, is possible only with a finite error.   To see this, we need only choose a spectral density $f(\lambda)$ with an exponential zero, but such that this zero does not cause the integral (8.4) to diverge.   For example, we can set

$$f(\lambda) = f_1(\lambda) \exp \left\{ -\frac{1}{|\lambda - \lambda_0|^\alpha} \right\} \qquad (0 < \alpha < 1),$$

where $f_1(\lambda)$ is a bounded continuous function, which is strictly positive.   It is clear that in this case, the integral (8.44) will diverge for arbitrary integers $m, k_1, k_2, \ldots, k_m$ and for arbitrary complex numbers $\alpha_1, \alpha_2, \ldots, \alpha_m$, whereas the integral (8.4) will converge.   This example shows that in the general case, the mean square interpolation error for unknown elements $\xi(t)$, $\xi(t + 1)$, $\xi(t + 2), \ldots, \xi(t + m)$ does not approach the mean square extrapolation error one step ahead as $m \to \infty$.

As already noted, in the case where the integral (8.44) does not diverge for any choice of the numbers $\alpha_1, \alpha_2, \ldots, \alpha_m$, determination of the maximum accuracy of linear interpolation when the values of a finite number of elements of the sequence $\xi(t)$ are unknown reduces to finding the minimum of the function $R(\alpha_1, \alpha_2, \ldots, \alpha_m)$.   Since this function is bilinear in the variables $\alpha_k$, when we differentiate $R(\alpha_1, \alpha_2, \ldots, \alpha_m)$ to obtain conditions for an extremum, we obtain a system of $m$ linear equations in the $\alpha_k$.   We observe that the problem of finding the minimum of the integral (8.44) resembles quite closely the problem of finding the minimum of the integral (8.2).   [It will be recalled that finding the best linear approximation to $\xi(t)$ in terms of the known values of a finite number of elements of the sequence reduces to the problem of finding the minimum of (8.2).]   The resemblance is especially striking when the sequence has a spectral density, for then finding the best approximation to $\xi(t)$ when the values of $\xi(t + k_1)$, $\xi(t + k_2), \ldots, \xi(t + k_m)$ are known reduces to finding the minimum of the integral

$$S(\alpha_1, \alpha_2, \ldots, \alpha_m) = \int_{-\pi}^{\pi} \left| 1 + \sum_{j=1}^{m} \alpha_j e^{ik_j\lambda} \right|^2 f(\lambda) \, d\lambda, \qquad (8.46)$$

and the mean square error of this approximation is given by the formula

$$\sigma^2_{(k_1, k_2, \ldots, k_m)} = \min_{\alpha_j} S(\alpha_1, \alpha_2, \ldots, \alpha_m). \qquad (8.47)$$

We see that in the case where $f(\lambda)$ is strictly positive, finding the mean square interpolation error reduces to solving a simple system of $m$ linear equations. However, if $f(\lambda)$ has zeros in the interval $[-\pi, \pi]$, special considerations are needed, as shown by the following examples.

***Example 2.1.*** Let

$$B(\tau) = \begin{cases} 1 & \text{for} \quad \tau = 0, \\ \dfrac{1}{2} & \text{for} \quad \tau = \pm 1, \\ 0 & \text{for} \quad |\tau| > 1, \end{cases} \qquad f(\lambda) = \frac{1}{\pi} \cos^2 \frac{\lambda}{2} \qquad (8.48)$$

[cf. formulas (8.17) and (8.18)].   Then, the integral (8.42) obviously diverges, so that we can choose a linear combination of the elements $\xi(t')$, $t' \neq t$, which approximates $\xi(t)$ arbitrarily closely.

Suppose now that the values $\xi(t)$ and $\xi(t + n)$ are unknown.   Then, in order for the corresponding integral (8.44) to converge, we have to choose the coefficient $\alpha$ in such a way that $1 + \alpha e^{in\lambda}$ vanishes at the points $\lambda = \pm \pi$.

Thus, we must set $\alpha = (-1)^{n+1}$, so that

$$\frac{|1 + \alpha e^{in\lambda}|^2}{\cos^2(\lambda/2)} = 4\left|\frac{1 - (-1)^n e^{in\lambda}}{1 + e^{i\lambda}}\right|^2$$

$$= 4|1 - e^{i\lambda} + e^{2i\lambda} - \cdots + (-1)^{n-1}e^{i(n-1)\lambda}|^2$$

and

$$\int_{-\pi}^{\pi} |1 + \alpha e^{in\lambda}|^2 \frac{d\lambda}{f(\lambda)} = 4\pi \int_{-\pi}^{\pi} |1 - e^{i\lambda} + \cdots + (-1)^{n-1}e^{i(n-1)\lambda}|^2 \, d\lambda$$

$$= 8n\pi^2.$$

It follows that

$$\sigma_{(n)}^2 = \frac{4\pi^2}{8n\pi^2} = \frac{1}{2n}. \tag{8.49}$$

Next, we consider the case where the values of the elements $\xi(t)$, $\xi(t + 1)$, $\xi(t + 2), \ldots, \xi(t + n)$ are unknown. Then, in order for the integral (8.44′) to converge, it is necessary that

$$1 + \sum_{k=1}^{n} \alpha_k e^{ik\lambda} = 0 \quad \text{for} \quad \lambda = \pm\pi,$$

i.e., we must be able to represent the polynomial in $e^{i\lambda}$ appearing on the left in the form

$$(1 + e^{i\lambda})(1 + \beta_1 e^{i\lambda} + \cdots + \beta_{n-1} e^{i(n-1)\lambda}),$$

and hence

$$\int_{-\pi}^{\pi} \left|1 + \sum_{k=1}^{n} \alpha_k e^{ik\lambda}\right|^2 \frac{d\lambda}{f(\lambda)}$$

$$= 4\pi \int_{-\pi}^{\pi} |1 + \beta_1 e^{i\lambda} + \cdots + \beta_{n-1} e^{i(n-1)\lambda}|^2 \, d\lambda$$

$$= 8\pi^2(1 + |\beta_1|^2 + \cdots + |\beta_{n-1}|^2).$$

It is clear that this expression takes its minimum value for

$$\beta_1 = \beta_2 = \ldots = \beta_{n-1} = 0,$$

which means that

$$\sigma_{(1, 2, \ldots, n)}^2 = \frac{4\pi^2}{8\pi^2} = \frac{1}{2}. \tag{8.50}$$

We see that (8.50) does not depend on $n$, so that the greatest accuracy attainable when the values of the elements $\xi(t)$, $\xi(t + 1)$, $\xi(t + 2), \ldots, \xi(t + n)$ are unknown ("omitted") is the same as when only the values of $\xi(t)$ and $\xi(t + 1)$ are unknown. This is understandable, since according to formulas (8.49) and (8.20), in the present case, the mean square interpolation error when the values of $\xi(t)$ and $\xi(t + 1)$ are unknown is the same as the mean square extrapolation error.

Finally, we consider the case where the values of the elements $\xi(t)$, $\xi(t + 1)$ and $\xi(t - 1)$ are unknown. Then, from the condition

$$1 + \alpha_1 e^{-i\lambda} + \alpha_2 e^{i\lambda} = 0 \quad \text{for} \quad \lambda = \pm\pi,$$

we find that

$$1 - \alpha_1 - \alpha_2 = 0, \qquad \alpha_2 = 1 - \alpha_1,$$

and hence

$$\min_{\alpha_1, \alpha_2} \int_{-\pi}^{\pi} \frac{|\alpha_1 e^{-i\lambda} + 1 + \alpha_2 e^{i\lambda}|^2}{(1/\pi)\cos^2(\lambda/2)} \, d\lambda$$

$$= \min_{\alpha} \pi \int_{-\pi}^{\pi} \frac{|\alpha e^{-i\lambda} + 1 + (1 - \alpha)e^{i\lambda}|^2}{\cos^2(\lambda/2)} \, d\lambda$$

$$= \min_{\alpha} 4\pi \int_{-\pi}^{\pi} |\alpha e^{i\lambda} + (1 - \alpha)|^2 \, d\lambda = \min_{\alpha} 8\pi^2 (|\alpha|^2 + |1 - \alpha|^2).$$

It is easily verified that $|\alpha|^2 + |1 - \alpha|^2$ takes its minimum value for $\alpha = 1/2$. Since this minimum value equals $1/2$, we have

$$\sigma_{(1, -1)}^2 = \frac{4\pi^2}{4\pi^2} = 1. \tag{8.51}$$

Thus, $\sigma_{(1, -1)}^2 = B(0)$, so that if we do not know the values of $\xi(t + 1)$ and $\xi(t - 1)$, we can say nothing at all about the value of $\xi(t)$.

The above results can easily be understood if we recall from Example 2, p. 30, that the sequence $\xi(t)$ with the correlation function (8.48) can be written as a moving average

$$\xi(t) = \frac{1}{\sqrt{2}} [\eta(t) + \eta(t - 1)]$$

with two terms, where $\eta(t)$ is a sequence of uncorrelated random variables. See also the paper K18, where explicit formulas are given for this case (when the values of a finite number of elements of the sequence are known).

**Example 2.2.** Next, we consider the more general case of a stationary sequence with the correlation function (8.17), where the spectral density is given by formula (8.18). This density has zeros of order two at the points

$$\lambda = \pm\frac{2\pi}{m}, \pm\frac{4\pi}{m}, \cdots, \pm\frac{2\left[\frac{m}{2}\right]\pi}{m}. \tag{8.52}$$

In this case, of course, the integral (8.42) diverges, so that $\xi(t)$ can be approximated arbitrarily closely by a linear combination of the other elements of the sequence.

Suppose now that in addition to $\xi(t)$, we know the value of one other element of the sequence, say, the element $\xi(t + n)$. Then, the corresponding integral (8.44) cannot converge unless there exists a value of $\alpha$ for which the equation

$$1 + \alpha e^{in\lambda} = 0$$

holds for all the values of $\lambda$ given by (8.52). It is easy to see that for $m > 2$, this is possible only if $e^{2n\pi i/m} = 1$, i.e., only if $n$ is a multiple of $m$. Thus we have

$$\sigma^2_{(n)} = 0, \quad \text{if} \quad m > 2 \text{ and } n \text{ is not a multiple of } m. \tag{8.53}$$

On the other hand, if the values of $\xi(t)$ and $\xi(t + m)$ are unknown, the denominator of (8.45) becomes

$$2\pi m \int_{-\pi}^{\pi} \frac{|1 - e^{im\lambda}|^2 \sin^2(\lambda/2)}{\sin^2(m\lambda/2)} \, d\lambda = 8\pi m \int_{-\pi}^{\pi} \sin^2(\lambda/2) \, d\lambda = 8\pi^2 m.$$

It follows that

$$\sigma^2_{(m)} = \frac{4\pi^2}{8\pi^2 m} = \frac{1}{2m}. \tag{8.53'}$$

Similarly, we can prove the more general formula

$$\sigma^2_{(km)} = \frac{1}{2km}. \tag{8.53''}$$

**37.3. The general problem of interpolating stationary random functions.** Both of the problems considered above, namely, interpolation of stationary random sequences, and extrapolation of stationary random sequences and processes, can be regarded as special cases of a single general problem. If we confine ourselves to the problem of finding the mean square error, this problem can be formulated as follows: Suppose that the set of all values of the argument of a stationary random function $\xi(t)$ [i.e., the set of all real numbers or of all integers] is divided into two subsets $K$ and $K'$, and suppose that $t$ belongs to $K$. Consider all possible finite sums

$$L_n = \sum_{j=1}^{n} \alpha_j \xi(t'_j), \tag{8.54}$$

where the $t'_j$ belong to $K'$ and the $\alpha_j$ are arbitrary complex numbers ($j = 1, 2, \ldots, n$). For each of the sums (8.54), form the mathematical expectation

$$\sigma^2(L_n) = E|\xi(t) - L_n|^2.$$

Then, given the correlation function $B(\tau)$, it is required to find the greatest lower bound

$$\sigma^2_{(K)} = \inf_{L_n} \sigma^2(L_n) \tag{8.55}$$

of the mathematical expectations $\sigma^2(L_n)$ for all permissible sums $L_n$; in particular, it is required to find the conditions under which this greatest lower bound equals zero.

Obviously, the extrapolation problem corresponds to the case where the set $K'$ consists of all numbers less than a given fixed number $t_0$, while the interpolation problem considered above corresponds to the case where the set $K$ is finite (and, in particular, consists of a single number). Moreover, we have also considered simple cases where the set $K'$ is finite; then the solution of the problem reduces to the solution of a system of linear equations (see Sec. 20).

In the paper Y2, another case is considered, i.e., the case where $t$ ranges over all real values [so that $\xi(t)$ is a stationary process], while $K'$ is just the set of all integers. For example, such a problem arises when all the observations of the quantity $\xi(t)$ are made at equally spaced instants of time, and it is required to use the data so obtained to "reconstruct" the value of $\xi(t)$ at some instant of time which is not a multiple of the interval between successive observations. It turns out that in this case, a simple formula can be obtained for the mean square interpolation error. In particular, by using this formula, it is easy to see under what conditions the quantity $\xi(t)$ [where $t$ is not an integer] can be approximated arbitrarily closely by a linear combination of the quantities

$$\ldots, \xi(-2), \xi(-1), \xi(0), \xi(1), \xi(2), \ldots$$

In the special case of a process $\xi(t)$ with a discrete spectrum, it is found that such perfect approximation is possible for all $t$, provided that there do not exist two points of the spectrum whose distance apart equals a multiple of $2\pi$; of course, this last result could have been obtained without considering the general case.

Similarly, it turns out that in the case where the spectral density $f(\lambda)$ vanishes outside of a finite frequency interval $\Lambda$ (i.e., when the process is *band-limited*), perfect approximation is possible if and only if the length of $\Lambda$ does not exceed $2\pi$. This implies that if $f(\lambda) = 0$ for $|\lambda| > W$, then for any $t$, $\xi(t)$ can be expressed as a linear combination of the quantities

$$\ldots, \xi\left(-\frac{2\pi}{W}\right), \xi\left(-\frac{\pi}{W}\right), \xi(0), \xi\left(\frac{\pi}{W}\right), \xi\left(\frac{2\pi}{W}\right), \ldots$$

From the general results of Y2 (see also L7), it is not hard to show that this linear combination is just

$$\xi(t) = \sum_{n=-\infty}^{\infty} \frac{\sin(Wt - n\pi)}{Wt - n\pi} \, \xi\left(\frac{n\pi}{W}\right). \tag{8.56}$$

Formula (8.56) is often called the *sampling theorem*, and is usually associated with the names of Kotelnikov and Shannon, who made extensive use of it in communication theory applications (see e.g., M2, Chap. 4, and W1, Sec. 19).

In the case where $t$ ranges over all real values [so that $\xi(t)$ is a process], while $K'$ is a finite interval of the real axis, conditions under which the mean square interpolation error $\sigma^2_{(K)}$ is identically zero can be derived from the results given in Krein's note (K22), which is devoted to the problem of continuing spirals (helical arcs) in Hilbert space (see also K21). The same case is discussed in the important paper Z1, which contains some results concerning the size of the corresponding mean square extrapolation error for processes $\xi(t)$ with rational spectral densities. The paper Z1 also discusses the more general case where $\xi(t)$ contains not only a stationary random component, but also a term which is a polynomial in the parameter $t$; moreover, this paper considers the filtering problem as well as the extrapolation problem.[12] A special problem of practical interest, from among those mentioned in Z1, had already been analyzed in detail earlier in C6. In the papers C6 and Z1, it is not claimed that the methods of proof are mathematically rigorous.

In the author's paper Y4, it is shown how to extend the methods of Chaps. 6 and 7, in a completely rigorous fashion, to solve the problem of finding the best linear approximation to the value of $\xi(t + \tau)$, where $\xi(t)$ is a stationary random process with a rational spectral density, in terms of the values of $\xi(t)$ on the finite interval $[-T, 0]$, say, or in terms of the values on $[-T, 0]$ of another process $\zeta(t)$, which is stationarily correlated with $\xi(t)$. (One can have either $\tau > 0$ or $\tau < -T$.) In particular, the corresponding generalizations of Example 4 of Sec. 31 and Example 2 of Sec. 34 are given. A variety of related problems, which are very germane to the subject matter of this book (but more specialized), are also discussed. Moreover, in the paper Y6, it is shown how to extend this "finite-time theory" to the case of multidimensional stationary processes with rational spectral densities, and to the case where $K'$ consists of several intervals (one or two of which may be semi-infinite).

**37.4. Smoothing.** By *smoothing* of a sequence $\xi(t)$ is meant going over to a new sequence $\Xi(t)$, obtained from the original sequence by forming the moving average

$$\Xi(t) = \sum_{k=-n}^{n} a_k \xi(t - k). \tag{8.57}$$

[For simplicity, we have written a finite sum in (8.57), but if certain convergence conditions are met, we can also write $n = \infty$.] If $\xi(t)$ is a stationary random sequence, the same is true of $\Xi(t)$; moreover, the correlation function and the spectral distribution function of the new sequence $\Xi(t)$ are related to the corresponding characteristics of $\xi(t)$ as follows:

$$B_{\Xi\Xi}(\tau) = \sum_{k,l} a_k \bar{a}_l B_{\xi\xi}(\tau - k + l),$$

$$F_{\Xi\Xi}(\lambda) = \int_{-\pi}^{\lambda} \left| \sum_{k} a_k e^{-ik\lambda} \right|^2 dF_{\xi\xi}(\lambda).$$

---

[12] The material in Z1 is presented in Chap. 8 of the book S12. For the case of discrete time (i.e., random sequences), the material in the monograph H1 is rather closely related to the problem considered in Z1.

In practice, smoothing is widely used to filter sequences, in order to diminish the effects of measurement errors and other high-frequency (i.e., weakly correlated) disturbances. For example, if

$$\xi(t) = x(t) + \eta(t),$$

where $x(t)$ is a definite function of time, or a random sequence with a slowly decreasing correlation function, while $\eta(t)$ is a sequence of uncorrelated random variables, then, after smoothing, the mean square of the second term is greatly decreased (as a rule), while at the same time, if the smoothing weights are properly chosen, the first term is either altered only rather slightly, or else may be appreciably altered, but in such a way that the change can easily be taken into account. (See e.g., Example 2 of Sec. 27.) This explains the origin of the term "smoothing." (However, it should be noted that the terms *smoothing* and *filtering* are sometimes used as synonyms, e.g., in the book W4.)

It is important to note that application of smoothing converts a sequence of uncorrelated random variables into a sequence which is no longer uncorrelated. In fact, if $dF_{\xi\xi}(\lambda)/d\lambda = \text{const}$, the spectral density of $\Xi(t)$ will have a maximum coinciding with the maximum of the function

$$\left| \sum_k a_k e^{-ik\lambda} \right|^2.$$

Long before the appearance of the spectral theory of random functions, it was shown by Slutski (S7, S8) that by repeatedly applying the smoothing operation to a sequence $\eta(t)$ of uncorrelated random variables, and by suitably choosing the smoothing weights, one can obtain a new sequence $\xi(t)$ such that during any given finite time interval, the values of any realization of $\xi(t)$ will differ by an arbitrarily small amount, and with probability arbitrarily close to unity, from the values of a function of the form

$$A \sin (\lambda t + \varphi),$$

where $A$, $\varphi$ and $\lambda$ are constants. This is the so-called "sinusoidal limit theorem," a more general formulation of which is given by Romanovski (R3, R4). In its day, this theorem played an important role, since it helped to show that the presence of a spectrum, even a spectrum with a sharp maximum, does not mean that the process under consideration is actually due to anything which is periodically varying. Today, we know that the presence of a spectrum is simply an automatic consequence of statistical stationarity, and by using the associated spectral theory of stationary random processes, we can very easily prove the old results of Slutski and Romanovski. Moreover, we can also obtain a whole series of new results pertaining to the smoothing of stationary time series (see e.g., M3, M4, N1 [13]).

---

[13] The continuous case is also considered in N1.

*Appendix* I

# GENERALIZED
# RANDOM PROCESSES[1]

Throughout the present book, we have used the fact that if $\xi(t)$ is a stationary random process for which

$$\mathbf{E}\xi(t) = 0, \qquad \mathbf{E}\xi(t)\overline{\xi(s)} = B(t - s) = B(\tau),$$

then

$$B(\tau) = \int_{-\infty}^{\infty} e^{i\lambda\tau} \, dF(\lambda), \tag{I.1}$$

where the function $F(\lambda)$, called the *spectral distribution function*, is non-decreasing and bounded [cf. formula (2.35)]. Conversely, every non-decreasing bounded function $F(\lambda)$ is the spectral distribution function of some stationary random process (cf. p. 47). In the case where $F(\lambda)$ is absolutely continuous, with derivative $f(\lambda)$, formula (I.1) reduces to

$$B(\tau) = \int_{-\infty}^{\infty} e^{i\lambda\tau} f(\lambda) \, d\lambda \tag{I.2}$$

[cf. formula (2.66)], where $f(\lambda)$ is called the *spectral density*.

As already noted in Sec. 14, it is often convenient to introduce stationary "white noise," with constant spectral density, i.e.,

$$f(\lambda) = \text{const.} \tag{I.3}$$

---

[1] References in "letter-number form" (see Translator's Preface) are to items in the main Bibliography, starting on p. 217. Numerical references (in brackets) are to items in the Supplementary Bibliography, starting on p. 224.

Since (I.3) corresponds to an unbounded spectral distribution function and to an infinite average noise power, it is clear that white noise is meaningful only as a mathematical idealization.[2]  The utility of this idealization stems from the fact that the frequency response of any actual physical measuring device is effectively zero outside some finite frequency interval $\Lambda$.  Consequently, when a random process $\xi(t)$ whose spectral density $f(\lambda)$ is effectively constant over $\Lambda$ is applied to such a device, there is no need to describe in detail how $f(\lambda)$ falls off at high frequencies (as it must), and one can regard $\xi(t)$ as being white noise, with constant spectral density $f(0) = f_0$ (cf. p. 63).

Moreover, by the same token, it is intuitively clear that the concept of the "value $\xi(t)$ of the process $\xi$ at the time $t$" is also a mathematical idealization. The point is that in practice one must always use some physical device to measure $\xi(t)$, and since the device always has "inertia" (or "memory"), corresponding to its nonzero "time constant," the input process $\xi(t)$ will inevitably be subjected to some time averaging.  This suggests introducing a new definition of a random process; it turns out that with a suitable definition, the case of white noise is no "worse" than the case of ordinary random processes (see below).

If we restrict ourselves to linear measuring devices, then the device is completely characterized by its *weighting function* or *impulse response* $\varphi(t)$ [cf. p. 42], and the result of measuring $\xi(t)$ is just the quantity[3]

$$\xi(\varphi) = \int_{-\infty}^{\infty} \xi(t)\varphi(t)\, dt. \tag{I.4}$$

In fact, in practice a physicist can only deal with quantities like (I.4), rather than with the process $\xi(t)$ itself.  These considerations suggest the following new definition of a random process: By a *generalized (random) process*, we mean a linear functional $\xi(\varphi)$ whose values are random variables.[4]  In other words, for every function $\varphi(t)$ in some function space $\mathscr{D}$, the quantity $\xi(\varphi)$ is a random variable, and moreover

$$\xi(\alpha_1\varphi_1 + \alpha_2\varphi_2) = \alpha_1\xi(\varphi_1) + \alpha_2\xi(\varphi_2), \tag{I.5}$$

where $\alpha_1$, $\alpha_2$ are arbitrary (real) numbers, and $\varphi_1$, $\varphi_2$ are arbitrary elements of $\mathscr{D}$.  The condition (I.5) is completely natural if we think of $\mathscr{D}$ as the space of weighting functions corresponding to all possible linear measuring devices, and the $\xi(\varphi)$ as the results obtained when various devices are used to measure an ordinary random process $\xi(t)$.  The word *generalized* appearing in the above definition is explained by the fact that any *ordinary* random process

---

[2] It will be recalled that white noise is a "purely random" process, in the sense of p. 64.

[3] For simplicity, measurements made with the same device at different times, corresponding to weighting functions $\varphi_1(t)$ and $\varphi_2(t) = \varphi_1(t + t_0)$, say, will be regarded as being made with different devices.

[4] Cf. the definition of an "ordinary" random process given on pp. 9, 11.

$\xi(t)$ gives rise to a generalized random process $\xi(\varphi)$ of a special kind, which is related to $\xi(t)$ by the formula (I.4) and is said to "have the point values $\xi(t)$." However, there also exist generalized processes $\xi(\varphi)$ which are not defined by the integral formula (I.4) for any $\xi(t)$ at all. This kind of generalized process is said to "have no point values," despite the fact that by repeatedly measuring such a process with the same linear device, one obtains sample values of a random variable with a perfectly definite probability distribution.

To specify a generalized random process, we parallel the treatment given on pp. 9, 10 for ordinary random processes. Thus, we shall regard the generalized random process $\xi(\varphi)$ as being specified if for any finite set of functions $\varphi_1(t), \varphi_2(t), \ldots, \varphi_n(t)$ in $\mathscr{D}$, we are given the distribution function

$$F_{\varphi_1, \varphi_2, \ldots, \varphi_n}(x_1, x_2, \ldots, x_n) = \mathbf{P}\{\xi(\varphi_1) < x_1, \xi(\varphi_2) < x_2, \ldots, \xi(\varphi_n) < x_n\} \quad (\text{I.6})$$

of the $n$-dimensional random variable

$$\boldsymbol{\xi} = (\xi(\varphi_1), \xi(\varphi_2), \ldots, \xi(\varphi_n)) \quad (\text{I.7})$$

[cf. (1.3)]. As in the case of ordinary random processes, the system of distribution functions (I.6) must satisfy the symmetry and compatibility conditions which are the obvious generalizations of (1.4) and (1.5). However, since in the case of the $n$-dimensional random variable (I.7), we can uniquely reconstruct the distribution function (I.6) from a knowledge of all the probabilities

$$P\{\alpha_1 \xi(\varphi_1) + \alpha_2 \xi(\varphi_2) + \cdots + \alpha_n \xi(\varphi_n) < x\}$$

for arbitrary $\alpha_1, \alpha_2, \ldots, \alpha_n$ and $x$ (see e.g., [7]),[5] a generalized random process can actually be specified by giving only the one-dimensional distribution of $\xi(\varphi)$ for arbitrary $\varphi(t)$ in $\mathscr{D}$.

From the standpoint of correlation theory, a generalized random process $\xi(\varphi)$ is characterized by its *mean value functional*

$$m(\varphi) = \mathbf{E}\xi(\varphi),$$

which is linear in $\varphi$, and by its *correlation functional*

$$B(\varphi_1, \varphi_2) = \mathbf{E}\xi(\varphi_1)\overline{\xi(\varphi_2)}, \quad (\text{I.8})$$

which is bilinear in $\varphi_1$ and $\varphi_2$. For simplicity, we shall henceforth assume that $m(\varphi) \equiv 0$.

We must now discuss how the space $\mathscr{D}$ is chosen. It follows from (I.5) that a natural requirement to impose on $\mathscr{D}$ is that it be linear, i.e., that $\alpha_1 \varphi_1 + \alpha_2 \varphi_2$ belong to $\mathscr{D}$ if $\alpha_1, \alpha_2$ are arbitrary constants and $\varphi_1, \varphi_2$ are

---

[5] See also [18], where explicit formulas are given for reconstructing the distribution functions (I.6).

arbitrary elements of $\mathscr{D}$.  We still have a wide choice of spaces $\mathscr{D}$, and corresponding to this freedom of choice, there are various kinds of generalized random processes.  Naturally, the larger the space $\mathscr{D}$, the narrower the corresponding generalized random processes, in the following sense: If $\mathscr{D}$ is contained in $\mathscr{D}'$, then all random processes $\xi(\varphi)$ defined on $\mathscr{D}'$ are still defined on $\mathscr{D}$, but in general there are random processes defined on $\mathscr{D}$ which cannot be defined on the larger space $\mathscr{D}'$.  Perhaps the most general space $\mathscr{D}$ for which generalized processes can be considered is the space $\mathscr{D}_0$ of piecewise continuous functions vanishing outside a finite interval of the axis $-\infty < t < \infty$.  Then, given any finite interval $\Delta t$, we can define a random interval function $\xi(\Delta t) = \xi(\varphi_{\Delta t})$, where $\varphi_{\Delta t}(t)$ is the function equal to 1 on $\Delta t$ and 0 elsewhere.  This leads us to the class of random interval functions $\xi(\Delta t)$ satisfying conditions (a) and (b) on p. 37, but in general not condition (c).  This class already contains processes $\xi(\varphi)$ for which no $\xi(t)$ satisfying (I.4) can be found, i.e., processes which have no point values.

One can also consider the smaller function spaces $\mathscr{D}_n$ consisting of functions $\varphi(t)$ which are differentiable at least $n$ times and vanish outside a finite interval (it is also stipulated that $\varphi$ and its first $n$ derivatives $\varphi'$, $\varphi''$, ..., $\varphi^{(n)}$ all vanish at the end points of the interval).  Gelfand and Itô, who independently introduced the concept of a generalized random process (G2, I1), followed the work of Schwartz [31] and chose $\mathscr{D}$ to be the space $\mathscr{D}_\infty$ of infinitely differentiable functions which vanish, together with all their derivatives, outside a finite interval (and also at its end points).  Finally, following the book by Gelfand and Shilov [9], one might even choose $\mathscr{D}$ to be a space of entire analytic functions of some class which fall off sufficiently rapidly at infinity (see e.g., [24]).[6]

It should be noted that generalized processes have certain very simple properties.  For example, if the functions $\varphi(t)$ are differentiable, we can always define the *derivative* $\xi'(\varphi)$ of a generalized process $\xi(\varphi)$ by means of the formula

$$\xi'(\varphi) = -\xi(\varphi'). \tag{I.9}$$

It is clear from (I.4) that if our generalized process has point values $\xi(t)$ and if $\xi(t)$ is differentiable in the sense of Sec. 5, then the appropriate definition of $\xi'(\varphi)$ is

$$\xi'(\varphi) \equiv \int_{-\infty}^{\infty} \xi'(t)\varphi(t)\, dt = -\int_{-\infty}^{\infty} \xi(t)\varphi'(t)\, dt,$$

which justifies (I.9).  However, unlike ordinary processes, generalized processes are *always* differentiable.  Therefore an ordinary process $\xi(t)$ may also be regarded as always differentiable, except that its derivative may be a

---

[6] Except for the trivial function which vanishes identically, the functions of $\mathscr{D}_\infty$ are all nonanalytic, since any analytic function vanishing outside an interval must vanish identically.

generalized process instead of an ordinary process. This shows the great convenience of the space $\mathscr{D}_\infty$, since when $\mathscr{D} = \mathscr{D}_\infty$, all generalized processes are *infinitely* differentiable.

It is natural to say that the generalized process $\xi(\varphi)$ is *stationary in the strict sense* if for every $\tau$, the random variable $\xi(T_\tau\varphi)$ has the same distribution as $\xi(\varphi)$. Here, $\varphi(t)$ is an arbitrary element of $\mathscr{D}$, and $T_\tau$ is the *translation* (or *shift*) *operator*, defined by

$$T_\tau\varphi(t) = \varphi(t + \tau).$$

Similarly, we say that the generalized process $\xi(\varphi)$ is *stationary in the wide sense* if it has a mean value functional $m(\varphi)$ and a correlation functional $B(\varphi_1, \varphi_2)$ such that for every $\tau$,

$$m(T_\tau\varphi) = m(\varphi),$$
$$B(T_\tau\varphi_1, T_\tau\varphi_2) = B(\varphi_1, \varphi_2),$$

where $\varphi$, $\varphi_1$ and $\varphi_2$ are arbitrary elements of $\mathscr{D}$ (cf. p. 15). In this connection, the following *spectral representation theorem* (first proved by Gelfand and Itô) is of basic importance: If $\xi(\varphi)$ is a generalized random process which is stationary in the wide sense, with $m(\varphi) = 0$, then $\xi(\varphi)$ can be represented in the form

$$\xi(\varphi) = \int_{-\infty}^{\infty} \tilde{\varphi}(\lambda)Z(d\lambda), \tag{I.10}$$

where

$$\tilde{\varphi}(\lambda) = \int_{-\infty}^{\infty} e^{i\lambda t}\,\varphi(t)\,dt.$$

Here, $Z(\Delta\lambda)$ is a random interval function with uncorrelated increments, i.e., $Z(\Delta\lambda)$ satisfies conditions (a) to (c) on p. 37, and the integral (I.10) is defined in the same way as the integral (2.37).

As in Part 1, we write

$$F(\Delta\lambda) = \mathbf{E}|Z(\Delta\lambda)|^2, \qquad F(\lambda) = F([-\infty, \lambda]),$$

and call $F(\lambda)$ the *spectral distribution function* of the generalized process $\xi(\varphi)$, but now the condition

$$\int_{-\infty}^{\infty} dF(\lambda) < \infty \tag{I.11}$$

need no longer hold [cf. formula (2.64)], although the integral

$$B(\varphi_1, \varphi_2) = \int_{-\infty}^{\infty} \tilde{\varphi}_1(\lambda)\overline{\tilde{\varphi}_2(\lambda)}\,dF(\lambda), \tag{I.12}$$

which defines the correlation functional of the process, always has meaning (as an ordinary improper Stieltjes integral). In the special case where (I.11) holds (and only in this case), the generalized process $\xi(\varphi)$ is given by formula (I.4), where $\xi(t)$ has the spectral representation (2.37), and then formula (I.12) is equivalent to formula (2.63). White noise is a typical generalized process which cannot be represented in the form (I.4) [i.e., which has no point values], since, as already noted, white noise does not satisfy the condition (I.11).

Naturally, the conditions characterizing the class $\mathscr{F}$ of spectral distribution functions $F(\lambda)$ of the processes $\xi(\varphi)$ depend on the choice of the space $\mathscr{D}$. If $\mathscr{D} = \mathscr{D}_\infty$, then $\mathscr{F}$ is the class of nondecreasing functions $F(\lambda)$ such that

$$\int_{-\infty}^{\infty} \frac{dF(\lambda)}{(1 + \lambda^2)^m} < \infty \qquad (I.13)$$

for some integer $m \geqslant 0$. Thus, in this case, $F(\lambda)$ can diverge at infinity, but no faster than some power of $\lambda$. For the derivative $\xi'(\varphi)$ of the process $\xi(\varphi)$, we have to replace $Z(d\lambda)$ by $i\lambda Z(d\lambda)$, and $F(\lambda)$ by

$$F^{[1]}(\lambda) = \int_{-\infty}^{\lambda} \lambda^2 \, dF(\lambda)$$

[cf. formula (2.83)]. It is clear that $F^{[1]}(\lambda)$ will always satisfy (I.13) if $F(\lambda)$ does; this simply corresponds to the fact that the derivative $\xi'(\varphi)$ always exists. In the case where $\mathscr{D} = \mathscr{D}_0$, $\mathscr{F}$ is the class of nondecreasing functions $F(\lambda)$ satisfying a much more stringent condition than (I.12), i.e.,

$$\int_{-\infty}^{\infty} \frac{\sin^2 a\lambda}{\lambda^2} \, dF(\lambda) < \infty$$

for any $a$. Similarly, by choosing $\mathscr{D}$ to be a space of entire analytic functions, one obtains stationary generalized processes with exponentially increasing spectral distribution functions (see [24]).

Using stationary generalized processes, we can give very simple derivations of all the results of Sec. 18 concerning ordinary processes with stationary increments. Moreover, we can go a step further and introduce the concept of a *generalized process with stationary increments*. In fact, since generalized processes are always differentiable for suitable $\mathscr{D}$, a (generalized) process with stationary increments can be defined simply as a process whose derivative is stationary.[7] Taking still another step in this direction, we can define a *(generalized) process with stationary increments of order n* as a process whose $n$th derivative (which is usually a generalized process) is stationary. Since an $n$-fold integral is defined only to within an arbitrary polynomial of degree $n$, from the very outset it is natural to consider a generalized process with stationary increments of order $n$ to be a random functional $\xi(\varphi)$ defined only on the subspace $\mathscr{D}^{(n)}$ of functions $\varphi(t)$ in $\mathscr{D}$ which satisfy the conditions

$$\int_{-\infty}^{\infty} \varphi(t) \, dt = 0, \qquad \int_{-\infty}^{\infty} t\varphi(t) \, dt = 0, \ldots, \qquad \int_{-\infty}^{\infty} t^n \varphi(t) \, dt = 0. \quad (I.14)$$

---

[7] Thus, if we use generalized processes, the theory of random processes with stationary increments is in principle no more complicated than the theory of random *sequences* with stationary increments. See the pertinent remarks on p. 93.

It is not hard to verify that such a process $\xi(\varphi)$, obtained by $n$-fold integration, again has a spectral representation of the form (I.10), but now instead of the condition (I.13), the function $F(\lambda)$ satisfies the more general condition[8]

$$\int_{-\infty}^{-\varepsilon} \frac{dF(\lambda)}{\lambda^{2m}} + \int_{-\varepsilon}^{\varepsilon} \lambda^{2n}\, dF(\lambda) + \int_{\varepsilon}^{\infty} \frac{dF(\lambda)}{\lambda^{2m}} < \infty, \qquad (I.15)$$

where $\varepsilon > 0$ is arbitrary. The convergence of the corresponding integral (I.10) at the point $\lambda = 0$ is guaranteed by the fact that (I.14) implies

$$\tilde{\varphi}(0) = \tilde{\varphi}'(0) = \cdots = \tilde{\varphi}^{(n-1)}(0) = 0.$$

As before, the correlation functional is given by formula (I.12). Thus, in the case of generalized processes with stationary increments of order $n$, $F(\lambda)$ can diverge not only at infinity, but also at the origin. On the other hand, in the case of an ordinary (nongeneralized) process with stationary increments, $F(\lambda)$ can diverge only at the origin, i.e., in this case we have to set $m = 0$ in (I.15) [cf. formula (3.56)]. Moreover, an ordinary process $\xi(t)$ with stationary increments of order $n$ has the spectral representation[9]

$$\xi(t) = \int_{-\infty}^{\infty} \left[ e^{it\lambda} - 1 - it\lambda - \cdots - \frac{(it\lambda)^{n-1}}{(n-1)!} \right] dZ(\lambda) + \xi_0 + \xi_1 t + \cdots + \xi_n t^n$$

which generalizes (3.58), and there is a corresponding generalization of (3.59) for a suitably defined $n$th order structure function.

The results just alluded to, concerning the spectral theory of ordinary processes with stationary increments of order $n$, were first obtained by Yaglom and Pinsker [33] (see also Y3 and [26]). Simple proofs of these results, based on the theory of generalized random processes, were pointed out independently by Gelfand (G2) and Itô (I1). Similarly, one can construct a theory of generalized homogeneous random fields in $k$-dimensional Euclidean space, and a theory of generalized random fields with homogeneous "differences" (increments) of any order $n$ (Y5 and [10]).[10] In fact, all the concepts discussed in Chap. 3 in connection with ordinary random fields (e.g., homogeneous and isotropic random fields, multidimensional random fields, etc.) have their counterparts in the theory of generalized random fields. An excellent general reference on this subject is the book by Gelfand and Vilenkin [10].

---

[8] Here, we assume that $\mathcal{D} = \mathcal{D}_\infty$.

[9] The quantities $\xi_0, \xi_1, \ldots, \xi_n$ are "constant" random variables, as in (3.58).

[10] When $n = 1$, the generalized random field is said to be *locally homogeneous* (cf. p. 93).

# SOME RECENT DEVELOPMENTS[1]

In this Appendix, we indicate some recent work in the theory of stationary random functions. Our task is made simpler by the fact that in the last few years a number of monographs and survey articles have appeared, devoted to various aspects of the subject [3, 11, 15, 16, 25, 30], and the reader interested in details may refer to these sources.

The first important work in prediction theory may be attributed to Szegö, who solved the basic minimum problem, and applied it to the theory of functions which are analytic in the unit circle. Szegö's techniques have recently proved effective in dealing with various statistical problems arising in the theory of stationary sequences, and are reported in [11] and G5. Again, Beurling [2] was led from problems involving isometric linear transformations in Hilbert space to corresponding problems involving spaces of analytic functions. This work has come to be recognized as belonging to the foundations of the theory of stationary sequences, and its techniques have been used in the theory. Therefore, it seemed natural to turn to prediction theory in dealing with problems involving generalizations of the theory of analytic functions of one variable. This was first done in [14] by allowing the time parameter to vary over more general sets than the integers, e.g., the set of points in the plane both of whose coordinates are integers. It was realized

---

[1] This appendix was written by D. B. Lowdenslager. References in "letter-number form" (see Translator's Preface) are to items in the main Bibliography, starting on p. 217. Numerical references (in brackets) are to items in the Supplementary Bibliography, starting on p. 224.

by Bochner [4] that these techniques essentially involve certain types of alge-
bras of functions on a general space. A book by Hoffman [16] covers this
work, starting from elementary facts of analysis. In addition to investigations
along these lines, Bochner [3] has explored the relationship between general
harmonic analysis and the foundations of the theory of stationary processes.

The extrapolation theory of multidimensional (or vector-valued) stationary
random functions has been developed by a number of authors, working independently. Thus, Masani and Wiener [20], Rozanov [29], and Helson and
Lowdenslager [14] have developed a structure theory for multidimensional
stationary sequences which are of *full rank* in the sense that if $\xi_j(n)$ denotes
the $j$th component of the sequence at time $n$, no linear combination

$$\sum_j a_j \xi_j(0) \tag{II.1}$$

is a limit of finite linear combinations of the form

$$\sum_j \sum_{k<0} b_j(k) \xi_j(k),$$

where the $b_j(k)$ are arbitrary complex numbers. Necessary and sufficient
conditions for a sequence to be of full rank are given in each of the papers
cited, and while the methods and proofs differ, the final results are almost
identical. This part of the theory is summarized in [15] and [30]. The
basic fact is that the mean square error is given in terms of a positive definite
matrix whose determinant may be simply computed from the spectrum.[2]
The problem of computing the actual matrix itself is unsolved, and seems to
be much harder. One of the basic difficulties is that the analytic functions
involved in these problems are *matrix-valued*, and hence may not commute;
this fact spoils many computations which are very simple in the scalar case.
The facts about processes can be deduced from the corresponding facts
about sequences by means of the device used by Akhiezer (A1, App. B)
and Doob (D6, Chap. 12, Sec. 5). The decomposition of the process or
sequence into regular and singular parts, due originally to Wold (cf. W7), and
analogous to that given by equation (8.6) on p. 186, is easily described in
terms of the spectrum. A number of somewhat different problems for
processes have been examined by Cramér [6].

The case where the random sequence is not of full rank offers a number of
complications, new to the multidimensional case. In the first place, the
extent to which "perfect prediction" is possible varies, and depends on
how many independent linear combinations (II.1) can be perfectly approximated by linear combinations of the $\xi_j(n)$ for $n < 0$. The *Wold decomposition* into mutually uncorrelated regular and singular parts (cf. Sec. 36) is
imperfectly understood in terms of the spectrum. Masani and Wiener [21],
Matveyev [22, 23], and Helson and Lowdenslager [14, 15] have developed

---

[2] I.e., the spectral distribution matrix (see p. 80).

structure theories and methods of attacking these problems, but the connection between the various methods is not clear.

The work of Beurling mentioned above has been taken up by Lax [19] and Halmos [13]. The scope of their method is wider than the problems actually dealt with, and lies at the heart of the Wold decomposition. Potapov [27] has investigated factorizations of matrix-valued functions analytic in the unit circle, generalizing the factorizations used by Beurling in his work. Devinatz [8] has investigated the factorization problem central to the discussion of sequences of full rank, in the case where the matrices are replaced by bounded linear operators in a Hilbert space.

The computational difficulties stemming from non-commutativity of matrix-valued functions make it considerably harder to solve concrete problems of multidimensional time series analysis. Masani and Wiener [20] have discussed approximation techniques for solving prediction problems. Yaglom (Y6) has examined the case where the spectrum is given by a rational matrix-valued function. The difficulties just mentioned have also hindered the solution of problems related to prediction, e.g., interpolation and filtering.

Finally, we mention certain interesting developments in the theory of one-dimensional stationary random processes. Chover [5] has studied the case where the best linear prediction of a process $\xi(t)$, in terms of its values on an interval $[a, b]$, can be represented in the form

$$\xi(t) = \int_a^b \xi(s) \, dF(s),$$

where $F$ is a function of bounded variation. This is the case in Example 1, p. 153, but not in Example 2, p. 155. In Example 1, the best linear prediction of $\xi(t)$ turns out to be a multiple of $\xi(b)$, and it has been shown by Hájek [12] that this is a reasonably good prediction whenever the correlation function is convex. Karhunen [17] has studied the problem of interpolating a stationary random process in an interval $a \leqslant t \leqslant b$, in terms of its values outside this interval.[3] One of the most challenging problems is that of determining the explicit form of the prediction formula when it is known that perfect prediction is possible. Rosenblatt [28] has solved this problem for a special case, using methods developed by Szegö (alluded to earlier). Belyaev [1] has proved that when the correlation function of a separable process is entire, so are the sample functions, with probability 1. This is the case in Examples 2.2 to 2.4 of Sec. 36, as already noted. Belyaev's paper also contains a rigorous discussion of the sampling theorem discussed on p. 204. Shapiro and Silverman [32] have examined the relationship between a stationary random process $\xi(t)$ and the random sequences obtained by *random* sampling of $\xi(t)$, and have shown under what conditions the spectrum of the sampled sequence uniquely determines that of the underlying process.

---

[3] This problem has also been studied by Yaglom (Y4).

# BIBLIOGRAPHY[1]

A1 Akhiezer, N. I., *Theory of Approximation*, translated by C. J. Hyman, Frederick Ungar Publishing Co., New York (1956).

A2 Akhiezer, N. I. and I. M. Glazman, *Theory of Linear Operators in Hilbert Space*, *Vol. 1*, translated by M. Nestell, Frederick Ungar Publishing Co., New York (1961).

B1 Bachelier, L., *Calcul des Probabilités*, Gauthier-Villars, Paris (1912).

B2 Bartlett, M. S., *An Introduction to Stochastic Processes*, Cambridge University Press, New York (1955).

B3 Batchelor, G. K., *The Theory of Homogeneous Turbulence*, Cambridge University Press, New York (1953).

B4 Blanc-Lapierre, A., *Sur certaines Fonctions aléatoires stationnaires; Applications à l'Étude des Fluctuations dues à la Structure électronique de l'Électricité*, Masson et Cie., Paris (1945).

B5 Blanc-Lapierre, A. and R. Brard, *Les fonctions aléatoires et la loi des grands nombres*, Bull. Soc. Math. France, **74**, 102 (1946).

B6 Blanc-Lapierre, A. and R. Fortet, *Sur une propriété fondamentale des fonctions de corrélation*, C. R. Acad. Sci., **224**, 786 (1947).

B7 Blanc-Lapierre, A. and R. Fortet, *Les fonctions aléatoires de plusieurs variables*, Revue Sci., **85**, 419 (1947).

B8 Blanc-Lapierre, A. and R. Fortet, *Théorie des Fonctions aléatoires*, Masson et Cie., Paris (1953).

B9 Bochner, S., *Monotone Funktionen, Stieltjes Integrale und harmonische Analyse*, Math. Ann., **108**, 378 (1933). English translation as supplement to S. Bochner: *Lectures on Fourier Integrals*, translated by M. Tenenbaum and H. Pollard, Princeton University Press, Princeton, N.J. (1959).

---

[1] In the interest of simplicity, the titles of all Russian books, and of all papers printed in Russian journals, are given in English. (In the case of papers, the original language is not always Russian, and poor English has occasionally been corrected.) (*Translator.*)

B10 Bode, H. W. and C. E. Shannon, *A simplified derivation of linear least square smoothing and prediction theory*, Proc. IRE, **38**, 417 (1950).

B11 Brooks, F. E. and H. W. Smith, *A computer for correlation functions*, Rev. Sci. Instrum., **23**, 121 (1952).

B12 Bunimovich, V. I., *The fluctuation process as an oscillation with random amplitude and phase*, J. Tekh. Fiz., **19**, 1231 (1949).

B13 Bunimovich, V. I., *Fluctuation Processes in Radio Receivers*, Sovietskoye Radio, Moscow (1951).

B14 Bunimovich, V. I. and M. A. Leontovich, *On the distribution of the number of large deviations in electrical fluctuations*, Dokl. Akad. Nauk SSSR, **53**, 21 (1946).

C1 Cauer, W., *Synthesis of Linear Communication Networks*, Vols. *1 and 2*, second edition, edited by W. Klein and F. M. Pelz, translated by G. E. Knausenberger and J. N. Warfield, McGraw-Hill Book Co., Inc., New York (1958).

C2 Chandrasekhar, S., *Stochastic problems in physics and astronomy*, Rev. Mod. Phys., **15**, 1 (1943). Reprinted in *Selected Papers on Noise and Stochastic Processes*, edited by N. Wax, Dover Publications, Inc., New York (1954), p. 3.

C3 Cramér, H., *On the theory of stationary random processes*, Ann. of Math., **41**, 215 (1940).

C4 Cramér, H., *On harmonic analysis in certain functional spaces*, Ark. Mat. Astr. Fys., **28B**, no. 12 (1942).

C5 Cramér, H., *Mathematical Methods of Statistics*, Princeton University Press, Princeton, N.J. (1946).

C6 Cunningham, L. B. C. and W. R. B. Hynd, *Random processes in problems of air warfare*, Suppl. J. Roy. Stat. Soc., **82**, 62 (1946).

D1 Davenport, W. B., Jr., R. A. Johnson and D. Middleton, *Statistical errors in measurements on random time functions*, J. Appl. Phys., **23**, 377 (1952).

D2 Davenport, W. B., Jr. and W. L. Root, *An Introduction to the Theory of Random Signals and Noise*, McGraw-Hill Book Co., Inc., New York (1958).

D3 Doob, J. L., *The law of large numbers for continuous stochastic processes*, Duke Math. J., **6**, 290 (1940).

D4 Doob, J. L., *The elementary Gaussian process*, Ann. Math. Stat., **15**, 229 (1944).

D5 Doob, J. L., *Time series and harmonic analysis*, Proceedings of the Berkeley Symposium on Mathematical Statistics and Probability, University of California Press, Berkeley and Los Angeles (1949), p. 303.

D6 Doob, J. L., *Stochastic Processes*, John Wiley and Sons, Inc., New York (1953).

E1 Einstein, A., *Investigations on the Theory of Brownian Movement*, edited by R. Fürth, translated by A. D. Cowper, Dover Publications, Inc., New York (1956).

F1 Fan, K., *Les Fonctions définies-positives et les Fonctions complètement monotones*, Mémorial des Sciences mathématiques, fascicule 114, Gauthier-Villars, Paris (1950).

F2 Foster, G. A. R., *Some instruments for the analysis of time series and their application to textile research*, Suppl. J. Roy. Stat. Soc., **8**, 42 (1946).

G1 Gelfand, I. M., *Spherical functions on symmetric Riemannian spaces*, Dokl. Akad. Nauk SSSR, **70**, 5 (1950).

G2 Gelfand, I. M., *Generalized random processes*, Dokl. Akad. Nauk SSSR, **100**, 853 (1955).

G3 Gnedenko, B. V., *The Theory of Probability*, translated by B. D. Seckler, Chelsea Publishing Co., New York (1962).

G4 Grenander, U., *Stochastic processes and statistical inference*, Ark. Mat., **1**, 195 (1950).

G5 Grenander, U. and M. Rosenblatt, *Statistical Analysis of Stationary Time Series*, John Wiley and Sons, Inc., New York (1957).

H1 Hald, A., *The Decomposition of a series of Observations Composed of a Trend, a Periodic Movement and a Stochastic Variable*, Thesis, University of Copenhagen (1948).

H2 Hanner, O., *Deterministic and non-deterministic stationary random processes*, Ark. Mat., **1**, 261 (1950).

I1 Itô, K., *Stationary random distributions*, Mem. Coll. Sci. Univ. Kyoto, Ser. A, **28**, 211 (1954).

J1 Jackson, L. C., *Wave Filters*, third edition, John Wiley and Sons, Inc., New York (1950).

J2 James, H. M., N. B. Nichols and R. S. Phillips, *Theory of Servomechanisms*, vol. 25 of the Massachusetts Institute of Technology Radiation Laboratory Series, McGraw-Hill Book Co., Inc., New York (1947).

K1 Kampé de Fériet, J., *Analyse harmonique des fonctions aléatoires d'ordre 2 définies sur un groupe abélien localement compact*, C. R. Acad. Sci., **226**, 868 (1948).

K2 Karhunen, K., *Über lineare Methoden in der Wahrscheinlichkeitsrechnung*, Ann. Acad. Sci. Fennicae, Ser. A. I. Math.-Phys., no. 37 (1947).

K3 Karhunen, K., *Über die Struktur stationärer zufälliger Funktionen*, Ark. Mat. **1**, 141 (1950).

K4 Keller, L., *Die Periodographie als statistisches Problem*, Beitr. Phys. frei. Atmosph., **19**, 173 (1932).

K5 Kerchner, R. M. and G. F. Corcoran, *Alternating-Current Circuits*, fourth edition, John Wiley and Sons, Inc., New York (1960).

K6 Khinchin, A. Y., *Korrelationstheorie der stationären stochastischen Prozesse*, Math. Ann., **109**, 604 (1934).

K7 Khinchin, A. Y., *The theory of damped spontaneous effects*, Izv. Akad. Nauk SSSR, Ser. Mat., **3**, 313 (1938).

K8 Kolmogorov, A. N., *Über die analytischen Methoden in der Wahrscheinlichkeitsrechnung*, Math. Ann., **104**, 415 (1931).

K9 Kolmogorov, A. N., *A simplified proof of the Birkhoff-Khinchin ergodic theorem*, Usp. Mat. Nauk, no. 5, 52 (1938).

K10 Kolmogorov, A. N., *Curves in Hilbert space which are invariant with respect to a one-parameter group of motions*, Dokl. Akad. Nauk SSSR, **26**, 6 (1940).

K11  Kolmogorov, A. N., *Wiener spirals and some other interesting curves in Hilbert space*, Dokl. Akad. Nauk SSSR, **26**, 115 (1940).

K12  Kolmogorov, A. N., *Stationary sequences in Hilbert space*, Bul. Moscow State Univ., **2**, no. 6 (1941).

K13  Kolmogorov, A. N., *Interpolation and extrapolation of stationary random sequences*, Izv. Akad. Nauk SSSR, Ser. Mat., **5**, 3 (1941).

K14  Kolmogorov, A. N., *The local structure of turbulence in incompressible viscous fluid for very large Reynolds numbers*, Dokl. Akad. Nauk SSSR, **30**, 301 (1941). Reprinted in *Turbulence: Classic Papers on Statistical Theory*, edited by S. K. Friedlander and L. Topping, Interscience Publishers, Inc., New York (1961), p. 151.

K15  Kolmogorov, A. N., *Dissipation of energy in locally isotropic turbulence*, Dokl. Akad. Nauk SSSR, **32**, 16 (1941). Reprinted in collection cited in Ref. K14, p. 159.

K16  Kolmogorov, A. N., *Statistical theory of oscillations with a continuous spectrum*, published in *Jubilee Collection, Part 1*, Izd. Akad. Nauk SSSR, Moscow (1947), p. 242.

K17  Kolmogorov, A. N., *Foundations of the Theory of Probability*, second English edition, translation edited by N. Morrison, with an added bibliography by A. T. Bharucha-Reid, Chelsea Publishing Co., New York (1956).

K18  Kozulyaev, P. A., *On problems of interpolation and extrapolation of stationary sequences*, Dokl. Akad. Nauk SSSR, **30**, 13 (1941).

K19  Kozulyaev, P. A., *On the theory of extrapolation of stationary sequences*, Uch. Zap. Moscow State Univ., **3**, no. 146, 59 (1950).

K20  Krechmer, S. I., *Investigation of microfluctuations of the temperature field in the atmosphere*, Dokl. Akad. Nauk SSSR, **84**, 55 (1952).

K21  Krein, M. G., *On the problem of continuing Hermitian-positive continuous functions*, Dokl. Akad. Nauk SSSR, **26**, 17 (1940).

K22  Krein, M. G., *On the problem of continuing helical arcs in Hilbert space*, Dokl. Akad. Nauk SSSR, **45**, 147 (1944).

K23  Krein, M. G., *On a generalization of some investigations by G. Szegö, V. I. Smirnov and A. N. Kolmogorov*, Dokl. Akad. Nauk SSSR (N.S.), **46**, 91 (1945).

K24  Krein, M. G., *On an extrapolation problem of A. N. Kolmogorov*, Dokl. Akad. Nauk SSSR (N.S.), **46**, 306 (1945).

K25  Krein, M. G., *Hermitian-positive kernels in homogeneous spaces*, Ukrain. Mat. Zh., **1**, no. 4, 64 (1949); **2**, no. 1, 10 (1950).

L1  Landau, L. D. and E. M. Lifshitz, *Statistical Physics*, translated by E. Peierls and R. F. Peierls, Addison-Wesley Publishing Co., Inc., Reading, Mass. (1958).

L2  Laning, J. H., Jr. and R. H. Battin, *Random Processes in Automatic Control*, McGraw-Hill Book Co., Inc., New York (1956).

L3  Lawson, J. L. and G. E. Uhlenbeck, *Threshold Signals*, vol. 24 of the Massachusetts Institute of Technology Radiation Laboratory Series, McGraw-Hill Book Co., Inc., New York (1950).

L4  Leontovich, M. A., *Statistical Physics*, Gos Izd. Tekh.-Teor. Lit., Moscow (1944).

L5 Lévy, P., *Processus stochastiques et Mouvement brownien*, with a supplement by M. Loève, Gauthier-Villars, Paris (1948).

L6 Lévy, P., *Le Mouvement brownien*, Mémorial des Sciences mathématiques, fascicule 126, Gauthier-Villars, Paris (1954).

L7 Lloyd, S. P., *A sampling theorem for stationary (wide sense) stochastic processes*, Trans. Amer. Math. Soc., **92**, 1 (1959).

L8 Loève, M., *Probability Theory*, second edition, D. Van Nostrand Co., Inc., Princeton, N. J. (1960).

M1 Maruyama, G., *The harmonic analysis of stationary stochastic processes*, Mem. Fac. Sci. Kyushu Univ. A, **4**, 45 (1949).

M2 Middleton, D., *An Introduction to Statistical Communication Theory*, McGraw-Hill Book Co., Inc., New York (1960).

M3 Moran, P. A. P., *The spectral theory of discrete stochastic processes*, Biometrica, **36**, 63 (1949).

M4 Moran, P. A. P., *The oscillatory behaviour of moving averages*, Proc. Camb., Phil. Soc., **46**, 272 (1950).

N1 Nagabhushanam, K., *The primary process of a smoothing relation*, Ark. Mat., **1**, 421 (1951).

O1 Obukhov, A. M., *On the distribution of energy in the spectrum of turbulent flow*, Izv. Akad. Nauk SSSR, Ser. Geograf. i Geofiz., nos. 4–5, 453 (1941).   German translation in *Sammelband zur Statistischen Theorie der Turbulenz*, Akademie-Verlag, Berlin (1958), p. 83.

O2 Obukhov, A. M., *Characteristics of the microstructure of the wind in the layer of the atmosphere near the earth*, Izv. Akad. Nauk SSSR, Ser. Geofiz., no. 3, 49 (1951).   German translation in collection cited in Ref. O1, p. 173.

O3 Obukhov, A. M., *Statistical description of continuous fields*, Trudy Geofiz. Inst. Akad. Nauk SSSR, **24**, 3 (1954).   German translation in collection cited in Ref. O1, p. 1.

O4 Obukhov, A. M. and A. M. Yaglom, *The microstructure of turbulent flow*, Prikl. Mat. Mekh., **15**, 3 (1951).   English translation as Tech. Memo. 1350, National Advisory Committee for Aeronautics, Washington (1953).   German translation in collection cited in Ref. O1, p. 97.

O5 Olson, H. F., *Dynamical Analogies*, second edition, D. Van Nostrand Co., Inc., Princeton, N. J. (1958).

R1 Raikov, D. A., *Harmonic analysis on commutative groups with Haar measure, and the theory of characters*, Trudy Mat. Inst. Steklov, no. 14 (1945).

R2 Rice, S. O., *Mathematical analysis of random noise*, Bell System Tech. J., **23**, 282 (1944); **24**, 46 (1945).   Reprinted in collection cited in Ref. C2, p. 133.

R3 Romanovski, V. I., *Sur la loi sinusoïdale limite*, Rend. Circ. Mat. Palermo, **56**, 82 (1932).

R4 Romanovski, V. I., *Sur une généralization de la loi sinusoïdale limite*, Rend. Circ. Mat. Palermo, **57**, 130 (1933).

R5 Rosenblatt, M., *Random Processes*, Oxford University Press, New York (1962).

S1 Schoenberg, I. J., *Metric spaces and completely monotone functions*, Ann. of Math., **39**, 811 (1938).

S2 Schoenberg, I. J., *Positive definite functions on spheres*, Duke Math. J., **9**, 96 (1942).

S3 Seiwell, H. R., *A new mechanical autocorrelator*, Rev. Sci. Instrum., **21**, 481 (1950).

S4 Silverman, R. A., *A matching theorem for locally stationary random processes*, Comm. Pure Appl. Math., **12**, 373 (1959).

S5 Silverman, R. A., *A note on the local structure of shot noise*, IRE Trans. Information Theory, **IT-6**, 548 (1960).

S6 Singleton, H. E., *A digital electronic correlator*, Proc. IRE, **38**, 1422 (1950).

S7 Slutski, E. E., *Sur un théorème limite relatif aux séries des quantités éventuelles*, C. R. Acad. Sci., **185**, 169 (1927).

S8 Slutski, E. E., *The summation of random causes as the source of cyclic processes*, Econometrica, **5**, 105 (1937).

S9 Slutski, E. E., *Sur les fonctions aléatoires presque periodiques et sur la décomposition des fonctions aléatoires stationnaires en composantes*, Actualités scientifiques et industrielles, no. 738, Hermann et Cie., Paris (1938), p. 33.

S10 Slutski, E. E., *Selected Works: Probability Theory, Mathematical Statistics*, Izv. Akad. Nauk SSSR, Moscow (1960).

S11 Solodovnikov, V. V., *Analysis and synthesis of servomechanisms and regulating systems subject to stationary random disturbances*, Izv. Akad. Nauk SSSR, Otd. Tekh. Nauk, **11**, 1648 (1950).

S12 Solodovnikov, V. V., *Introduction to the Statistical Dynamics of Automatic Control Systems*, translation edited by J. B. Thomas and L. A. Zadeh, Dover Publications, Inc., New York (1960).

T1 Tatarski, V. I., *Wave Propagation in a Turbulent Medium*, translated by R. A. Silverman, McGraw-Hill Book Co., Inc., New York (1961).

T2 Taylor, G. I., *The spectrum of turbulence*, Proc. Roy. Soc., **A164**, 476 (1938). Reprinted in collection cited in Ref. K14, p. 100.

T3 Titchmarsh, E. C., *Introduction to the Theory of Fourier Integrals*, second edition, Oxford University Press, New York (1948).

V1 Van der Ziel, A., *Noise*, Prentice-Hall, Inc., Englewood Cliffs, N. J. (1954).

V2 Von Neumann, J. and I. J. Schoenberg, *Fourier integrals and metric geometry*, Trans. Amer. Math. Soc., **50**, 226 (1941).

V3 Von Smoluchowski, M., *Abhandlungen über die Brownsche Bewegung und verwandte Erscheinungen*, Ostwalds Klassiker der exacten Wissenschaften. no. 207, Leipzig (1923).

W1 Wainstein, L. A. and V. D. Zubakov, *Extraction of Signals from Noise*, translated by R. A. Silverman, Prentice-Hall, Inc., Englewood Cliffs, N. J. (1962).

W2 Wang, M. C. and G. E. Uhlenbeck, *On the theory of the Brownian motion II*. Rev. Mod. Phys., **17**, 323 (1949). Reprinted in collection cited in Ref. C2, p. 113.

W3 Weil, A., *L'Intégration dans les Groupes topologiques et ses Applications*, Actualités scientifiques et industrielles, no. 869, Hermann et Cie., Paris (1940).

W4 Wiener, N., *Extrapolation, Interpolation, and Smoothing of Stationary Time Series*, MIT Technology Press and John Wiley and Sons, Inc., New York (1950).

W5 Wilks, S. S., *Mathematical Statistics*, Princeton University Press, Princeton, N. J. (1944).

W6 Wold, H., *On prediction in stationary time series*, Ann. Math. Stat., **19**, 558 (1948).

W7 Wold, H., *A Study in the Analysis of Stationary Time Series*, with an appendix by P. Whittle, second edition, Almqvist and Wiksell, Stockholm (1954).

Y1 Yaglom, A. M., *Homogeneous and isotropic turbulence in a viscous compressible fluid*, Izv. Akad. Nauk SSSR, Ser. Geograf. i Geofiz., **12**, 501 (1948). German translation in collection cited in Ref. O1, p. 43.

Y2 Yaglom, A. M., *On the problem of linear interpolation of stationary random sequences and processes*, Usp. Mat. Nauk, **4**, no. 4, 173 (1949).

Y3 Yaglom, A. M., *Correlation theory of processes with stationary random increments of order n*, Mat. Sb., **37**, 141 (1955). English translation in *American Mathematical Society Translations, Series 2, Vol. 8*, American Mathematical Society, Providence, R.I. (1958), p. 87.

Y4 Yaglom, A. M., *Extrapolation, interpolation and filtering of stationary random processes with rational spectral densities*, Trudy Mosk. Mat. Ob., **4**, 333 (1955).

Y5 Yaglom, A. M., *Some classes of random fields in n-dimensional space, related to stationary random processes*, Theory Prob. and Its Appl., English edition, **2**, 273 (1957).

Y6 Yaglom, A. M., *Effective solutions of linear approximation problems for multivariate stationary processes with a rational spectrum*, Theory Prob. and Its Appl., English edition, **5**, 239 (1960).

Y7 Yaglom, A. M., *Second-order homogeneous random fields*, Proceedings of the Fourth Berkeley Symposium on Mathematical Statistics and Probability, University of California Press, Berkeley and Los Angeles (1961), vol. 2, p. 593.

Y8 Yaglom, A. M., *Examples of optimum nonlinear extrapolation of stationary random processes*, Proceedings of the Sixth All-Union Conference on Probability Theory and Mathematical Statistics, Vilnius (1962), p. 273.

Z1 Zadeh, L. A. and J. R. Ragazzini, *An extension of Wiener's theory of prediction*, J. Appl. Phys., **21**, 645 (1950).

Z2 Zasukhin, V. N., *On the theory of multidimensional random processes*, Dokl. Akad. Nauk SSSR, **33**, 435 (1941).

# SUPPLEMENTARY
# BIBLIOGRAPHY

1 Belyaev, Y. K., *Analytic random processes*, Theory Prob. and Its Appl., English edition, **4**, 402 (1959).

2 Beurling, A., *On two problems concerning linear transformations in Hilbert space*, Acta Math., **81**, 239 (1949).

3 Bochner, S., *Harmonic Analysis and the Theory of Probability*, University of California Press, Berkeley and Los Angeles (1955).

4 Bochner, S., *Generalized conjugate and analytic functions without expansions*, Proc. Nat. Acad. Sci., **45**, 855 (1959).

5 Chover, J., *Conditions on the realization of prediction by measures*, Duke Math. J., **25**, 305 (1958).

6 Cramér, H., *On the structure of purely non-deterministic stochastic processes*, Ark. Mat., **4**, 249 (1960).

7 Cramér, H., *Random Variables and Probability Distributions*, second edition, Cambridge University Press, New York (1961).

8 Devinatz, A., *The factorization of operator valued functions*, Ann. of Math., **73**, 458 (1961).

9 Gelfand, I. M. and G. E. Shilov, *Generalized Functions, Vol. 2: Spaces of Basic and Generalized Functions*, Gos. Izd. Fiz.-Mat. Lit., Moscow (1958).

10 Gelfand, I. M. and N. Y. Vilenkin, *Generalized Functions, Vol. 4: Some Applications of Harmonic Analysis. Rigged Hilbert Spaces*, Gos. Izd. Fiz.-Mat. Lit., Moscow (1961).

11 Grenander, U. and G. Szegö, *Toeplitz Forms and Their Applications*, University of California Press, Berkeley and Los Angeles (1958).

12 Hájek, J., *Predicting a stationary process when the correlation function is convex*, Czech. Math. J., **8**, 150 (1958).

13 Halmos, P., *Shifts on Hilbert spaces*, J. Reine Angew. Math., **208**, 102 (1961).

14 Helson, H. and D. Lowdenslager, *Prediction theory and Fourier series in several variables, Part I*, Acta Math., **99**, 165 (1959); *Part II*, ibid., **106**, 175 (1962).

15 Helson, H. and D. Lowdenslager, *Vector-valued processes*, Proceedings of the Fourth Berkeley Symposium on Mathematical Statistics and Probability, University of California Press, Berkeley and Los Angeles (1961), p. 203.

16 Hoffman, K., *Banach Spaces of Analytic Functions*, Prentice-Hall, Inc., Englewood Cliffs, N.J. (1962).

17 Karhunen, K., *Zur Interpolation von stationären zufälligen Funktionen*, Ann. Acad. Sci. Fennicae, Ser. A. I. Math.-Phys., no. 142 (1952).

18 Khachaturov, A. A., *Determination of the values of a measure for a region in n-dimensional Euclidean space from its values for all half-spaces*, Usp. Mat. Nauk, vol. 9, no. 3, 205 (1954).

19 Lax, P., *Translation invariant subspaces*, Proceedings of the International Symposium on Linear Spaces, Jerusalem Academic Press and Pergamon Press, London (1961), p. 299.

20 Masani, P. and N. Wiener, *Prediction theory of multivariate stochastic processes, Part I, The regularity condition*, Acta Math., **98**, 111 (1957); *Part II, The linear predictor*, ibid., **99**, 93 (1958).

21 Masani, P. and N. Wiener, *On bivariate stationary processes and the factorization of matrix-valued functions*, Theory Prob. and Its Appl., English edition, **4**, 38 (1959).

22 Matveyev, R., *On singular multidimensional stationary processes*, Theory Prob. and its Appl., English edition, **5**, 33 (1960).

23 Matveyev, R., *On regular multidimensional stationary processes*, Teor. Veroyatnost. i Primenen., **6**, 164 (1961).

24 Onoyama, T., *Note on random distributions*, Mem. Fac. Sci. Kyushu Univ., Ser. A, **13**, 208 (1959).

25 Parzen, E., *An approach to time series analysis*, Ann. Math. Stat., **32**, 951 (1961).

26 Pinsker, M. S., *Theory of curves in Hilbert space with stationary increments of order n*, Izv. Akad. Nauk SSSR, Ser. Mat., **19**, 319 (1955).

27 Potapov, V. P., *The multiplicative structure of J-contractive matrix functions*, Trudy Mosk. Mat. Ob., **4**, 125 (1955). English translation in *American Mathematical Society Translations, Series 2, Vol. 15*, American Mathematical Society, Providence, R.I. (1960), p. 131.

28 Rosenblatt, M., *Some purely deterministic processes*, J. Math. Mech., **6**, 801, (1957).

29 Rozanov, Y. A., *Spectral theory of multidimensional stationary stochastic processes with discrete time*, Usp. Mat. Nauk, vol. 13, no. 2, 93 (1958). English translation in *Selected Translations in Mathematical Statistics and Probability, Vol. I*, American Mathematical Society, Providence, R.I. (1961), p. 253.

30 Rozanov, Y. A., *Spectral properties of multivariate stationary processes and boundary properties of analytic matrices*, Theory Prob. and Its Appl., English edition, **5**, 362 (1960).

31 Schwartz, L., *Théorie des Distributions*, second printing, Actualités scientifiques et industrielles Nos. 1245 and 1122, Hermann et Cie., Paris, vol. 1 (1957), vol. 2 (1959).

32 Shapiro, H. S. and R. A. Silverman, *Alias-free sampling of random noise*, J. Soc Indust. Appl. Math., **8**, 225 (1960).

33 Yaglom, A. M. and M. S. Pinsker, *Random processes with stationary increments of order n*, Dokl. Akad. Nauk SSSR, **90**, 731 (1953).

# NAME INDEX

## A

Akhiezer, N. I., 27, 104, 185, 189, 215, 217

## B

Bachelier, L., 4, 217
Bartlett, M. S., 39, 217
Batchelor, G. K., 16, 84, 93, 217
Battin, R. H., 5, 220
Belyaev, Y. K., 216, 224
Beurling, A., 214, 216, 224
Bharucha-Reid, A. T., 220
Birkhoff, G. D., 22
Blanc-Lapierre, A., 2, 22, 39, 46, 50, 84, 217
Bochner, S., 47; 84, 214, 215, 217, 224
Bode, H. W., 5, 218
Brard, R., 22, 217
Brooks, F. E., 19, 218
Bunimovich, V. I., 2, 16, 20, 42, 73, 218

## C

Cauer, W., 42, 218
Chandrasekhar, S., 2, 218
Chover, J., 216, 224
Corcoran, G. F., 42, 219
Cowper, A. D., 218
Cramér, H., 15, 16, 39, 80, 100, 209, 215, 218, 224
Cunningham, L. B. C., 205, 218

## D

Davenport, W. B., Jr., 2, 5, 19, 42, 218
Devinatz, A., 216, 224

## Doob

Doob, J. L., 6, 11, 20, 22, 39, 71, 76, 90, 92, 113, 116, 121, 122, 154, 162, 182, 185, 187, 189, 190, 191, 193, 215, 218

## E

Einstein, A., 2, 70, 218

## F

Fan, K., 85, 86, 218
Fortet, R., 39, 46, 50, 84, 217
Foster, G. A. R., 3, 218
Friedlander, S. K., 220
Fürth, R., 218

## G

Gelfand, I. M., 86, 93, 210, 211, 213, 219, 224
Glazman, I. M., 27, 104, 217
Gnedenko, B. V., 6, 15, 16, 20, 22, 24, 39, 44, 47, 219
Grenander, U., 21, 39, 214, 219, 224

## H

Hájek, J., 216, 224
Hald, A., 205, 219
Halmos, P., 216, 224
Hanner, O., 189, 219
Helson, H., 214, 215, 224
Hoffman, K., 214, 215, 224
Hyman, C. J., 217
Hynd, W. R. B., 205, 218

## I

Itô, K., 93, 210, 211, 213, 219

# SUBJECT INDEX

## A

Approximate differentiation, 195–197
Autocorrelation function, 14, 78
Averaging operator, 13, 22

## B

Bochner theorem, 47, 50–51, 56
  general form of, 84
  multidimensional, 83–84
Brownian motion, 2, 63–65, 69–70, 86, 155
  of a harmonic oscillator, 75, 92–93, 157

## C

Central limit theorem, 16
Chebyshev's inequality, 18
Compatibility condition, 10, 209
Continuity in the mean, 22, 27
Convergence:
  almost sure, 22
  in the mean (square), 17, 27
  in probability, 18
Correlation coefficient, 20
Correlation functions, 14 ff.
  centered, 20
  of complex stationary random functions, 25
  entire, 191, 216
  normalized, 19, 20
  properties of, 22–24
  spectral representation of, 21, 43–51, 55
Correlation functional, 209
Correlation matrix, 79
Correlation theory, 4, 14
Correlation time, 62, 175

## Covariance function, 14
  stationary, 14
Cross-correlation functions, 78, 79
Cross-spectral densities, 81
Cross-spectral distribution function, 80

## D

Damped oscillations, 71, 76
Decay time, 175
Distribution functions, 1, 9–10
  compatibility condition for, 10
  multidimensional, 1
  symmetry condition for, 10

## E

Einstein-Smoluchowski theory, 70
Elementary events, 10
Ergodic theorem, 16–22
Extrapolation (prediction), 4, 27, 97 ff.
  formula, 97
  general theory of, 182–195
  linear vs. nonlinear, 99–100, 144
  mean square error of, 4, 98, 145
  problem, 97
  related problems, 195–206
  of stationary random processes, 5, 144–166, 189–192
    complex variable formulation of, 150–153
    examples of, 153–161
    multidimensional case, 195, 215
    statement of problem for, 144–150
    with general rational spectral densities, 161–166
  of stationary random sequences, 97–125, 183–189
    examples of, 110–121, 187–189

# A CATALOGUE OF SELECTED DOVER BOOKS
## IN ALL FIELDS OF INTEREST

# A CATALOGUE OF SELECTED DOVER
# BOOKS IN ALL FIELDS OF INTEREST

CONDITIONED REFLEXES, Ivan P. Pavlov. Full translation of most complete statement of Pavlov's work; cerebral damage, conditioned reflex, experiments with dogs, sleep, similar topics of great importance. 430pp. 5⅜ x 8½. 60614-7 Pa. $4.50

NOTES ON NURSING: WHAT IT IS, AND WHAT IT IS NOT, Florence Nightingale. Outspoken writings by founder of modern nursing. When first published (1860) it played an important role in much needed revolution in nursing. Still stimulating. 140pp. 5⅜ x 8½. 22340-X Pa. $3.00

HARTER'S PICTURE ARCHIVE FOR COLLAGE AND ILLUSTRATION, Jim Harter. Over 300 authentic, rare 19th-century engravings selected by noted collagist for artists, designers, decoupeurs, etc. Machines, people, animals, etc., printed one side of page. 25 scene plates for backgrounds. 6 collages by Harter, Satty, Singer, Evans. Introduction. 192pp. 8⅞ x 11¾. 23659-5 Pa. $5.00

MANUAL OF TRADITIONAL WOOD CARVING, edited by Paul N. Hasluck. Possibly the best book in English on the craft of wood carving. Practical instructions, along with 1,146 working drawings and photographic illustrations. Formerly titled *Cassell's Wood Carving*. 576pp. 6½ x 9¼. 23489-4 Pa. $7.95

THE PRINCIPLES AND PRACTICE OF HAND OR SIMPLE TURNING, John Jacob Holtzapffel. Full coverage of basic lathe techniques—history and development, special apparatus, softwood turning, hardwood turning, metal turning. Many projects—billiard ball, works formed within a sphere, egg cups, ash trays, vases, jardiniers, others—included. 1881 edition. 800 illustrations. 592pp. 6⅛ x 9¼. 23365-0 Clothbd. $15.00

THE JOY OF HANDWEAVING, Osma Tod. Only book you need for hand weaving. Fundamentals, threads, weaves, plus numerous projects for small board-loom, two-harness, tapestry, laid-in, four-harness weaving and more. Over 160 illustrations. 2nd revised edition. 352pp. 6½ x 9¼. 23458-4 Pa. $6.00

THE BOOK OF WOOD CARVING, Charles Marshall Sayers. Still finest book for beginning student in. wood sculpture. Noted teacher, craftsman discusses fundamentals, technique; gives 34 designs, over 34 projects for panels, bookends, mirrors, etc. "Absolutely first-rate"—E. J. Tangerman. 33 photos. 118pp. 7¾ x 10⅝. 23654-4 Pa. $3.50

# CATALOGUE OF DOVER BOOKS

**GEOMETRY, RELATIVITY AND THE FOURTH DIMENSION,** Rudolf Rucker. Exposition of fourth dimension, means of visualization, concepts of relativity as Flatland characters continue adventures. Popular, easily followed yet accurate, profound. 141 illustrations. 133pp. 5⅜ x 8½.
23400-2 Pa. $2.75

**THE ORIGIN OF LIFE,** A. I. Oparin. Modern classic in biochemistry, the first rigorous examination of possible evolution of life from nitrocarbon compounds. Non-technical, easily followed. Total of 295pp. 5⅜ x 8½.
60213-3 Pa. $4.00

**PLANETS, STARS AND GALAXIES,** A. E. Fanning. Comprehensive introductory survey: the sun, solar system, stars, galaxies, universe, cosmology; quasars, radio stars, etc. 24pp. of photographs. 189pp. 5⅜ x 8½. (Available in U.S. only)
21680-2 Pa. $3.75

**THE THIRTEEN BOOKS OF EUCLID'S ELEMENTS,** translated with introduction and commentary by Sir Thomas L. Heath. Definitive edition. Textual and linguistic notes, mathematical analysis, 2500 years of critical commentary. Do not confuse with abridged school editions. Total of 1414pp. 5⅜ x 8½.
60088-2, 60089-0, 60090-4 Pa., Three-vol. set $18.50